E. CHAN
'6

HIGH PERFORM

High Performance Computer Imaging

Ihtisham Kabir

Greenwich
(74° w. long.)

For electronic browsing of this book, see:
http://www.brosebooks.com

The publisher offers discounts on this book when ordered in quantity.
For more information, please contact:

Special Sales Department
Manning Publications Co.
3 Lewis Street
Greenwich, CT 06830

Fax: (203) 661-9018
email: orders@manning.com

Recognizing the importance of preserving what has been written, it is the policy of Manning to have the books they publish printed on acid-free paper, and we exert our best efforts to that end.

Library of Congress Cataloging-in-Publication Data
Kabir, Ihtisham
High performance computer imaging / Ihtisham Kabir.
p. cm.
Includes bibliographical references and index.
ISBN 1-884777-26-0 (pbk.)
1. High performance computing. 2. Computer graphics.
3. Image processing—Digital techniques. I. Title.
QA76.88.K33 1996
006.6—dc20 96-28263
CIP

Manning Publications Co.
3 Lewis Street
Greenwich, CT 06830

Copyeditor: Robin Gee
Typesetter: Aaron C. Lyon
Cover designer: Leslie Haimes

Printed in the United States of America
1 2 3 4 5 6 7 8 9 10 – CR – 00 99 98 97 96

For my mother

my father

Sony

Ihsan

contents

preface

This book is intended for the technically proficient reader who wants to implement or use computer imaging techniques. It will be successful if the reader can use it as a bridge between the theory of image processing and the implementation of high performance imaging products.

Computer imaging, currently enjoying significant growth, is concerned with the acquisition, processing, and display of images using computers. From its roots in research laboratories, it has branched out into numerous industrial and consumer applications from remote sensing to machine vision to digital photography and multimedia. This growth has been nurtured by advances in imaging techniques, as well as by improvements in the performance of computer software and hardware. The appearance, in abundance, of powerful desktop computers with color display capabilities has brought computer imaging within easy reach of the end user.

However, those who want to implement or utilize this exciting new technology face an obstacle: the dearth of useful literature on its practice. For the uninitiated, the transition from the theory of image processing to its practice can be time-consuming and difficult. While several books address the theory exhaustively, few address the implementation details of common algorithms, and none cover engineering design issues such as performance and software reliability. This book offers a direct path to the practice of computer imaging by focusing on common algorithms and their implementation.

Having grown out of a course offered to practicing engineers in California's Silicon Valley, this book is the crystallization of the author's dozen years of experience in the design and implementation of computer imaging products. It answers many questions and problems that engineers face during the development of imaging products, and provides useful information for the user of those products.

As interest in computer imaging has grown, numerous image processing algorithms have been developed. While many of these are application specific and thus find limited use, some have proved robust enough to be applicable to a wide variety of situations. These commonly used algorithms are described in this book.

While high performance desktop computers have made imaging possible on the desktop, computational complexity remains a critical issue. Digital images are essentially very large arrays of numbers—therefore, processing them is computationally expensive. For execution at an acceptable speed, it is frequently necessary to implement an imaging algorithm in a manner that exploits, to the maximum possible extent, the architecture and instruction set of the hardware. This requires casting the algorithm in terms of the primitives with which the specific hardware being used performs best. On the other hand, naive and brute-force implementations usually result in suboptimal performance. In this book, the implementation of common imaging algorithms is covered in detail, emphasizing fast implementations whenever possible.

The key features of this book are:

- Exposition of current, commonly used computer imaging techniques
- Extensive information on efficient implementation of these techniques
- A software engineering framework for development of imaging software
- Numerous C programming examples
- Discussion of industry standard imaging and video compression algorithms, including CCITT Group 3 (fax), JPEG, and MPEG
- Tips on designing high performance imaging software tuned for special- and general-purpose computers
- Discussion of performance evaluation methods
- Exposition of the various types of imaging devices
- Coverage of modern color imaging including color spaces and accurate color reproduction
- Use of the Java programming language for image processing over the Internet
- Techniques for evaluating commercial imaging software and hardware products

This book is divided into three parts. The first part, Chapter 1, presents background and introductory material. Imaging systems are the focus of the second part, Chapters 2 through 4. Imaging algorithms and their implementation are the subjects of Chapters 5 through 9, which constitute the third part.

Chapter 1 introduces computer imaging. It defines the digital image and discusses some key applications and special operations needed in imaging.

Imaging devices are the subject of Chapters 2 and 3. In Chapter 2, various devices for the acquisition, storage, and display of digital images are discussed. Chapter 3 is devoted to pixel computation devices. Chapter 4 presents an approach to designing high-quality and robust imaging software.

The rest of the book is devoted to imaging algorithms. Efficient implementation is discussed with C programming examples. Chapter 5 is devoted to pointwise imaging operations, that is, those performed on a pixel-by-pixel basis. Examples include image copying, monadic and dyadic operations, histogram equalization, and contrast enhancement using a lookup table.

In Chapter 6, image filters are discussed. Linear filters using convolution, as well as non-linear filters such as median filters and morphological filters, are presented. The chapter also explores efficient implementation.

Color theory and its application to computer imaging is discussed in Chapter 7. Some useful color spaces are defined, and techniques are presented for efficient color space conversion and accurate color reproduction.

Chapter 8 presents algorithms useful for geometric processing and manipulation of digital images, including image scaling, rotation, and transposition. The ubiquitous affine transformation is also presented.

Image data compression is the subject of Chapter 9. Several techniques used in compression systems and details of the compression standards CCITT Group 3, JPEG, and MPEG are discussed.

It is hoped that this book will benefit the reader by helping him or her put the ideas of computer imaging into practice in a direct and efficient manner.

acknowledgments

A book is seldom the product of only one person's labor, and although this book bears my name, numerous individuals have contributed to its fruition. I want to take this opportunity to thank as many of them as possible.

The idea for this book came from my father, who, once I took it up, insisted that I complete it. I am grateful to him and to my late mother for their support and nurturing.

The person who assisted me most in this endeavor was my wife Sonia who gave me love, time, support, and encouragement. Without her help this book would not have been completed and I thank her for everything she has given me.

Bernie Hutchins and Cornell's DSP Laboratory first told me about image processing. The late Anil Jain taught me the basics of image processing. Bill Pratt opened the doors of industrial image processing for me and opened my eyes to high performance image processing. Steve Howell taught me much about reliable software. I am grateful to them for getting me started in image processing.

I have been fortunate to have had several expert colleagues who have reviewed the manuscript for this book at various stages and offered me insights and suggestions. They include John Watkins, Mike Hsieh, Bill Radke, Peter Farkas, John Recker, Robert Hoffman, David Berry, John Furlani, Yates Fletcher, Gerard Fenando, Aman Jabbi, and Walt Donovan. Tim van Hook taught me a good deal about performance and architecture. Mike Lavelle taught me about graphics hardware. Jasvinder Nijjar helped me in the laboratory, especially with the SX. Discussions with Lee Westover, Daniel Rice, and Alex Mou have also contributed to my understanding of imaging algorithms. Ray Roth taught me about optimization. I am grateful to all of them.

My thanks go to Munsi Haque, Surendra Ranganath, and Cliff Reader for their reviews and discussions regarding the form and content of this book.

Roger Day, Manager of Graphics Software at Sun, deserves special thanks for his support of this endeavor, as does Kipp Kramer, Director of Graphics and Imaging. They have created an environment where high performance imaging shines. Thanks are also due to my previous manager, Rob Mullis, for his encouragement of this work.

My editor Marjan Bace patiently worked with me throughout the development of this book. Aaron Lyon typeset the book, cleaned up the diagrams, and performed other tasks for the manuscript. I am grateful to them for their effort.

Finally, thanks to Hasan Z. Rahim for the inspiration, and Paul Winternitz for the friendship!

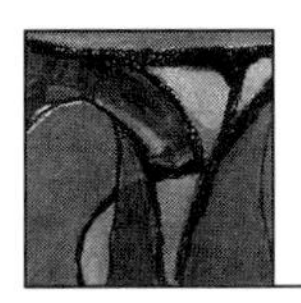

chapter 1

Introduction to computer imaging

1.1 Introduction

Pictures have been a powerful means of communication in human civilization. The sense of sight, one of the most precious aspects of human existence, enables us to communicate in myriad different ways using pictures. One hundred and fifty-five years ago, visual communication entered a new era with the invention of photography. Today, we are poised at the beginning of yet another revolution in visual communication, brought about by advances in technology, and exemplified by the advances in computer imaging.

Computer imaging is the acquisition, processing, and display of pictures using computers. During acquisition, an image is converted into a form readable by computer. An analog representation of an image is thus turned into a digital image. During processing, the digital image is manipulated using computers. The digital image is then displayed for the user to look at.

Thus, computer imaging enables the user to benefit from the powers of the computer to process and display pictures and photographs.

Often, there is confusion between computer imaging and computer graphics. The fundamental difference is that in the latter, the image is drawn, or created, using a computer, whereas in the former, an image from the real world is captured and often manipulated using a computer. The difference is somewhat analogous to the difference between a photographer and a painter. A painter creates images; a photographer captures them from the real world.

1.2 Origin and growth of computer imaging

The birth of computer imaging can be traced to a few research centers where pioneering work was done. These include the Massachusetts Institute of Technology (MIT), California Institute of Technology (Jet Propulsion Laboratory), University of Southern California, and the University of Maryland.

Some of the earliest work on image processing was done at MIT during the late 1950s and early 1960s. This includes Schreiber's early work on image coding [1], and Roberts's work on gradient edge detection [2].

The early days of NASA's space flights had a large influence on modern computer imaging, when large numbers of photographs of the moon were sent to the earth by the Ranger missions [3]. These images were sent to the California Institute of Technology's Jet Propulsion Laboratory (JPL) for processing. Early work on image enhancement and geometric processing was carried out here. One of the first image processing textbooks, *Digital Image Processing* [4], came out of this work.

At about the same time, the Image Processing Institute (IPI) at the University of Southern California was organized. This group concentrated on image enhancement and compression.

Some of the image enhancement work from JPL was contracted out to the IPI. In addition, the first vector-space formulation of image processing mathematics was done here [5]. Transform image coding (used in all image compression systems today) was also discovered at IPI. Out of this work came the encyclopedic book *Digital Image Processing* [6]. Subsequent invention of the Discrete Cosine transform by Rao [7] replaced the Fourier and Hadamard transforms in image compression; however, many of the general techniques developed here for transform compression are still used.

At the University of Maryland, another group was working on image processing; however, their early thrust was on image analysis and understanding. This work concentrated on extraction of information from images. Examples of these types of algorithms include segmentation of images into individual parts, and the extraction of features from those parts. The result of this work was the book *Digital Picture Processing* [8].

Another image processing reference book, *Fundamentals of Digital Image Processing* [9], grew out of the work being done at the University of California, Davis, during the late 1970s and early 1980s. Mathematical models for image processing and image compression are two areas of contribution from this work.

In the middle and late 1970s, image processing spread into more places as it became possible to perform image processing using minicomputers. During the middle of the 1970s, dedicated image processors became available. Commercial vendors of image processing hardware included Comtal, Vicom, De Anza, and International Imaging Systems.

Also during the 1970s, larger image processing software libraries became available. Examples of this kind of library include VICAR at JPL and DAISY at the University of California, Davis. These libraries typically included command-line support for executing commonly used image processing functions.

Up until the late 1970s, image processing was limited to dedicated processors or minicomputers. It was available to selected people working in industry or graduate schools. In the early 1980s, several vendors started selling image processing hardware as add-on cards for IBM PCs. Towards the middle of the 1980s, image processing hardware and software were also developed for the various desktop computers (Apple Macintosh, IBM PC-compatibles, and UNIX workstations). At that point, image processing became available for use by a large number of people.

In the late 1980s and early 1990s, as microprocessor speeds increased, memory sizes increased, and display technology became better in general, image processing moved closer to everyone's range. The advent of several user-friendly software products, such as Adobe Photoshop, also hastened the migration of image processing technology from the laboratory and factory to the consumer. Imaging software libraries, such as the Programmer's Imaging Kernel System (PIKS) and the XIL library from Sun Microsystems, have also brought new levels of sophistication to the user and programmer.

In the future, with increasingly better technology, faster processors, and advanced algorithms, we expect to see image processing in many more applications. In particular, we expect to see the convergence of several different technologies—including still and video image processing, computers, television, and telecommunications—that will substantially alter the role of visual communications in our lives.

1.3 Application areas

In this section we will discuss some application areas of computer imaging.

1.3.1 Technical image processing

Technical image processing was the earliest application area of digital imaging. In this area, image processing has been used as a research tool to solve technical problems [10].

Scientists in virtually all the natural sciences, including physics, chemistry, biology, materials science, astronomy, and geology use image processing as a tool.

Examples of image processing usage are the measurement of particle sizes in material science, the enhancement of images of celestial bodies for astronomy, and analyzing images of particle collisions in particle physics.

Users in technical image processing are technically proficient. Typically, they have had to develop many of their own algorithms for solving their problems.

Some commonly used functions in technical image processing include image histogram (Chapter 5), convolution and other neighborhood filters (Chapter 6), and morphological image processing (Chapter 6). A topic not covered in this book, but used in technical image processing, is Fourier domain image processing. The reader should see references 6 and 9, which cover this material.

1.3.2 Medical imaging

Image processing has found substantial use in medical imaging from the early days, because medical imaging technology has evolved with computer imaging as analog techniques have been replaced with digital ones. Several types of medical imaging data, including computed tomography, magnetic resonance imaging, and medical ultrasound, are gathered in digital form, making them specially suitable for computer processing.

In most medical imaging applications, the two important phases are image acquisition and image display. The manner in which images are acquired in medical imaging is called the *modality*. Detailed study of the different modalities in medical imaging are quite involved and beyond the scope of this book. The reader should see reference 11 for a treatment of these modalities.

Regardless of which modalities are used for acquiring the images, the same common functions are used in the display of these images. These functions include table lookup (Chapter 5), copying images (Chapter 5), convolution (Chapter 6), and geometric image resampling (for example, scaling an image) (Chapter 8).

1.3.3 Remote sensing

Computer imaging can be used to observe, study, and measure objects from a distance. This is the general idea behind remote sensing. Examples of remote sensing include the measurement of earth resources (for example, forests, water supplies, urbanization) and tracking the changes in these resources, as well as general geographical information gathering. Another area of application related to this is defense and intelligence gathering.

Image acquisition for remote sensing must be done from a satellite or an airplane, using specialized high-speed cameras. These images are then transmitted to a remote sensing image processing system, which performs the necessary work.

Some of the imaging algorithms useful for this application area are geometric modifications of images (Chapter 8) and color algorithms (Chapter 7), as well as basic point operations (Chapter 5) and neighborhood filters (Chapter 6).

1.3.4 Desktop publishing

Desktop publishing includes a wide range of functions, related to publishing, that are performed using desktop computers. These functions include word processing, page layout, and so forth.

Increasingly, image processing has become an integral part of desktop publishing. The most common example is to integrate images into a document; another is the production of photographic (or mostly photographic) books using the computer [12].

The imaging functions needed for desktop publishing include geometric image modifications (Chapter 8), blending (also known as compositing) (Chapter 5), and neighborhood filters such as convolution and median filters (Chapter 6).

1.3.5 Graphic arts

Another area, related to publishing, where image processing plays an important role is graphic arts. In this application, the goal is the design and production of images for advertisement as well as artistic purposes.

The users in this area tend to be more artistic than technical. However, with the advent of powerful, user-friendly software such as Adobe Photoshop and a variety of special effects *plug-ins*, which augment the capabilities of Photoshop, more and more graphic-arts users are turning to computer imaging.

Image processing functions used most in graphic arts are those that create special effects. This includes geometric modification algorithms (Chapter 8), convolution (Chapter 6), and table lookup functions (Chapter 5).

1.3.6 Document imaging

In document imaging, the goal is to process and store documents using computer imaging techniques. There are several reasons why imaging techniques are attractive in this area. First is the elimination of paper documents. Second is the efficiency achieved while processing documents in digital form.

In document imaging systems, paper documents are usually scanned in using a high-speed scanner. The digital images are typically binary (one bit). Images are usually stored in TIFF format, including fax Group 4 compression.

Work flow is important in many document imaging systems. In many real-life situations, several people may need to read and write into a document. For example, a business travel authorization form may need input from the traveler, his or her manager, accountants, and so on. A work-flow system controls the flow of the document and moves it on from one person to another, thus eliminating the need to pass the document by hand.

Some of the concepts used in document imaging include fax compression and geometric modifications, which are covered in Chapters 8 and 9, respectively. Items not covered include databases for document images.

1.3.7 Digital photography

Digital photography is another emerging area of application for computer imaging. Almost all the major camera manufacturers have announced digital cameras in 35 mm, medium format, and large format. These cameras are discussed in detail in Chapter 2.

While the prospect of going "filmless" must be quite alluring to many photographers, there are several hurdles on the road to widespread acceptance of digital cameras. These include cost ($10,000 or more for reasonable-quality pictures), bulkiness, and the somewhat cold and gritty nature of these images when compared to film images.

Almost all the image processing functions discussed in this book can be useful for digital photography.

1.3.8 Industrial inspection

Computer imaging is used also in industrial inspection. The goal in this area is to use computer imaging to inspect manufactured products reliably and effectively.

Often the performance requirements are very demanding in industrial inspection because the parts being inspected are frequently moving, for example, in an assembly line. Therefore, fast image processing hardware is required.

Some important functions used in this area include image histograms, image thresholding and other table lookup functions (Chapter 5), convolution to detect edges (Chapter 6), and morphological image processing (Chapter 6). Some of the topics not covered include BLOB analysis and segmentation and feature extraction. The reader should see reference 8 for these.

1.3.9 Video processing

Video is an important area of technology promising to become even more widespread with the convergence of telecommunications, computers, and television.

While video is a subject unto itself, image processing techniques are used in many video functions. In particular, image compression techniques are used in video compression algorithms used in turn in video teleconferencing. Thus, all of the basic compression building blocks described in Chapter 9 are useful for video processing.

Concepts from computer color are also important in video processing. In addition, video images often require filtering using neighborhood filters described in Chapter 6.

1.3.10 Consumer applications

Computer imaging has moved into several consumer application areas. These areas include fax and PhotoCD.

Fax has revolutionized the world of communications. The algorithms used in compressing fax images are discussed in Chapter 9.

PhotoCD is a new technology, introduced by Kodak, that enables the consumer to enter the world of computer imaging without substantial investment in image acquisition hardware. PhotoCD is discussed in more detail in Chapter 2.

Other emerging consumer applications requiring imaging technology include the digital video disk and digital television.

1.4 Notations and terminology

In most of the operations described in this book, input images are processed to produce output images. Input images are also referred to as *source* images, while output images are called *destination* images. A pixel in the *source image* is called a *source pixel;* one in the destination image is called a *destination pixel.*

Mathematical notations and equations are kept to a minimum. When needed, images are specified as two-dimensional functions. For example, the notations $f(x,y)$, and $g(x,y)$ are used to represent source and destination images, respectively. During mathematical derivations, we have often considered one-dimensional digital signals. These are represented as functions of one variable, for example, f(x).

1.5 Basic parameters of a digital image

A digital image is a digital representation of a continuous valued, or analog, image. The digital image is obtained from the analog image by sampling and quantization, which are described in detail in Chapter 2. In essence, the process consists of placing a two-dimensional grid over the analog image, and assigning a digital number (consisting of a fixed number of bits) to each square of the grid. The number assigned to a square, when properly interpreted, corresponds to the average brightness or gray level of the analog image in that square. Each square is called a *picture element*, or a *pixel.* Thus, a digital image is nothing but a two-dimensional array of pixels. A picture of this representation is shown in Figure 1.1.

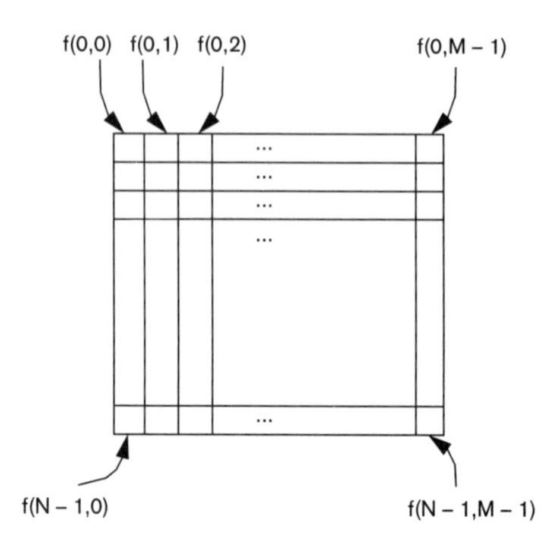

Figure 1.1 N × M digital image layout

When a digital image is stored in computer memory, several other parameters are needed to represent it. Important among these are *pixel depth* or *precision*, *image size*, and the *number of bands*.

The pixel depth is the number of bits used to digitally represent the pixel. This determines the dynamic range of the values in the image. Common pixel representations are in bits (0 or 1), unsigned bytes (0 to 255), signed bytes (–128 to 127), unsigned shorts (0 to 65535), signed shorts (–32768 to 32767), and floating point (large dynamic range).

The size of the image tells us the number of rows and columns in the image. Some common image sizes are 640 × 480, 512 × 512, and 1024 × 1024. In general, the higher the resolution, the closer the digital image in appearance to the analog image.

The number of bands is another important and commonly used image attribute. Each band represents the same image, but emphasizes certain frequency characteristics. For example, a color image consists of three bands (typically red, green, and blue), which together hold all the information required to construct the color image. In a more technical setting, remote sensing images may be made with a number of spectral filters, each measuring a certain aspect of the landscape, and the number of bands in this case can be larger than three. For a monochrome (gray-scale) image, the number of bands is one.

In addition to these parameters, there are other image attributes that are needed during the implementation of an imaging product. Some of these are illustrated in the following examples.

1.5.1 Image formats for API libraries

In computer software usage, *API* stands for *application programmer's interface*, that is, a software library of functions used for writing application programs.

Image formats for a low-level API library Our first example is a low-level API, that is, an imaging library written in hardware-specific terms for a particular piece of hardware, utilizing simple and basic data structures.

For such a library, the image may be described in terms of the following parameters: *pixel depth* (in bits or bytes per pixel), *width, height, linebytes, pixel_stride*, and a *pointer* to the first pixel. Note that it is not always necessary to include the number of bands in this description. This is because the higher-level software that calls this library can separate the bands if needed.

We have several new parameters here. *Linebytes* is the number of bytes between the same column positions in successive rows. It is not necessarily equal to the width of the image, since successive rows may be noncontiguous memory addresses. This is shown in Figure 1.2. The image is represented by the solid lines; however, it is a subset of a larger image, represented by the dashed lines. In computer imaging practice, it is the rows of the larger rectangle, not the image, that make up contiguous addresses in memory. Therefore, the linebytes parameter is needed in addition to the width of the image. (In many practical cases, width and linebytes are the same, of course.)

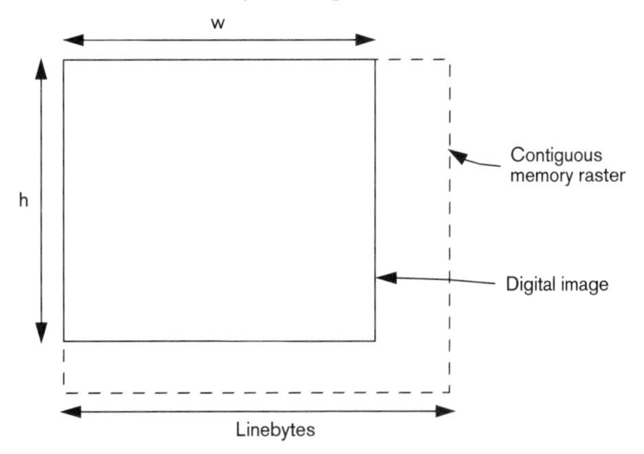

Figure 1.2 The need for linebytes

The second parameter of interest is *pixel_stride*. This is the number of bytes that one must add to the pointer of the current pixel to get to the next pixel of the same band. This number is not always trivially equal to the product of bytes per pixel and number of bands. The reason is the bands can be arranged in at least two different ways: *band-interleaved* (also known as *pixel-sequential*) or *band-sequential*. This is shown in Figure 1.3.

In the first case, where the pixels are band-interleaved, the pixel_stride is 3; in the second case, where the pixels are band-sequential, the pixel_stride is one.

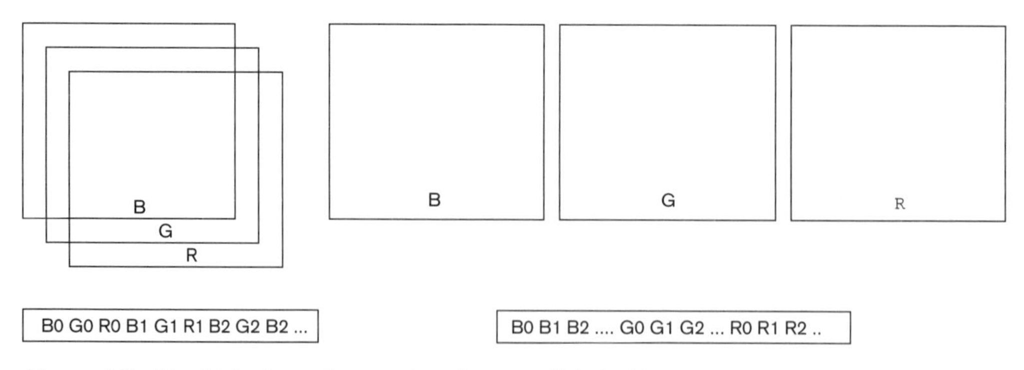

Figure 1.3 Band-interleaved versus band-sequential pixel layout

Finally, the third quantity of interest is the pointer to the first pixel. This is the first pixel of the image that is to be processed, and may not be the pixel at location (0,0) of the image. This allows for a nonzero origin to be defined for the image. Origins are described in the section below.

An example of a low-level API is the pci library developed for this book. The pci_image structure is given by:

```
typedef struct pci_image {
     unsigned char *data;
     unsigned int width;
     unisgned int height;
     unsigned int bands;
     unsigned int pixel_stride;
     unsigned int linebytes;
} pci_image;
```

Image formats for a high-level API library While a low-level image processing API library is usually tightly coupled to a particular piece of hardware, a high-level API library often attempts to maintain device independence so that it can be portable across different devices. In a high-level imaging API, the image processing functions are abstracted in such a way that they behave in the same manner independent of the device on which they are executed. The design of these kinds of APIs is discussed in more detail in Chapter 3.

The API often has different objects, including the image, which the application programmer can specify for use without having to deal with their internals. For example, an image in a high-level API may obtain a handle to an image by specifying the width, height, pixel depth, and the number of bands, and creating the image with these parameters. The actual image memory is device-dependent and may not be allocated until the image is actually used.

Additional attributes of an image used in a high-level API may include regions of interest and origins. A region of interest (ROI) is an area within an image. It can be a rectangular or arbitrary shape, and can be represented internally as a list of rectangles or as a bit mask. When an imaging operation is performed, the ROI of the source image(s) is intersected with the ROI of the destination image. The result is the area in the destination image that may be written. This is illustrated in Figure 1.4, where the shaded area in the destination is where pixels may be written. All other areas of the destination must be left untouched. From this shaded area, we can also derive the corresponding areas of the source image from which pixels are read.

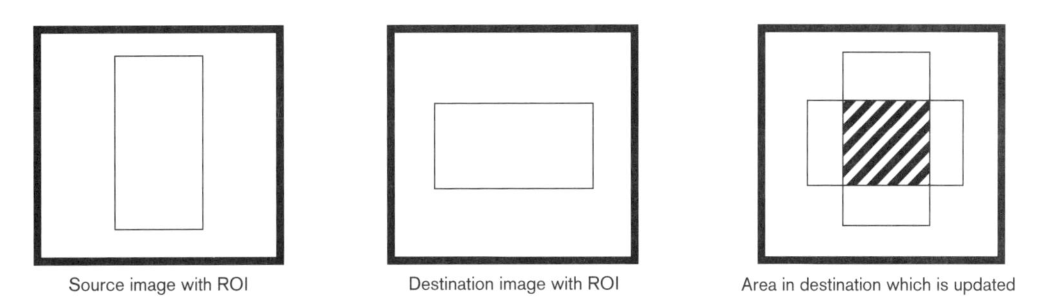

Figure 1.4 Intersection of ROIs between source and destination images

An image origin is an (x,y) coordinate pair associated with an image. Normally, the origin of an image is the upper left corner. The coordinate system is positive x increasing to the right and positive y increasing downward. When an operation is performed, the origins of all the source and destination images are aligned, and the images (or their ROIs) are intersected. This is shown in Figure 1.5.

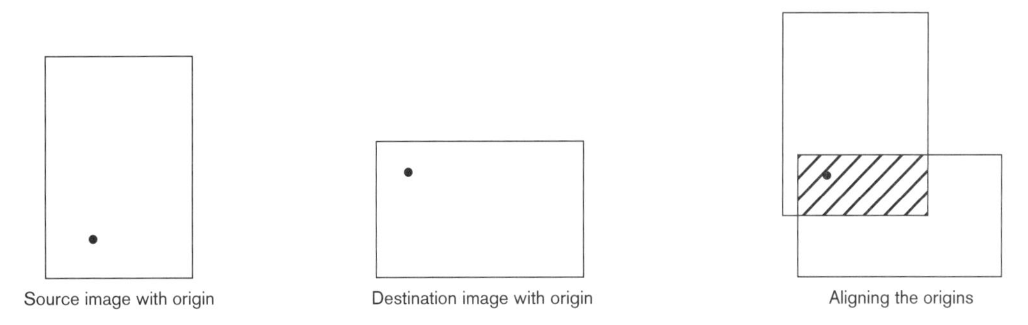

Figure 1.5 The use of image origins in an imaging operation

Again, an ROI may be a list of rectangles or an arbitrary area within an image. These attributes take on specific meanings during an operation. For example, if two images are to be added into a third image, the origins of the two source images must be aligned, and addition begins there. Then, the output pixels are written in alignment with the origin of the output image.

Regions of interest usually work as write masks in an image. In the case where there are regions of interest associated with the source and destination images, these regions are intersected to find the region of the destination where pixels can be written.

Finally, it should be mentioned that there are several other attributes that may be associated with images. These may include a color space, color lookup table for displaying the image, and various flags indicating the state of the image.

1.6 Pixel arithmetic

In this section we will look at the data types used to represent pixels and the arithmetic used for these pixels.

1.6.1 Pixel data types

As indicated earlier, pixels in digital images can be represented in several forms. The most important among these are 1-bit pixels, used to represent binary images; 8-, 16-, and 32-bit signed and unsigned pixels, used to represent gray-scale images; and 24-bit unsigned pixels, used to represent color images. Frequently, 24-bit pixels are actually represented by 32-bit memory words, where the top 8 bits are used as an alpha band (see alpha blending in Chapter 5), as an overlay plane, or left unused.

In all computer imaging operations, pixels may be treated just like normal numbers, and all the rules governing computer arithmetic apply to pixels as well. However, there are some operations specific to imaging that can substantially enhance the speed of imaging operations. While these operations can always be performed by using a sequence of C expressions, having them available as hardwired *specials* enables speedup of many imaging routines. These operations are described below.

1.6.2 Special operations for pixels: formatting, clipping, masking

Special operations dedicated for computer imaging include *pixel formatting*, *clipping*, and *masking*.

Pixel formatting is a term used for some operations that arrange the pixels in the processor's registers in the manner best-suited for processing. One type of pixel formatting is casting pixels into higher precision than is needed to maintain intermediate precision. Formatting is typically done in the registers of the processor that will process the images. It may be done while loading the pixels into the registers (such as in the SX pixel processor, described in Chapter 3), or it may be done after the pixels have already been loaded (in most microprocessors). While formatting, several operations, including sign extension and shifting may also be performed. Two examples of this type of formatting are shown in Figure 1.6.

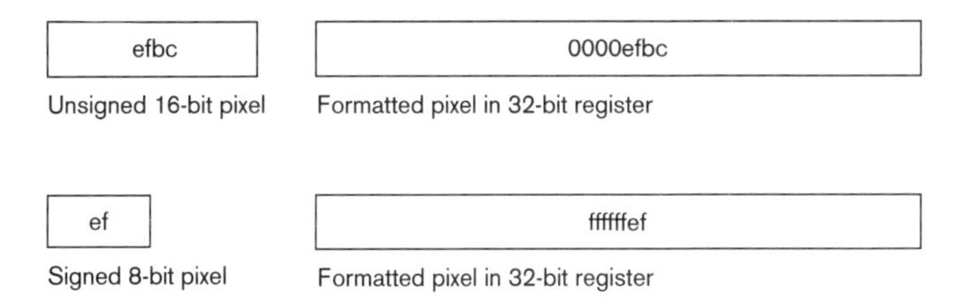

Figure 1.6 Pixel formatting

Another type of formatting consists of the hardware's ability to understand different bands in the digital image to be processed. This is necessary when processing color images, represented by 24 or 32 bits, which are composed of 8-bit bands. For such an image, if the hardware is capable of understanding the concept of bands (for example, by being able to load and store only the green band of the image, or being able to load and store subsequent bands into separate registers), the processing speed is greatly enhanced.

After the pixels are processed, the inverse operations of the formatting must also be performed. In addition, the resultant pixel value must be *clipped* before being written to the destination. Clipping simply clamps the resultant pixel between a minimum and a maximum value. For example, if two unsigned 8-bit source pixels, with values 250 and 251, are added to produce an unsigned 8-bit destination pixel, the result in an intermediate higher precision is 501. If only the lower 8 bits of this result are taken, we would have 245, not the expected 255. Whereas, if the result is clipped between 0 and 255, we would end up with 255. If it must be done explicitly, clipping requires two comparisons. Built-in hardware clipping, such as those available in the SX processor, or the FPACK instructions in the Visual Instruction Set (both are discussed in detail in Chapter 3), enable the programmer to bypass the two comparisons.

Another operation, specific to pixels, is *masking* of certain bits when the resultant pixels are written. Often, we have a situation where only certain bits of the destination pixel may be written. For example, in a 32-bit pixel, the upper 8 bits may be dedicated to an overlay plane or to an alpha channel (Chapter 5). These bits may not be corrupted. When a 24-bit pixel needs to be written to the lower 24 bits of a 32-bit pixel, read-modify-write must be performed instead of a single write. In other words, the 32-bit pixel must be read, its contents modified to include the desired 24 bits, and the results written. Often, imaging hardware performs this operation at no additional cost. An example of pixel masking with and without hardware assistance is shown in Figure 1.7.

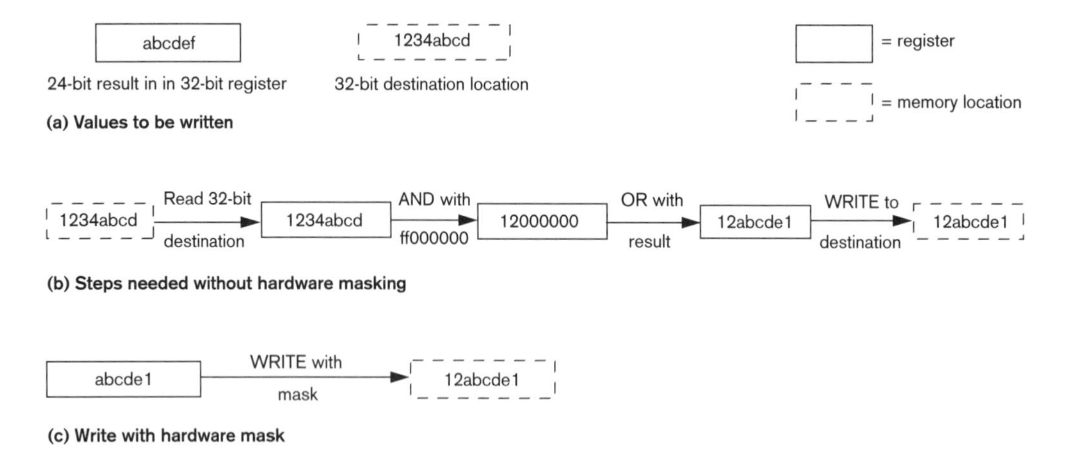

Figure 1.7 Steps needed for writing pixels with mask with and without hardware assistance

1.6.3 Fixed-point arithmetic

Fixed-point arithmetic is often needed when performing floating-point operations using an integer processor. Such processors are very common in image processing systems.

The idea behind fixed-point arithmetic is simple: Since all quantities must be represented as integers, any floating-point quantity must be scaled up appropriately. After the processing operation is completed, the resultant pixel value is scaled down by the same amount.

A classic example is convolution, where the kernel is often specified in floating point. Prior to performing convolution on an integer processor, the kernel values must be scaled up by the maximum amount possible (without causing an overflow in the final result). After the convolution is performed, the resultant pixel value is scaled down by the same amount before being written to the destination.

Note that the availability of a hardware-assisted shift during pixel store greatly accelerates fixed-point arithmetic. This shift is available, for example, in the SX processor.

Finally, we note that the errors caused by the use of fixed-point arithmetic in general signal processing operations have been studied exhaustively (reference 13, for example). In image processing, the same error analysis applies, for which reason this subject is not traditionally covered in image processing textbooks.

1.7 Examples of digital images

The images used in this book to illustrate the algorithms are shown in Figures 1.8 through 1.13. These images were scanned from color negatives and slides into a PhotoCD (discussed in Chapter 2). Then, the luminance component of the images was extracted. The resolution of Figure 1.8 is 512×768. The others are 768×512 pixels.

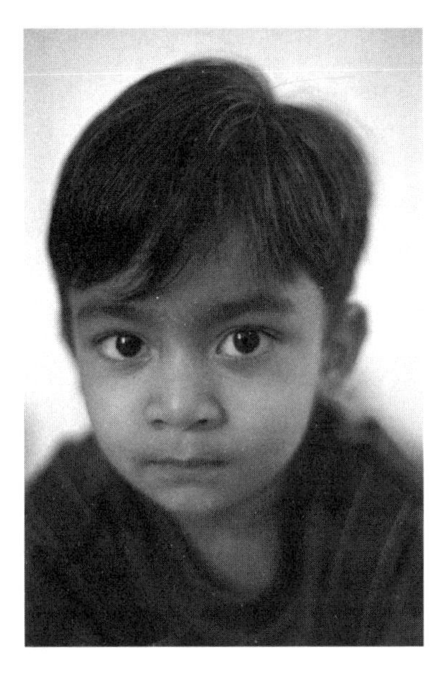

Figure 1.8 Ihsan

Figure 1.9 Museum

Figure 1.10 Bus

Figure 1.11 Boat

Figure 1.12 Snog Rock

Figure 1.13 Zabriskie

1.8 System considerations

In this section we will look at the requirements for a computer imaging system.

1.8.1 Storage requirements for digital images

There is a large storage requirement for digital images. The designer of any imaging system must address this issue. A "normal" digital image, that is, one that looks close to a photographic image in resolution, can contain about 1024×1024 pixels at 24 bits per pixel. This is 3 megabytes of memory required for simply one image.

There are two situations where this storage demand is further exacerbated. The first is the case of high-resolution images. In most graphic arts and color prepress applications, the resolution of the image must be higher than 1024×1024. Typically they can go up to 4096×4096 or more. This is a direct result of improvements in scanner and charge-coupled device (CCD) technology (discussed in Chapter 2). Graphic artists require the high-resolution images because they want the finished product to look photographic. The second situation where the storage demand is aggravated is when a series of images is acquired. This may happen, for example, in a medical imaging situation, where a heart beat is being monitored. Several digital images may be acquired every second.

Therefore, the storage of images must be a prime concern for the designer of the computer imaging system. Besides having the amount of storage necessitated by the application, the only other way to improve the situation is to use image data compression (Chapter 9). There are several standards for still and moving picture compression that enable the user to compress images by a substantial amount. In particular, if the user tolerates some loss of fidelity in the images, then lossy compression techniques can greatly help.

1.8.2 Performance requirements for imaging algorithms

The performance requirements for imaging algorithms are usually quite demanding. This is because images consist of a large number of pixels, each of which must be loaded into the processor, processed, and stored. Frequently, processing of one pixel may require loading several pixels into the processor. Using the algorithms and tuning methods described in Chapters 4 through 9, the performance requirements of various imaging algorithms can often be reduced.

Since there is no accepted benchmark for imaging operations, it is difficult to decide whether a particular hardware device will be able to meet the performance requirements of the application. To a large extent, this depends on the job at hand. For example, if an imaging system is required to perform a certain sequence of operations at 30 frames per second, then the hardware needs to be explicitly benchmarked for this sequence.

Dedicated image processing hardware often performs much better than general-purpose computers for certain image processing algorithms. However, if the application requires that a large number of different imaging algorithms be used, then a general-purpose processor is the best choice.

1.8.3 Bandwidth requirements

The bandwidth requirements of an imaging system are also very important. In general, the bandwidth of a computer system refers to the bandwidth of the bus through which data move around in the machine. For example, if images need to be moved to the display at real-time rates, then the bus connecting the display must support these rates. Similarly, any I/O device acquiring or printing images will be constrained by the bandwidth of the I/O bus.

1.8.4 Requirements for the display of digital images

Image displays have some particular requirements. Perhaps the most important among these are the image resolution and the pixel depth. In order to display images, they are loaded into a special type of memory called Video RAM (VRAM). The pixels of the VRAM are essentially passed through a bank of digital-to-analog (D/A) converters to create the analog signals that feed the display monitor. The size of the VRAM must be able to accommodate the maximum image resolution to be displayed at the desired pixel depth. In addition, the bus bandwidth of the display bus, and the speed of the D/A converters also plays an important role.

1.8.5 Software requirements

In any system, it is the software that interacts with the user. Therefore, the software of an imaging system must have the right amount of user friendliness in order for the system to be successful in the market. At the same time, the software must be efficient so that in the execution of image processing functions, the performance comes close to the hardware performance.

Much of this book deals with software algorithms for image processing. Chapter 4 is devoted entirely to the software design of an image processing system. In addition to the imaging software, other pieces of software that are important include the operating system, display system (for example, the X Windows system), and any relevant database and networking software necessary for an application.

1.9 Conclusion

In this chapter, we have examined the origins and development of computer imaging. We have also examined the digital image in some detail. We have looked at application areas of computer imaging, and discussed some key requirements of imaging systems.

1.10 References

1 W. R. Schreiber, C. F. Knapp, and N. D. Kay, "Synthetic Highs, an Experimental TV Bandwidth Reduction System," *Journal of SMPTE*, Vol. 68, August 1959, pp. 525–537.

2 L. G. Roberts, "Machine Perception of Three Dimensional Solids," *Optical and Electro-Optical Information Processing*, J. T. Tippett et al., eds., Cambridge, MA: MIT Press, 1965.

3 G. Baxes, *Digital Image Processing*, New York: Wiley, 1994.

4 K. Castleman, *Digital Image Processing*, Englewood Cliffs, NJ: Prentice Hall, 1978.

5 W. K. Pratt, "Vector Formulation of Two Dimensional Signal Processing Operations," *Computer Graphics and Image Processing*, Vol. 4, No. 1, March 1975, pp. 1–24.

6 W. K. Pratt, *Digital Image Processing*, New York: Wiley, 1978.

7 K. R. Rao, *The Discrete Cosine Transform*, New York: Academic Press, 1988.

8 A. Rosenfeld and A. C. Kak, *Digital Picture Processing,* New York: Academic Press, 1978.

9 A. K. Jain, *Fundamentals of Digital Image Processing*, Englewood Cliffs, NJ: Prentice Hall, 1988.

10 B. Jahne, *Digital Image Processing*, Berlin: Springer-Verlag, 1989.

11 A. Macovski, *Medical Image Processing,* Englewood Cliffs, NJ: Prentice Hall, 1983.

12 S. Johnson, *Making a Digital Book*, Pacifica, CA: Stephen Johnson Photography, 1993.

13 A. V. Oppenheim and R. Schafer, *Digital Signal Processing*, Englewood Cliffs, NJ: Prentice Hall, 1975.

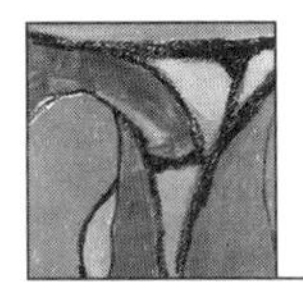

chapter 2

Imaging devices I: acquisition, display, and storage

2.1 Introduction

As we saw in Chapter 1, today's imaging systems are used in a wide variety of applications. An imaging system usually consists of several imaging devices that are connected to perform the tasks required for an imaging application. These devices may be responsible for physically creating the digital images, processing them, storing them, and displaying the images in either electronic or hardcopy forms.

The parts comprising a typical imaging system are shown in Figure 2.1. Images are acquired from the physical world by means of an acquisition device. This device can be a video digitizer (frame grabber), a scanner, a digital camera, or one of a number of other devices that are used in specialized applications of imaging (for example, an ultrasound scanner). After acquisition, the image data may be processed before being stored. Images can be read from storage into the processor for display or further processing. In many real-time imaging systems, the data can be sent directly from the acquisition device into the display subsystem.

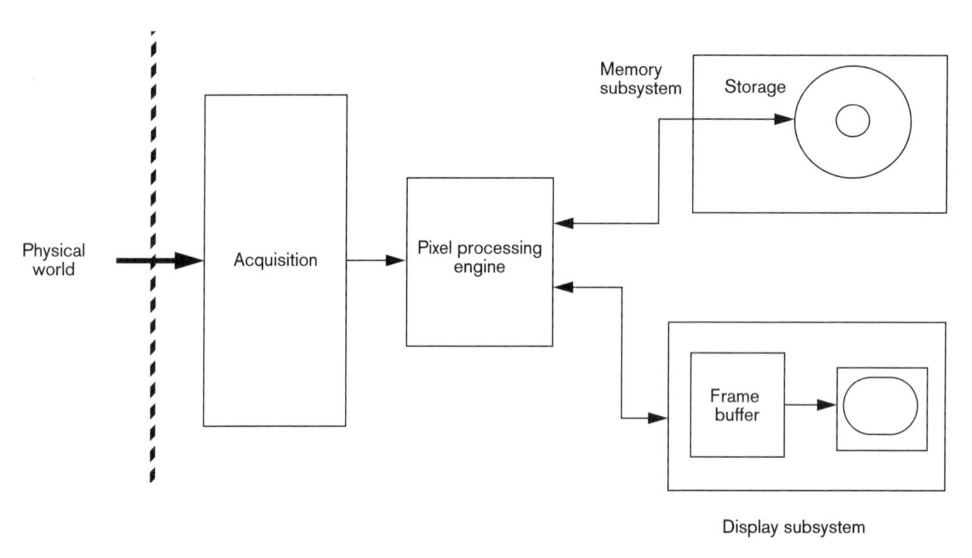

Figure 2.1 Parts of a digital imaging system

The purpose of this chapter is to examine some of the more common devices used in the acquisition, storage, and viewing of digital images. We shall examine the principles used in the operation of both input and output devices used in imaging systems, and discuss several popular devices from each class. The users and the designers of imaging systems will gain an understanding of the inherent technologies as well as the various features of these devices. The processing engine, which is responsible for pixel computation and actual image processing, is the subject of the next chapter.

In Section 2.2 we will take a closer look at the devices available for image acquisition. These devices are mostly based on the charge-coupled device (CCD) and include scanners, digital cameras, and frame grabbers. Image output, including display and hardcopy, is the subject of Section 2.3, where we will examine recent trends in the design of frame buffers and hardcopy output devices. In Section 2.4 we will look at the storage of images, including a discussion of the memory hierarchy of a computer. The recently introduced PhotoCD storage device will also be discussed.

2.2 *Image acquisition*

Image acquisition is the first step in any imaging operation. It is the process by which we obtain a digital image from an analog representation in the physical world. After acquisition, the digitized image can be stored away, or displayed, or processed using image processing techniques.

The conceptual steps used during image acquisition are shown in the block diagram of Figure 2.2. These steps are image formation, sampling, quantization, and scan conversion. We now take a detailed look at them in order to gain a deeper understanding of image acquisition.

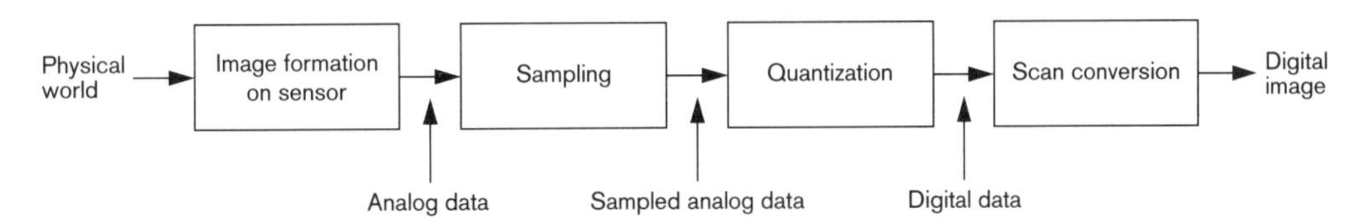

Figure 2.2 Steps in the image acquisition process

During image formation, a scene from the real world is imaged into the sensor. This scene is often—but not always—a two-dimensional projection of a three-dimensional scene. An example is the image formed by the lens system of a camera. Other examples include the ultrasound image formed by the echoes received by an ultrasound transducer emitting ultrasound pulses, or the x-ray projections used to form cross-sectional images of three-dimensional objects using computed tomography.

In the second step of the acquisition process, the analog data are *sampled.* This process is shown for a one-dimensional signal in Figure 2.3. Sampling is equivalent to multiplying the analog function by a comb function, called the *sampling function.*

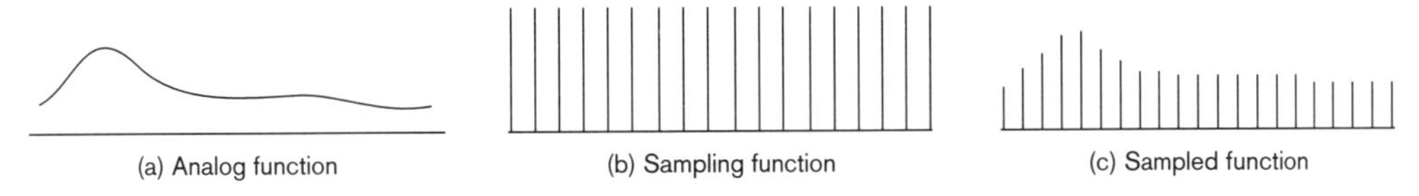

Figure 2.3 Sampling a one-dimensional analog function

The density of the samples in the sampling function (labeled *b* in the Figure 2.3), also known as the *sampling frequency*, is important. If the samples are too far apart, then rapid changes in the analog signal will not be reflected in the sampled signal. This is shown in Figure 2.4. The loss of high-frequency information from the signal is called *aliasing*. The *Sampling Theorem* [1] tells us that in order to avoid aliasing, the sampling frequency must be at least twice the maximum frequency present in the analog signal.

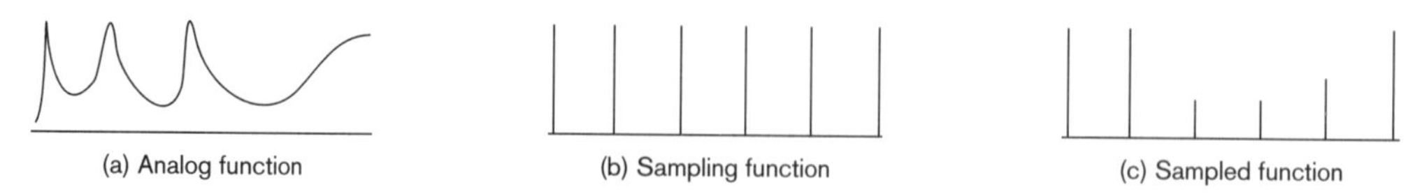

Figure 2.4 Undersampling a function

Since images are functions in two dimensions, the principles of sampling one-dimensional signals must be carried over to two dimensions. Note that frequency in a two-dimensional signal (such as an image) really means *spatial* frequency and refers to the rate of change of intensity along space. Therefore, in an image, areas of constant brightness have low frequency, whereas areas where there are a lot of changes in brightness, such as edges, have higher frequencies.

To sample a two-dimensional signal, we multiply it with a two-dimensional comb function, as shown in Figure 2.5. As in one-dimensional sampling, the two-dimensional version of the sampling theorem must be obeyed in order to avoid aliasing. Often, the analog image formed by the imaging system is sharp and contrasty, leading to high-frequency components. If the sampling frequency is inadequate, then the high-frequency components of the signal must be attenuated so that the sampling theorem is satisfied. This may be done by throwing the lens slightly out of focus or by employing an antialias filter, which is essentially a low-pass filter.

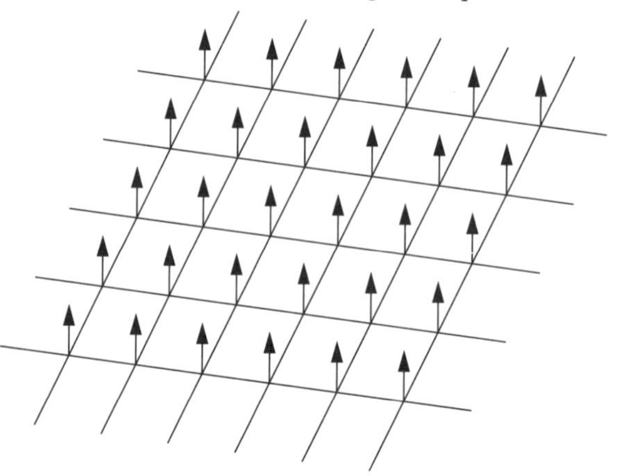

Figure 2.5 Two-dimensional comb function

The third step in image acquisition is *quantization*. The samples obtained after sampling are continuous-valued, with infinite precision. However, for a digital representation of the image, we need to discretize these samples into digital words with finite length. Quantization turns the samples into discrete-valued quantities. Thus, the continuous-valued samples become, for example, discrete-valued 8-bit pixels with values between 0 and 255. Quantization is an irreversible process. That is, once the samples are quantized, the original continuous-valued samples cannot be reconstructed.

Because the continuous-valued samples are essentially truncated to a finite number of bits during quantization, errors are introduced into the image. The design of optimal quantizers [2] focuses on minimizing these errors.

In practice, quantization is the result of mapping the sampled values using a function such as that shown in Figure 2.6, which shows a *uniform* quantizer. The output values are equally spaced, and the range of input values is divided equally into the number of output values. In real imaging systems, a uniform quantizer is quite common.

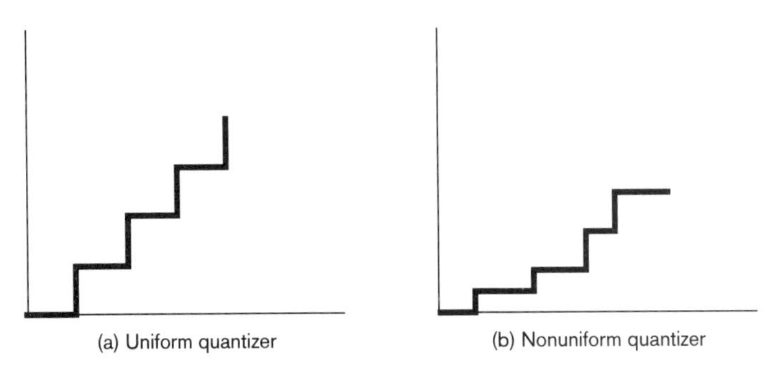

Figure 2.6 Quantization functions

Quantizers can also be *nonuniform* (labeled *b* in Figure 2.6). Note that both the horizontal and vertical spacings can be irregular. Nonuniform quantizers can be used to compensate for nonlinearities in the image sensor (including the human visual system, which is much more sensitive to changes in low-luminance scenes). While it is difficult to build a nonuniform quantizer, the same effect can be achieved by passing the samples through a nonuniform function (such as a logarithmic filter) and using a uniform quantizer on the result.

The fourth step in the digitization process is *scan conversion*. The result of sampling and quantization is a set of pixels. However, the data coming out of the quantizer are usually a monolithic stream of pixels. Scan conversion assigns (x,y) coordinate values to each pixel.

Scan conversion is a trivial exercise when the analog function being sampled is a two-dimensional, rectangular function, such as a photograph. However, it can be more complicated in imaging systems that create two-dimensional projections of three-dimensional data. For example, in an ultrasound acquisition system, a rotating transducer transmits a signal and listens for the reflection at discrete intervals over the angle of rotation. In such a system, the transducer's angular velocity and position must be translated into an (x,y) coordinate for each data element obtained. Often, special-purpose hardware is needed for scan conversion if real-time display is necessary.

Thus far we have examined the steps required, in principle, to acquire an image. We are now ready to look at the way practical acquisition devices work.

2.2.1 *Acquisition devices*

There are several classes of image acquisition devices. These devices are responsible for image formation, sampling, quantization, and scan conversion discussed in the above section. While the methods used for image formation may vary from device to device, the vast majority of acquisition devices rely on photosensitive materials to generate electrical signals corresponding to the light signals making up the image. In this section we will begin with the CCD (used in many acquisition devices) and proceed to scanners, digital cameras, and video digitizers.

Charge-coupled devices At the heart of many acquisition devices is the charge-coupled device (CCD). The CCD is a silicon device sensitive to light. It is used for converting light information into electricity. The basic device is shown in Figure 2.7. The charge generated by light striking the photosensitive surface is accumulated in a potential well by the electrode voltage.

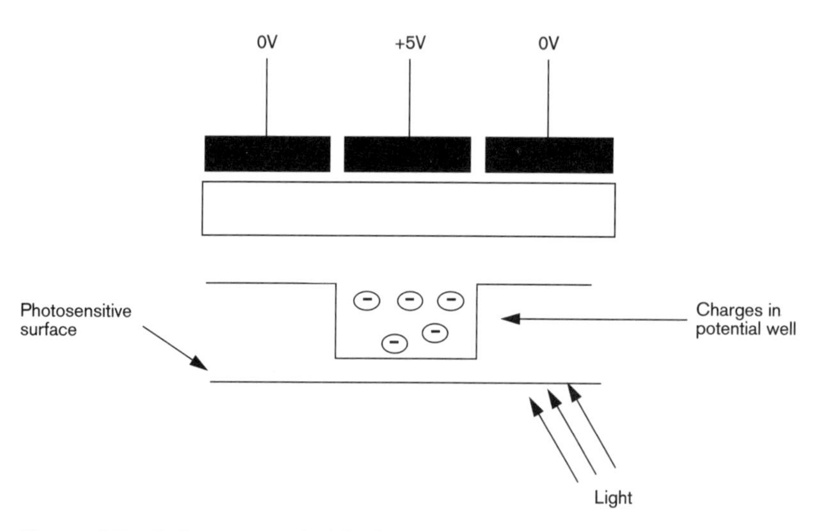

Figure 2.7 A charge-coupled device

A typical CCD consists of a two-dimensional array of these devices. Up to several hundred thousand can be accommodated, one for each pixel. The charges are read out by a pulse signal applied to the electrodes. This is sometimes called a *bucket brigade* device because of the similarity with a line of firemen moving buckets of water. The operation of one row of the CCD is shown in Figure 2.8. In two dimensions, another pulse signal reads the voltages along columns after they are read out from the rows.

Next, we examine some acquisition devices based on the CCD. For use in these devices, the CCDs are usually connected to an A/D converter and timing circuitry allowing the pixels to be read out in digital form at the sampling rate.

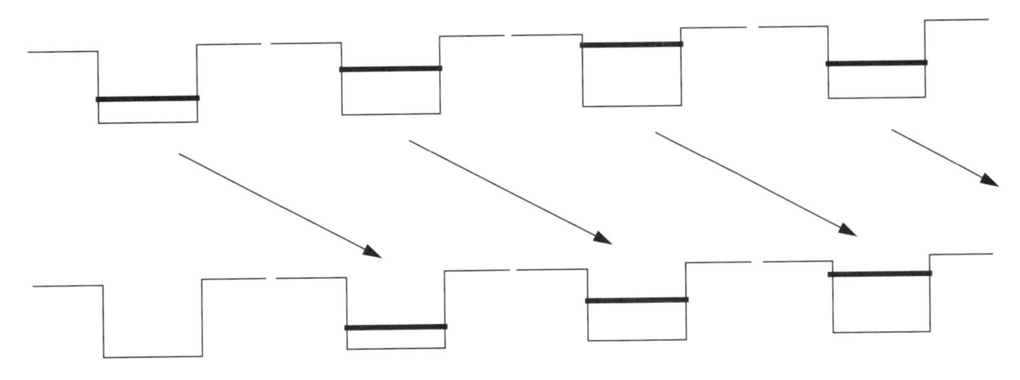

Figure 2.8 Bucket brigade device operation of a CCD array

Scanners A scanner scans a two-dimensional image (for example, a photograph, slide, or negative), turning it into a digital image. Scanners are probably the most common image acquisition devices in use today. They are useful for generating digital versions of photographs that have already been made with a camera, as well as for digitizing line drawings and text. Scanners are used extensively in many areas of imaging applications, including desktop publishing, color prepress, and document imaging.

At its heart, a scanner normally has a one-dimensional CCD strip, which can generate one line of a digital image. Therefore, to create a two-dimensional digital image, each line of the photograph or slide must be projected into this CCD. If color images are to be digitized, then either three passes over the photograph must be made (one with a different filter), or a color CCD composed of three CCD strips must be used.

While there are many different types of scanners, three types have found common usage today. These are the *flatbed* scanner, the *slide* scanner, and the *handheld* scanner.

For the nonprofessional user, whose needs are not exacting, flatbed scanners perhaps offer the best compromise between quality and price. The block diagram of a flatbed scanner is shown in Figure 2.9. The light source moves across the page, and is reflected and captured by a CCD strip.

The second popular type of scanner is the slide scanner. This device also uses a CCD and a moving light source, but it scans a photographic negative or a slide, rather than a print. Light is projected through the negative or slide into the CCD, creating the digital image.

In general, digital images made from slides or negatives tend to have better fidelity than those generated from photographic prints. This is simply because the digital image made from the print is actually a copy of a copy. The occasional user of digital images may find it expensive to purchase a slide scanner; however, a convenient and inexpensive system of transferring slides

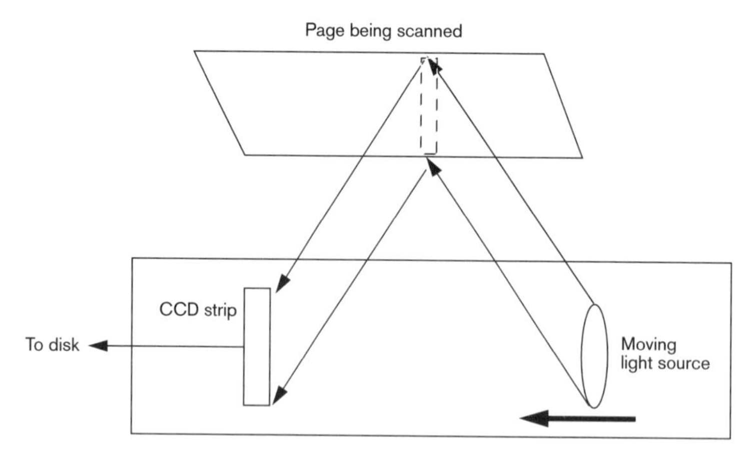

Figure 2.9 Operation of a flatbed scanner

or negatives directly into digital images is offered by the Kodak PhotoCD system, which is discussed later in this chapter.

Handheld scanners are also popular for digitizing images. These are low-cost devices, which must be dragged over the image by the user. In cases where the width of the scanner is smaller than the width of the image, the final digital image may need to be assembled by software.

In addition to the above types of scanners, we should also mention that there are several other more exotic types of scanners. The *drum scanner* yields high-quality images, but is expensive and difficult to operate. The *sheetfed scanner* is useful for document imaging applications where many pages need to be scanned rapidly and scanning speed is of paramount importance. *Overhead scanners* have a camera and light source mounted on top of a scanning bed, creating a bird's-eye view of the object—which can be two- or three-dimensional—being scanned.

Several characteristics of scanners are important to the user. Perhaps the most important is the *resolution* of the digitized image. The scanning control software bundled with the scanner allows the user to select the resolution of the scan. For flatbed and handheld scanners, resolution is specified in dots per inch (dpi) (for example, 600 dpi). The resolution of slide scanners is specified in pixels (for example, 1024 × 1500 pixels). Often one needs to distinguish between the true resolution of the scanner and the maximum resolution as quoted by the manufacturer, since the latter can be based on pixels interpolated between pixels scanned at the true resolution. In other words, the scanned image is zoomed using one of the interpolation algorithms discussed in Chapter 8.

The number of *bits per pixel* is another important characteristic. This determines the dynamic range of the digitized image. Currently available scanners can scan between 4 and 12 bits per pixel.

For color scanners, *color fidelity* is important. That is, the colors of the digitized image must look similar to the original image, without changes in color or added artifacts. Other important considerations are the price, the speed of scanning, maximum size of the paper that can be scanned, and the file formats supported by the associated software.

These scanner types are compared in Table 2.1. Low cost is less than $500, moderate cost is $500–$2,000, and high cost is more than $2,000.

Table 2.1 Comparison of scanner types

Type	Resolution	Image quality	Cost	Example
Handheld	400 dpi	Fair	Low	Logitech Scanman
Flatbed	600 dpi	Good	Moderate	HP Scanjet 4C
Slide scanner	2000 dpi	Excellent	High	Kodak RFS 2035 Plus Film Scanner

A program controlling the function of the scanner is usually in the form of a graphical user interface (GUI) to adjust the scanner controls. The programmer needs to learn the input/output interface that is being used to connect the scanner to the computer. A commonly used interface is the Small Computer Systems Interface (SCSI). Control software for a SCSI scanner translates the user's GUI choices into SCSI instructions that are sent to the scanner. The technical manual for the scanner usually has a description of the I/O interface and how the interface commands are translated into scanner functions.

Digital cameras While scanners are used to digitize images that have already been created by a camera or some other means, digital cameras are used to digitize the image at the time of image creation. Using a CCD coupled with an analog-to-digital converter, digital cameras directly create digital images of the scene imaged by the camera's lens. Thus the intermediate step of developing and printing the film is eliminated.

Use of a digital camera simplifies the sequence of steps in image acquisition. If an application requires the creation of a large number of digital images—for the creation of a merchandise catalog, for example—then digital cameras are a convenient substitute for the more traditional approaches, which require film to be processed before being scanned using a scanner. Digital cameras also simplify some of the systems integration issues of an imaging system. For example, the user can integrate the camera software with image database software. In addition, the need for film and print storage is eliminated.

There are, however, a number of differences between CCD and film that translate into important differences between digital camera images and film camera images. CCD has much lower resolution than film. Hence, the photographer is unable to zoom in on a small part of the image and still retain reasonable detail in the image. In addition, film cameras offer more control over the contrast of the final image. For example, when an image is captured on film, the exposure and film development times can be adjusted to match the dynamic range of the film to the dynamic range of the scene using a technique called the Zone System. In digital cameras, very little control like this is available. In addition, because of the finer grain, images captured on film have a warmer feel, while CCD images have a grittier, colder quality to them [3].

There are many different types of digital cameras on the market. In this chapter, we will be concerned with two main classes of digital cameras: *two-dimensional CCD* and *linear CCD* cameras. The difference between these cameras is that the two-dimensional CCD camera uses a CCD array to generate the image, while the linear CCD camera uses a one-dimensional strip of CCD much like a flatbed scanner.

Two-dimensional CCD cameras come in many price and quality categories. At the low end are cameras resembling instamatic cameras, producing images about the same resolution as TV images. An example of this type of camera is the Apple Quicktake. In the middle range are 35-mm single-lens reflex (SLR) cameras that have been converted to digital cameras. These cameras, including the Kodak DCS200, can take reasonably high-resolution images of up to 6 megabytes per image. At the higher end are cameras such as the Leaf Back camera, which can take even higher resolution images. However, this quality comes at high prices and slow acquisition speeds.

Linear CCD cameras need a motor mechanism to cover the two-dimensional image. Their main advantage is that the resolution can be high because only a one-dimensional CCD strip is needed. However, the scanning speed can be slow, and this type of camera can only be used in a studio situation. These camera types are compared in Table 2.2.

Table 2.2 Comparison of digital cameras

Type	Resolution	Image quality	Speed	Cost	Example
2-d CCD instamatic	Low–medium	Fair	Fast	Low	Apple Quicktake
2-d CCD SLR	Medium–high	Good	Medium–fast	Moderate	Kodak DCS 200
2-d CCD high-end	High	Excellent	Slow	High	Leaf Back
1-d linear CCD	High	Good–excellent	Slow	Medium	Leaf Lumina

While choosing a digital camera, several other considerations need to be made. Primary among these is image storage. Most of these cameras come with some form of on-board memory to store the digitized images. Often the memory is in the form of nonvolatile flash memory in a PCM CIA card format, which can store several images. Obviously, the portability of the camera is largely determined by how many images it can store. Color fidelity as well as the ability to resolve color textures are important qualities of these cameras.

It should be mentioned here that there are also some other, more exotic types of digital cameras on the market. One of these is the Kontron *stepping chip* camera, which uses a two-dimensional CCD that is smaller than the final image size. The image is acquired in tiles, and software is used to mosaic the tiles to create the final image.

Another type of camera, a "smart camera," was recently announced by Sierra Digital Imaging Inc [4]. The novelty of this camera is that a SPARC microprocessor and an image processing

DSP have been integrated into the camera.This allows enhancement, processing, and compression of the images to take place within the camera.

From the programmer's point of view, control software written for digital cameras must use the interface (e.g., SCSI) that is provided in much the same way as scanners. In addition, for smart cameras, software must be written for on-board processors to process the image.

Video digitizers While scanners and digital cameras acquire digital images from two-dimensional pictures and three-dimensional scenes, video digitizers or frame grabbers, which are another source of digital images, acquire them from an analog video input. These devices "grab" a static picture from the moving video picture sequence and turn it into a digital image. Hence they are sometimes called "frame grabbers."

Moving video can be thought of as a sequence of pictures or *frames*. Each frame can, in turn, be composed of two *fields*, whose horizontal *scan lines* interlace each other to make up the frame.

The video signal is actually composed of two types of information: picture signals and timing signals. The picture part of the signal consists of the frame and field brightness (also called *luminance*) information. The timing part consists of synchronization information, including signals to begin a new scan line and a new field or frame.

In color video signals, color information (also called *chrominance*), is mixed in with the luminance. This kind of color video is called *composite video*, to distinguish it from *component video*, where the red, green, and blue channels are separated. The color information is usually represented in the YUV color space, which is discussed in detail in Chapter 7.

For a particular video *standard*, the timing portions and the picture portions are precisely defined. For monochrome video, RS 170 is the prevailing standard. For color video, there are several standards, including the National Television Standards Committee (NTSC) and Phase Alternating Line (PAL). The reader should see reference 5 for a discussion of these standards.

All of this information must be taken into account when a frame grabber is designed. Some important considerations in the design of a frame grabber are the choice between color and monochrome images, the video standards to be supported, and the resolution of the images to be grabbed.

There are several ways to implement a frame grabber. A detailed description of the construction of a monochrome frame grabber is given in reference 6. In this design, the sync signal is essentially stripped from the picture signal. Timing information from the sync signal is then used to drive the clock of an A/D converter, which digitizes the picture information.

A high-level block diagram of a color video digitizer (VideoPix) is shown in Figure 2.10. In this design, the input is composite video, and the output is a digitized image representing the separate color components of the image.

The input signal is first conditioned using some analog circuitry. This is necessary, for example, for low-pass filtering (antialias filtering) the signal to eliminate aliasing. Then, the signal is sampled. The sampled signal is passed into a color decoder chip. This decoder separates the sync signal from the picture information and separates the chrominance (U and V) from the

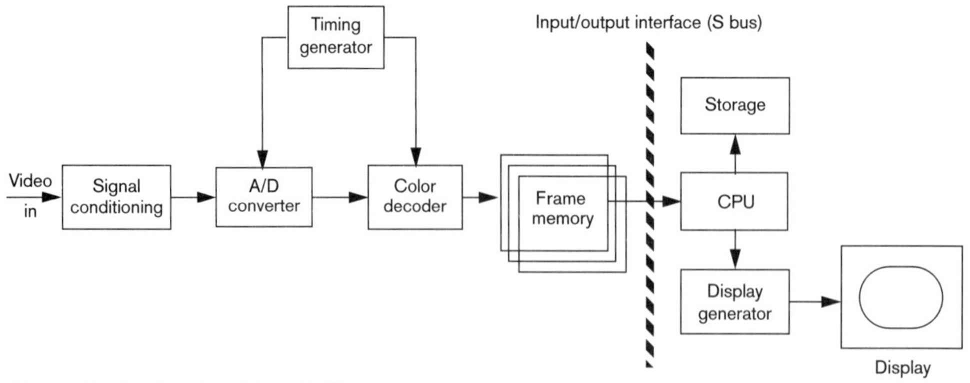

Figure 2.10 A color vIdeo digitizer

luminance (*Y*) information. The *Y*, *U*, and *V* samples are then stored in the buffer memory of the digitizer.

The digitized image often needs to be buffered, since the bandwidth of the I/O bus of the computer may not be sufficient to support a direct data transfer from the A/D converter into the CPU. After the image is transferred, it is stored in the memory hierarchy of the computer or displayed using a display controller. Often, acquisition devices allow direct memory access (DMA) between the buffer memory and the computer's memory.

A logical extension of the video digitizer is a video capture device. This kind of device allows the capture of live video for multimedia applications. Examples of this type of application are video teleconferencing and distance learning. In such cases, the captured video must be transmitted over a transmission channel, for example, a computer network or a digital telephone (ISDN) line. The video data then need to be compressed to fit in the bandwidth available from this channel.

An example of a video capture device is the SunVideo card, which captures, digitizes, and compresses NTSC or PAL video from a variety of analog video sources. A block diagram of this device is shown in Figure 2.11.

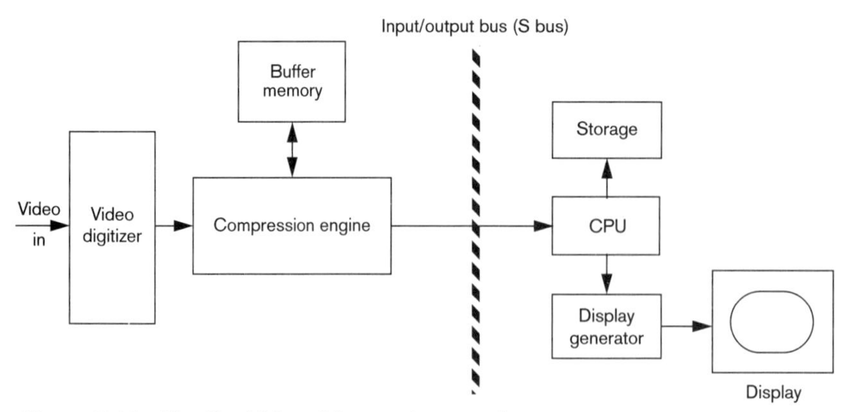

Figure 2.11 The SunVideo video capture card

The input signal into the device is digitized by a video digitizer. This digitizer is very similar in principle to the frame grabber of Figure 2.10. The digitized video stream is then passed to a compression engine (C-Cube CL 4000), which can compress the stream according to a number of compression standards including JPEG, MPEG1, and H.261. (The JPEG and MPEG standards are discussed in detail in Chapter 9.) A buffer memory is needed by the compression engine for storing intermediate results.

Note that in order to display "live video," the SunVideo card must rely on the display subsystem of the computer since it does not have its own display generator. Some video capture cards, such as the Parallax, have their own display subsystem. This enables them to bypass the various system busses and bottlenecks associated with them and display high-quality live video.

Other acquisition methods The above methods of acquiring images are by no means exhaustive. Several specialized areas of imaging employ special techniques for gathering image data. These areas include medical imaging [7], seismic imaging [8], and imaging for materials science [9].

2.3 Image viewing

Having looked at devices that acquire images, we are now ready to look at devices used for viewing digital images. These devices fall into two broad categories: *frame buffer displays* and *hardcopy devices*. Frame buffer displays, which are ubiquitous because they are an integral part of today's desktop computers, are used to display digital images on a computer monitor. Hardcopy devices, which are needed whenever a print copy of a digital image must be made, are used to make copies of the digital image on paper, film, or other media.

2.3.1 Frame buffer displays

Devices that display digital images on a computer monitor are called *frame buffer displays*, or *display generators* (or *display* in short). The frame buffer display converts each pixel of the image into a spot of brightness (color or gray) on the monitor. Along with microprocessors, memory, and hard disks, displays are an integral part of today's desktop computers.

There are several characteristics of frame buffer displays that are important from the perspective of computer imaging. The *resolution* of the frame buffer specifies the maximum number of pixels that can be drawn horizontally and vertically. The *color depth* specifies how many colors can be displayed by each pixel. The *refresh rate* specifies how rapidly the image is drawn on the monitor (faster to avoid flicker). The *amount of memory* available in the frame buffer to store images is also an important consideration.

In general, there is a trade-off between the resolution and the pixel depth. For example, display controllers for document imaging, which require 1 or 4 bits per pixel, but higher resolution, can go up to 2048 × 2048 resolution, whereas ordinary color displays usually go up to resolutions of about 1024 × 768 (in PCs) or 1600 × 1280 (in UNIX workstations).

For displaying color images, there are two main subclasses of frame buffers: *true color* and *pseudocolor*. True color frame buffers rely on separate red, green, and blue components of each pixel. If each component is represented by one byte, then up to 2^{24} colors can be displayed. Pseudocolor frame buffers, on the other hand, can display one of n possible colors where n (typically 16 to 256) is usually much smaller than the number of possible colors (typically 256 to 2^{24}) that may be present in the image. Therefore, pseudocolor frame buffers must use color quantization techniques during image display. Color quantization is discussed in detail in Chapter 7.

Manufacturers of frame buffers for IBM-compatible PCs have defined several standards, based on the above parameters (plus others), to which their frame buffers conform. These standards include VGA (Video Graphics Adapter), SVGA (Super Video Graphics Adapter), and VESA. No such standardization exists on Macintosh or UNIX workstations.

Figure 2.12, based on reference 10, shows a simplified block diagram of the functional parts of a frame buffer display. We will use this diagram to shed more light on the characteristics of these displays.

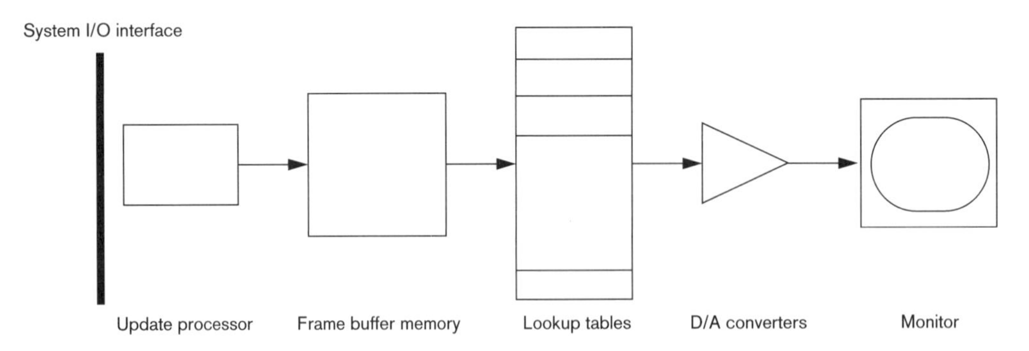

Figure 2.12 Parts of a display device

The frame buffer display is usually connected by an input/output interface (for example, PCI bus, S bus, or VL bus) to the computer system. The device consists of an update processor, frame buffer memory, a set of lookup tables, and a bank of D/A converters.

The update processor is responsible for reading and writing pixels from the frame buffer memory. This is done when the image being displayed is updated, for example, after the completion of an image processing operation. In many frame buffer displays, the update processor can also accelerate some basic imaging and graphics functions. These functions include copying blocks of pixels from one location to another, alpha blending, raster operations (ROP), and various two-dimensional graphics operations. More advanced frame buffer processors can also accelerate some imaging and three-dimensional graphics operations.

The changes made by the update processor often require a type of operation called *read-modify-write*. In this operation, pixels are read from the memory, modified, and then written back. Read-modify-write can be an expensive operation, and often the display system needs hardware acceleration to speed it up.

The frame buffer memory contains all the pixels being displayed. For display, the memory contents are read at a constant rate by the video generator and converted by the video generating circuitry into analog signals driving a monitor. Simultaneously, the memory can be written at an irregular rate by the update processor when new image data are available. Therefore, this memory needs to have an *update port* and a *display port*, making it *dual-ported*. This type of memory is also called Video RAM (VRAM).

The pixels of the frame buffer memory are read by a video generating circuitry. They are first passed through a set of lookup tables. These tables vary between different types and models of frame buffer displays. They are used for various reasons. For example, when the frame buffer is used to display a window system, the lookup tables are used for allocating a certain set of colors for images and reserving others for window system needs. Another example of the use of these tables is in gamma correction (described in detail in Chapter 7), where the nonlinearity of the input-output function of the monitor is compensated for. Finally, for pseudocolor displays, these tables can be used for dithering the color images.

The lookup table output is fed into a bank of A/D converters. These converters generate the voltage signal into component video signals (*R*,*G*, and *B*), which are input to the monitor.

The hardware design of frame buffer displays continue to advance at a rapid rate. Examples of some frame buffers with image processing functionalities can be found in Chapter 3. Frame buffer displays can be programmed by mapping the display (memory, control registers, lookup tables) into the address space of the program.

2.3.2 Hardcopy devices

Hardcopy devices enable the user to make a copy of a digital image on a print medium, such as paper or photograph. Several classes of hardcopy devices are used in computer imaging. These include color printers and film recorders.

Once considered exotic and expensive, color printers are rapidly finding widespread use today. The types of color printers used most commonly today include *thermal transfer*, *dye sublimation*, *inkjet*, and *color laser* printers.

Thermal transfer printers use heat-sensitive paper to transfer digital images onto paper. A printhead containing an array of heaters presses into the paper. The color image is converted into a pattern of tiny dots in which the density of dots is directly proportional to the density of the color, using a technique called *dithering*. A dot that needs to transfer a color is heated, and it presses against a ribbon with ink, which liquefies onto the paper. Several passes are made through the paper, one for each primary color (for example, cyan, magenta, and yellow.) Thermal transfer prints are quick and inexpensive; however, the color quality is often poor.

Dye sublimation printers also rely on heat to transfer digital images to paper. Among all the classes of color printers, this class most closely approaches photographic-quality printing. This is because the dots on the printhead can be heated to varying temperatures, and the temperature determines the amount of ink transferred to the paper. Thus, the output has the appearance of a continuous-tone print. The disadvantage of a dye sublimation printer is that it is expensive to make each print.

Inkjet printers use a high-speed jet of ink to write to the paper. There are two types of inkjet technology: continuous and drop-on-demand [11]. The former uses pressure to build up a high-speed stream of ink droplets, generating over 100,000 drops per second. Each drop must either hit the paper or be captured on a receptacle for recycling. This is accomplished by applying electrostatic charges to the unneeded drops, which are attracted to a charged receptacle. Drop-on-demand technology, on the other hand, creates as many drops as needed for the image, and every drop of ink is used on the paper. Inkjet printers offer an all-round compromise, making medium-quality prints at moderate prices.

A laser printer uses toner—a heat-sensitive, powdered ink—to transfer digital images to paper. Electrostatic attraction is used to deposit the toner on the paper. First, a laser "writes" the image on a positively charged drum inside the printer, converting the image areas into neutral charges. Next, the toner is attracted to the neutral areas by rotating the drum through positively charged toner powder. After this, the drum is rolled over positively charged paper, transferring the appropriate patterns of powder into the paper. Finally, the powder is fused to the paper by passing the paper through hot rollers before it exits the printer. Color laser printers can make reasonable-quality hardcopy images at a price slightly higher than inkjet printers.

While color printers make hardcopy prints that can be used immediately, film recorders make high-resolution copies of digital images on photographic film or slides. These can then be developed and photographic copies of the pictures made.

A film recorder is essentially a video monitor and a camera inside a lightproof housing. The input to the monitor comes from the display controller of the imaging system. In older film recorders, called analog film recorders, the film inside the camera was exposed manually. When the recorder was activated, three exposures, through red, green, and blue filters, were made. In newer recorders, called digital film recorders, mechanical control can be exercised over the way in which pixels are "written" to the film, and very high resolution images (up to 4096×4096) can be downloaded to the recorder and exposed to film or slide. Unfortunately, the price paid for the high quality is high—film recorders run to several thousand dollars. Having looked at various image display devices, we now proceed to look at image storage.

2.4 Image storage

Image storage is an important part of imaging systems because images are large chunks of data. In this section we will examine methods of image storage and discuss a detailed example of image storage using Kodak PhotoCD.

2.4.1 Methods of storage—memory hierarchy

The need to store images may arise during any phase of imaging. For example, the image may need to be stored immediately after digitization, or after some processing, or after it has been displayed. In order to address image storage, we need to understand the memory hierarchy of a computer system.

Figure 2.13 shows a typical memory hierarchy. The position of the memory relative to the processor of the computer is shown for each type of memory.

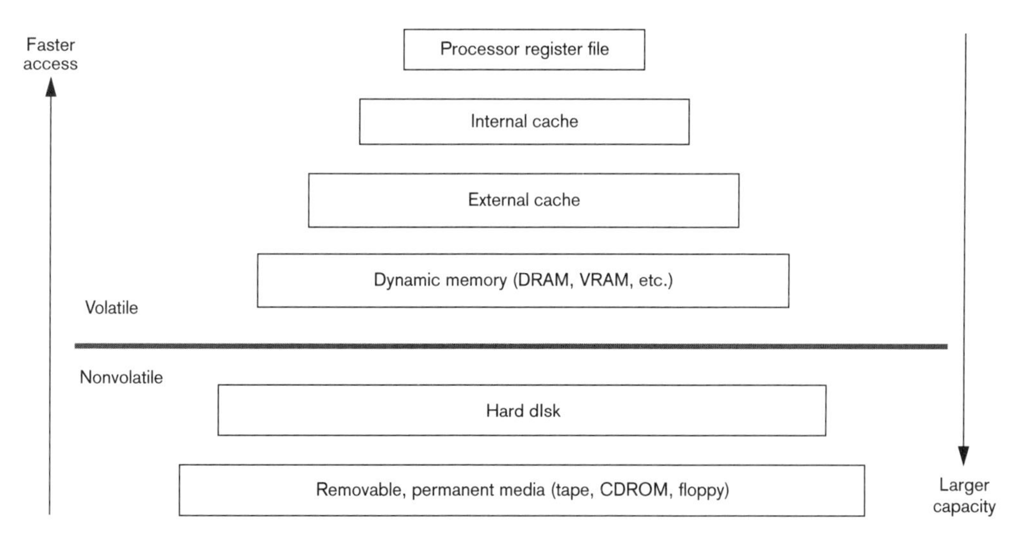

Figure 2.13 Memory hierarchy of a computer system

In principle, the image can be stored in any one of these memory types. The main trade-offs to be considered while making this choice are *access time* (the time it takes to read and write the pixels), *capacity* (the number of pixels that can be stored), *volatility* (the retention of stored data when power is turned off), and *portability* of the storage device.

In the permanent, nonvolatile storage category are several storage devices that can be used to store data. The most important of these is the hard disk, which stores a large amount of data.

Removable devices, such as tapes and CDs, are also available for storing larger amounts of data permanently.

The structure of the image file stored in these devices follows one of several well-known file formats. These file formats, such as GIF, TIFF, JPEG, PhotoCD, and PNG, contain a *header* followed by the image data. The header contains global information about the image such as image size, pixel type, and so forth. Following the header, there may be a lookup table associated with displaying the image data. The final portion of the file contains the image data laid out in some predefined fashion. In some cases, such as JPEG (discussed in Chapter 9), the data are compressed, and significant work must be done to decompress the data. Detailed descriptions of image file formats have been covered in great detail in a number of texts, including reference 12, and are therefore not covered in this book.

Once a program begins to execute, much of it is placed from the hard disk into *main program memory*, which is usually fast dynamic RAM (DRAM). Images are read into the DRAM from a more permanent storage medium or from a digitizer source. In case the image or program is larger than available main memory, relevant parts are swapped in and out by the operating system.

The next fastest access time is provided by the internal and external *caches* of the processor. The idea behind caches is the principle of *locality of reference*, which states that if a certain memory location is accessed by a program, then the probability that the memory close to that location will also be accessed is high. Therefore, performance can be improved by loading the internal cache with the contents of a particular memory location and its neighbors.

The fastest access time is available from the *register file* of the processor. From the programmer's point of view, pixels are read into the register file when they are going to be processed. The number of registers available in most processors is extremely limited.

In Figure 2.14 we show the flow of a typical imaging program and data movement between different types of storage.

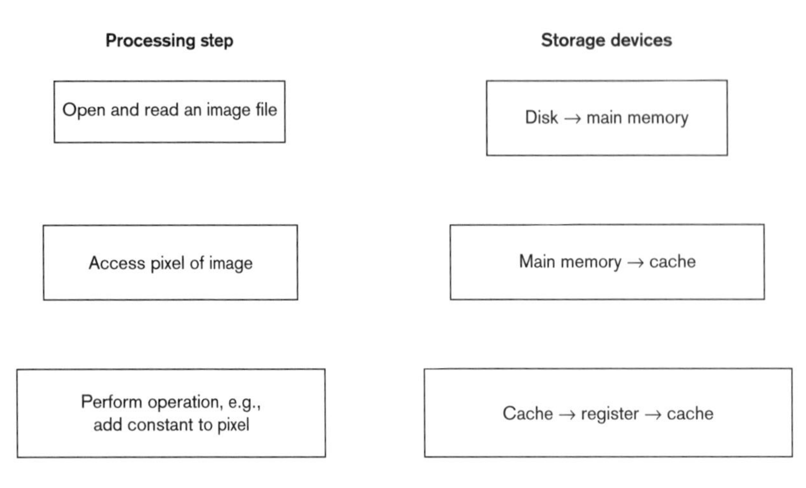

Figure 2.14 Uses of different types of image memory

When an image data file is opened and read from the program, the data are moved between the nonvolatile device, such as the hard disk or the floppy drive, into the program memory. Alternatively, in a real-time acquisition system, the data could be fed into main memory directly from the acquisition device through DMA.

At some point after this, parts of the image may be loaded into the cache system of the computer. This happens when the program references a memory location corresponding to a pixel in the image. When an operation is actually performed on this pixel, the data are loaded into the processor register file from cache and written back into cache after the operation is done.

Design of memory systems has been studied widely, and is outside the scope of this book. See, for example, reference 13 for details. In this section, we shall proceed to look at an example of image storage, dedicated for commercial imaging, which has made a significant impact on the prevalent use of digital imaging. This is Kodak PhotoCD.

2.4.2 Kodak PhotoCD—a system for storing photographic-quality images

The PhotoCD system was introduced by Eastman Kodak Company in 1991. Its goal is to enable the acquisition and storage of photographic-quality color images in a convenient format that is usable across multiple platforms in a variety of applications. PhotoCD makes high-quality digital images more accessible to the everyday computer user.

The scenario for using PhotoCD is like this: The photographer takes his or her pictures using color film. When dropping off the film at the processing laboratory, the photographer requests that the images be put into a PhotoCD. The laboratory then processes the film in the usual way. To place the negatives in a PhotoCD, a PhotoCD workstation is used to scan the negatives using a high resolution scanner. The digital image is then written out into a write-once-read-many (WORM) compact disk. This disk can now be read in any CDROM drive using the appropriate software.

There are several varieties of PhotoCD disks available to the user. The most common is the PhotoCD Master disk, which stores approximately 100 images from 35-mm negatives or slides. Other formats include the Pro PhotoCD Master, which can incorporate higher-resolution images for 120 and 4 × 5 film, and the Portfolio and Catalog formats, which trade off resolution for number of images.

The ImagePac The basic unit of image storage in a PhotoCD is called an *ImagePac*. Images are stored in a PhotoCD in several resolutions, which together make up the ImagePac. The scanning resolution is at 2048 × 3072 pixels and this is called the *16Base* or *Base*16* image. The other resolutions, shown in Table 2.3 [14], make up the rest of the ImagePac files.

Table 2.3 ImagePac resolution of images

ImagePac component	Resolution	Example use
Base/16	128 × 192	Index images, thumbnail
Base/4	256 × 384	Image browsing
Base	512 × 768	Monitor viewing
4Base	1024 × 1536	HDTV, small prints
16Base	2048 × 3072	Hardcopy prints
64Base	4096 × 6144	Large proofs and prints

The details of the actual file format used in a PhotoCD are proprietary to Kodak. However, manufacturers that wish to integrate support for PhotoCD technology in their product can license this technology from Kodak.

From the image processing point of view, there are two items of interest in PhotoCD technology. These are the color spaces and the compression scheme.

Compression in PhotoCD In a PhotoCD Master disk, every image is represented in five resolutions from Base/16 to 16Base. The initial scan is a 16Base resolution. The 16Base is decimated into 4Base using the algorithm shown in Figure 2.15. The image is subsampled in horizontal and vertical directions by 2×; this is the 4Base image. The subsampled image is then zoomed back up to the original resolution using an interpolated zoom, and the error between the original image and the zoomed image is computed.

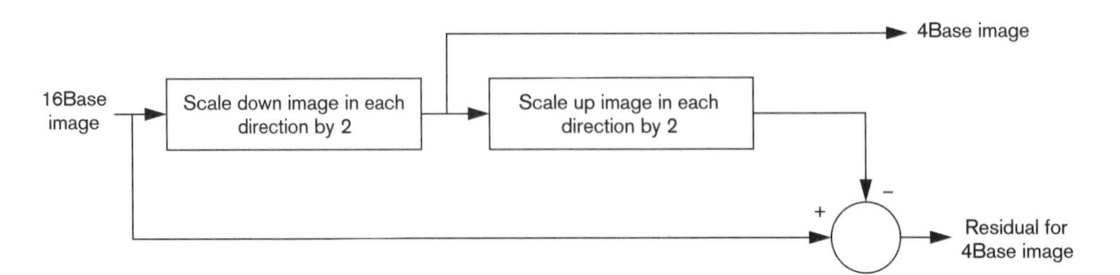

Figure 2.15 The compression of 16Base image to 4Base image + residue

The 4Base is decimated into Base resolution using the same algorithm. When the Base image is decimated into Base/4 and Base/16, the residue image is not computed.

Using this scheme, the average ImagePac size is about 4.5 MB (compared to a straight storage requirement of 18 MB for R, G, and B components of the 2048 × 3072 scan). Also, this format allows the user ready access to lower resolutions.

In summary, the PhotoCD image storage system gives an inexpensive way to scan high-quality digital images for the everyday user. This technology enables and encourages the continued usage of silver-halide photographic material in the digital age.

Color in PhotoCD The images created on a PhotoCD have beautiful color. This is because a proprietary color space, developed specifically for PhotoCD, is used. This color space, called PhotoYCC, is described in detail in Chapter 7. After being scanned in RGB color space, the image is converted into the PhotoYCC color space before being stored.

2.5 *Conclusion and further reading*

In this chapter we have examined several devices used in the various stages of image processing, including image acquisition, storage, and display devices. In the next chapter we will examine devices dedicated to pixel computation for image processing.

Since the technology described in this section changes rapidly, the best source for information is manufacturers' catalogs and trade journals. To probe further on how to build an acquisition device, the reader should see reference 6. For a discussion of image display, the reader should see reference 10. In reference 11 there is a discussion of various printer technologies.

2.6 *References*

1 A. V. Oppenheim and R. Schafer, *Digital Signal Processing*, Englewood Cliffs, NJ: Prentice Hall, 1975.

2 A. K. Jain, *Fundamentals of Digital Image Processing*, Englewood Cliffs, NJ: Prentice Hall, 1985.

3 P. Siegel, "In Search of the Perfect Digital Camera," *Imaging*, March 1995.

4 "RISC Gets Shot in Digital Camera," *Electronic Engineering Times*, March 6, 1995.

5 *Color Television*, D. Fink, ed., New York: Society of Motion Picture and Television Engineers, 1970.

6 C. Lindley, *Practical Image Processing in C*, Englewood Cliffs, NJ: Prentice Hall, 1983.

7 A. Macovski, *Medical Imaging Systems,* Englewood Cliffs, NJ: Prentice Hall, 1983.

8 Robinson and Treitel, *Geophysical Signal Analysis*, Englewood Cliffs, NJ: Prentice Hall, 1980.

9 J. C. Russ, *Image Processing Handbook*, Boca Raton, FL: CRC Press, 1992.

10 R. F. Sproull, "Frame Buffer Display Architectures," *Tutorial: Computer Graphics Hardware: Image Generation and Display*, Reghbati and Lee, eds., Los Alamitos, CA: IEEE Computer Society Press.

11 L. G. Thorell and W. G. Smith, *Using Computer Color Effectively*, Englewood-Cliffs, NJ: Prentice Hall, 1990.

12 C. W. Brown and B. Shepherd, *Graphics File Formats*, Greenwich, CT: Manning, 1995. J. D. Murray, *Encyclopedia of Graphics File Formats*, Sebastopol, CA: O'Reilly and Associates, Inc., 1994.

13 J. Hennessy and D. Patterson, *Computer Architecture: A Quantitative Approach*, San Mateo, CA: Morgan Kaufmann, 1988.

14 G. Baxes, *Digital Image Processing*, New York: Wiley, 1995.

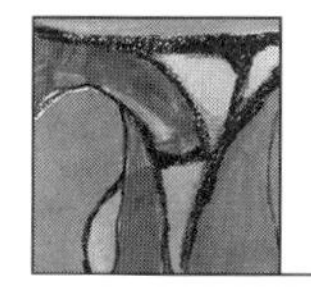

chapter 3

Imaging devices II: the processing engine

3.1 Introduction

In the previous chapter we looked at three components of an imaging system: image acquisition, image display, and image storage. The pixel computation or processing engine is the fourth component of the system. The processing engine is responsible for the computation of all pixels, and thus the execution of all the algorithms described in Chapters 5 through 9 of this book. In this chapter we will discuss various processing engines used for pixel computation.

The processing engine can be categorized into two broad classes: *special-purpose imaging hardware* and the *general-purpose microprocessor.* Until the late 1980s, dedicated hardware (which was often expensive) was the only choice available to the user for whom speed of processing was important. However, recent years have seen faster microprocessors, which have enabled the user to perform many imaging functions on the desktop computer. At the same time, the cost of specialized imaging hardware has decreased, while functionality and performance have improved. Therefore, while the general-purpose microprocessor has appropriated some of the more traditional imaging tasks from specialized hardware, other markets have been opened up by new and improved hardware solutions to imaging problems.

In this chapter we will present pixel computation issues, which we hope will be relevant to the user and designer of imaging systems. To provide a perspective on the evolution of imaging hardware, we examine several representative samples from different periods. While we cannot possibly do an exhaustive survey of all the pixel computation devices built for image processing, we hope to touch on some important trends, thus establishing a frame of reference for the reader interested in the capabilities of processing engines.

We also take a detailed look at two programmable imaging devices: the SX and the Visual Instruction Set (VIS). It is hoped that examining the steps involved in programming these devices will help the reader who is faced with the task of designing similar (or more advanced) devices or of programming them.

The discussion of specific hardware devices in a book can be risky because they change so rapidly; however, our goal of observing trends and understanding the complexities of designing imaging hardware will be well-served by these examples.

We continue in Section 3.2 with a look at special-purpose imaging hardware. This discussion includes dedicated imaging hardware and programmable imaging hardware, as well as imaging acceleration on the frame buffer. Section 3.3 is concerned with the use of general-purpose microprocessors for imaging. In Sections 3.4 and 3.5 we will take a detailed look at programmable pixel computation using the SX and the VIS.

3.2 *Special-purpose imaging hardware*

Since imaging is so compute-intensive, numerous types of hardware have been designed to accelerate imaging. Many such experimental hardware processors have been reported in the academic world [1]. In the late 1970s and early 1980s, commercially available hardware included stand-alone image processors that executed under the control of a host computer. Examples of these are the image processors from Comtal [2] and Vicom [3]. During the mid-1980s, image processing boards, which plug into bus slots (such as the AT bus of IBM-compatible PCs or the VME bus of UNIX workstations) became popular. In the 1990s, two distinct trends can be discerned: cheaper and faster ASICs (and thus add-on cards), which incorporate useful image processing and related functions, and the integration of acceleration capabilities into faster microprocessors and computer systems enabling them to do many general-purpose imaging and video functions at a high speed.

In this section, we discuss several examples of specialized imaging hardware. We have distinguished between three classes of imaging hardware: *dedicated*, *programmable*, and *frame buffer* hardware.

Dedicated hardware is designed to perform specific imaging operations. These operations can be convolution, lookup table and other point operations, morphological operations, and statistical data extraction (such as histogram) from images, as well as related video operations such as JPEG or MPEG compression and decompression.

Programmable hardware, as the name suggests, can be programmed to perform a potentially wider variety of imaging operations. The advantage of programmable hardware lies in its flexibility. Usually, more imaging operations can be implemented on programmable hardware than on dedicated hardware. Dedicated hardware, on the other hand, tends to be faster, and easier to program.

When an arbitrarily complex image processing operation—that is, one requiring a sequence of primitive operations on the image—is being performed, programmable hardware has a distinct advantage over dedicated hardware solutions. It can read the required pixels into its registers, perform the entire operation on the pixels, and then write out the results. Dedicated hardware, on the other hand, requires that the complex operation be decomposed in terms of the primitive operations that the hardware can accelerate. Then, each primitive operation must be executed separately on the image, adding significantly to the overhead. A comparison of the two types of approach is shown in Table 3.1.

Table 3.1 Comparison of the pros and cons of dedicated and programmable imaging hardware

Dedicated hardware	Programmable hardware
Usually faster	Usually slower
Easier to program	Large programming effort
Smaller set of functions	Large set of functions

Table 3.1 Comparison of the pros and cons of dedicated and programmable imaging hardware (continued)

Dedicated hardware	Programmable hardware
Composite operations need multiple passes over image or multiple hardware pieces	Composite operations can be programmed in a single pass over image
More expensive	Less costly than dedicated hardware

The third class of image processing hardware we examine is the frame buffer. The basic functionality of the frame buffer—to display an image—has been discussed in Chapter 2. In this chapter we will examine the image processing capabilities of some frame buffers.

3.2.1 Dedicated imaging hardware

In this section we examine several dedicated image processors from the past and the present. These include the Vicom, an image processing system of the 1980s; two image processing boards from Data Translation and Imaging Technology, Inc.; and an ASIC chip to assist in simple imaging tasks from Media Computer Technologies.

Vicom image processor The Vicom was a member of the earliest breed of stand-alone image processors. It was designed to implement convolution and point operations at real-time rates. The key innovation in the design of the Vicom was the use of hardware multipliers to perform 3×3 convolution.

A block diagram of the Vicom is shown in Figure 3.1. The pipeline controller effectively controlled the flow of the pixels through the processors. The neighborhood processor was mainly used for performing convolution; it had nine multipliers needed for a 3×3 convolution. The point processor had an ALU and a lookup table for performing arithmetic operations on the image. The processors, controller, and image memory were connected by a high-speed image bus.

The user interacted with the Vicom by typing commands from the keyboard, which were sent to the pipeline controller from the 68000 host computer. A command-line interpreter was provided for performing basic image processing operations. An extensive library of image processing functions was also provided for users who needed to write application programs.

The main advantage of the Vicom was being able to perform 3×3 convolutions on a 512×512, 16-bit image at 30 frames per second. In addition, using a Singular Value Decomposition/Small Generating Kernel algorithm, which is discussed in Chapter 6, larger convolution kernels could be decomposed in terms of a sequence of 3×3 convolutions. For example, a 5×5 convolution kernel could be decomposed in terms of two (or four) 3×3 kernels depending on the number of singular values the 5×5 kernel had. This enabled the Vicom to perform convolution with large kernels at a high speed.

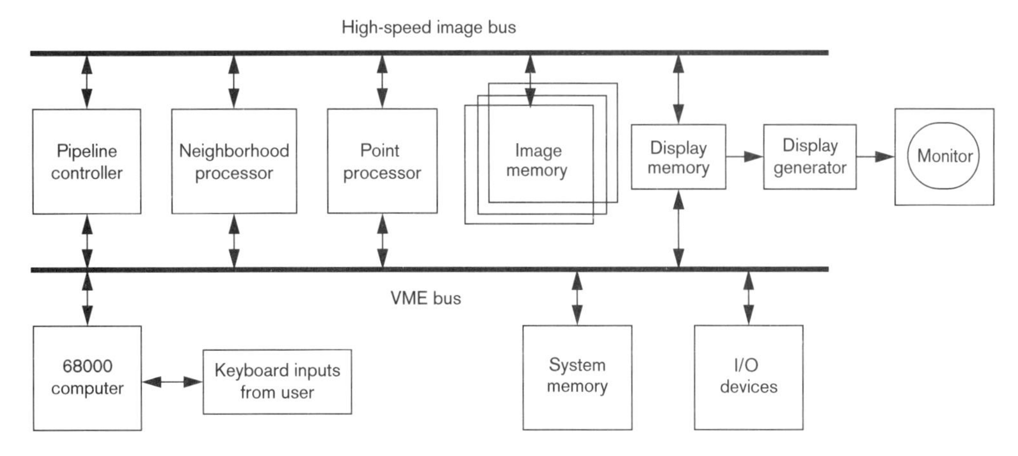

Figure 3.1 Block diagram of the Vicom hardware image processor

The speed of other accelerated operations, such as monadic and dyadic functions, and lookup table operations was also 30 frames per second. Imaging functions that could not gain any benefit from the hardware acceleration were executed on the 68000-based computer at significantly slower speeds.

Table-driven morphology was executed as a composite operation of convolution followed by a table lookup (Chapter 9). It ran at approximately 15 frames per second. The Sobel edge detector and Wallis statistical differencing operator (for enhancing edges) were also cast as a sequence of primitive operations and ran at five to ten frames per second.

While its ability to perform convolution at 30 frames per second was revolutionary for its time, the main drawback of the Vicom was its cost and size. It was eventually overshadowed by cheaper and faster image processing boards and chips. Some Vicom performance results, based on 512 × 512 images, are tabulated in Table 3.2. Note that Vicom pixels were 16-bits wide.

Table 3.2 Vicom performance

Function	Megapixels/sec
3 × 3 convolution	7.864
5 × 5 convolution	1.966–3.932
Copy	7.864
Table lookup	7.864
Add, subtract	7.864
3 × 3 median (unaccelerated)	~ 0.0524

Data Translation Model 2868 image processor The Vicom image processor was an earlier stand-alone machine using a Motorola 68000 microprocessor as an on-board computer; later (VME) versions used a Sun workstation as a host. The Data Translation Image Processor [4] introduced in the late 1980s, on the other hand, was an AT bus card, which worked in conjunction with frame grabbers made by Data Translation. The intention was to provide low-cost acceleration on the PC for the most basic imaging functions.

In this processor, internal pixel precision was maintained at 16 bits. The board consisted of a 512 × 512 × 16-bit image memory, an ALU, a histogram generator, and conversion tables to convert from 8-bit to 16-bit pixels. A block diagram is shown in Figure 3.2.

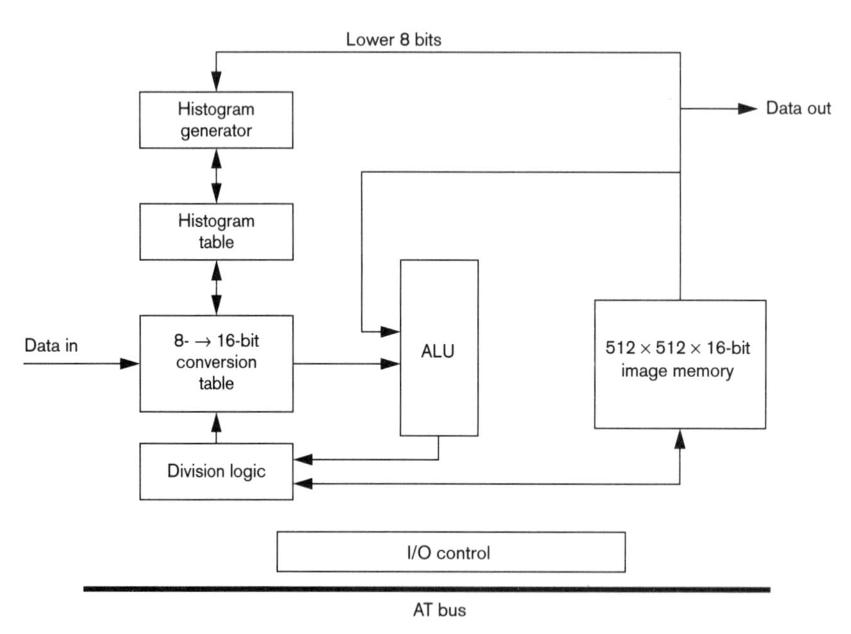

Figure 3.2 Data Translation Model 2868 processor

The ALU executed at 10 MHz, performing one addition or multiplication per cycle. It performed convolution, histograms, monadic and dyadic operations, and pan and zoom functions. The performance of the 2868 is shown in Table 3.3.

Table 3.3 Data Translation 2868 performance

Function	Megapixels/sec
Histogram	2.620
3 × 3 convolution	0.397
Add two images	2.016
Average two images	0.690
7 × 7 convolution	0.133

The main strengths of the 2868 were its ability to provide moderate acceleration of imaging functions, and its compatibility, through the DT-Connect I/O interface, with the rest of the Data Translation offerings, which included various acquisition devices. For its day, the performance of the product was quite respectable; however, this level of performance can be achieved by many microprocessors today.

Imaging Technology 150/40 imaging modules The Imaging Technology image processing modules [5], which can plug into VME, PCI, or VL bus slots (and into S bus and AT bus slots through bus translators) are designed for image analysis and machine vision applications such as industrial inspection and optical character recognition. The series consists of a set of mix-and-match modules, which include an Image Manager (IM), including a frame grabber and a display unit, and seven Computational Modules for image processing.

The Computational Modules are: the Convolver/Arithmetic Logic Unit (CM-CLU), Programmable Accelerator (CM-PA), Histogram/Feature Extraction Processor (CM-HF), Median and Morphological Processor (CM-MMP), Binary Correlator (CM-BC), and Memory Expansion (CM-ME). One of these can be attached to the Image Manager as a daughterboard. Thus, a basic configuration takes up one VME slot.

For compute-intensive applications where multiple Computational Modules are required, a Computational Module Controller (CM-CMC) is available to host up to three more CM-xx's. The basic and compute-intensive configurations are shown in Figure 3.3.

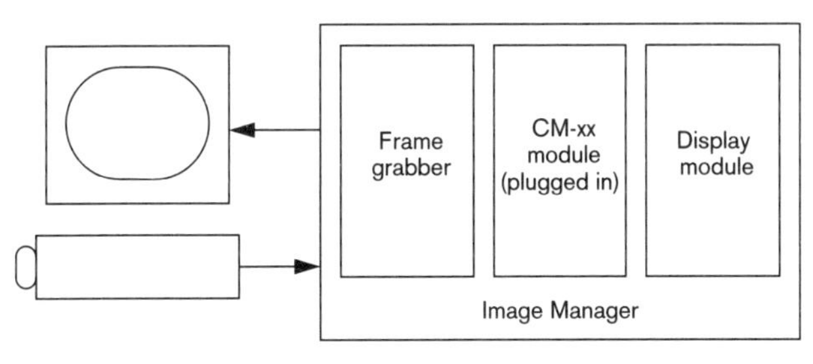

a) Basic configuration

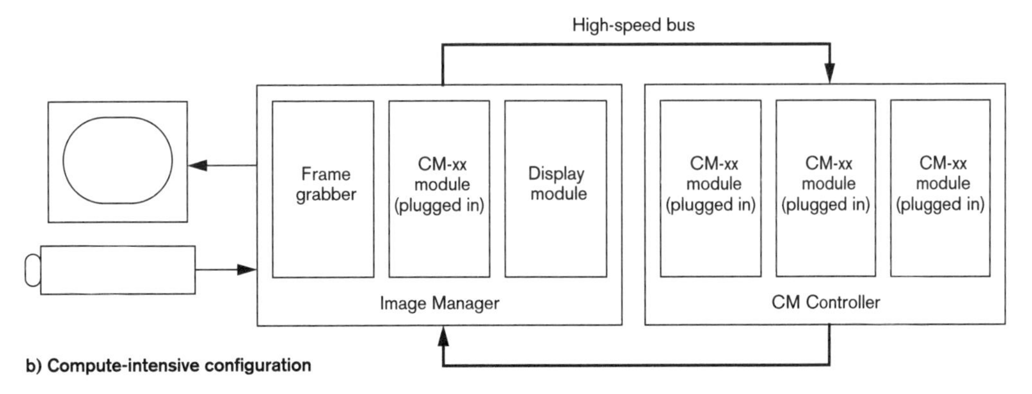

b) Compute-intensive configuration

Figure 3.3 Imaging Technology 150/40 modules

All the processing modules operate at 40 MHz. Some performance numbers are listed in Table 3.4 for 8-bit pixels. The main advantage of the IT 150/40 hardware is the impressive performance; however, the relatively high cost is a disadvantage.

Table 3.4 Imaging Technology 150/40 performance

Function	Megapixels/sec
4 × 4 convolution	34.950 (CM-CLU)
Dyadic	34.950 (CM-CLU)
Histogram	34.950 (CM-HF)
Binary correlation	17.475 (CM-BC)
Gray-scale erosion/dilation	34.950 (CM-MMP)

MCT Video Manager+ chip The Video Manager+ (VM+) chip from Media Computer Technology is a recent (1994) example of the use of imaging in low-end desktop computers. It shows the increasing usage of imaging and video on the consumer computer. The chip is intended for use in digital video PC boards, which provide desktop video display and capture for the IBM-PC multimedia and video applications market.

The VM+ provides a convenient interface between a digital video input and the computer display. It has built-in YUV-to-RGB color conversion, pixel-replicated and interpolated zooming, and mirroring (useful for rear-view projection), as well as a variety of video controls such as cropping and color keying. It is also able to handle chromatic subsampling ratios of 4:1:1 and 2:1:1 in the YUV space. A functional block diagram of the VM+ is shown in Figure 3.4.

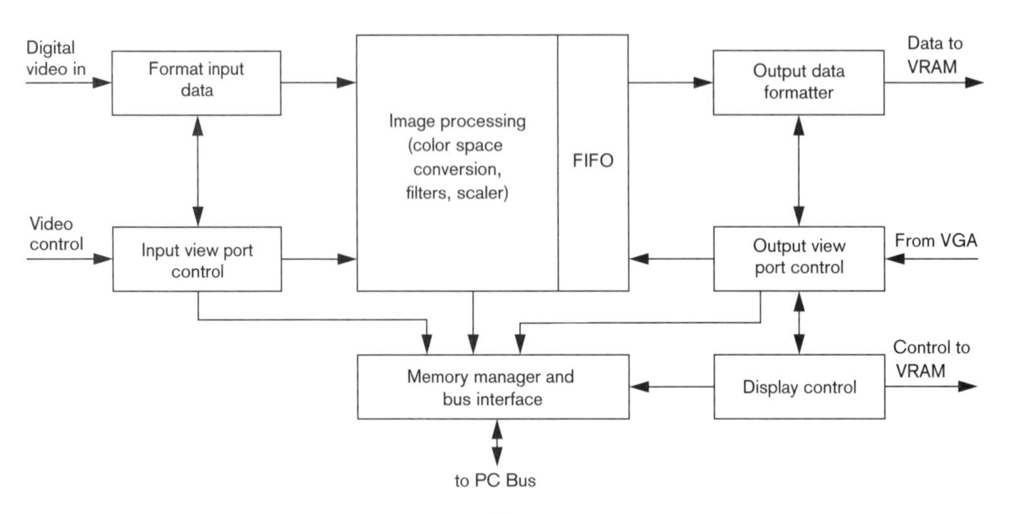

Figure 3.4 Functional block diagram of the VM+ chip

The block diagram of a possible digital video card using the Video Manager is shown in Figure 3.5.

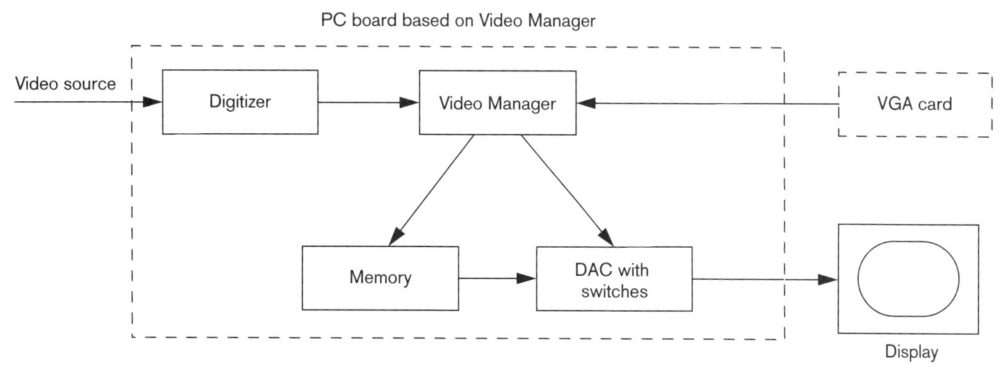

Figure 3.5 Example application of Video Manager+ chip

The card is capable of receiving video from several sources. It digitizes the video, performs any of the functions listed above, and then displays it, mixed with the VGA signal.

The main advantage of the VM+ is the very low cost. This is necessary for the consumer market, where boards based on chips such as this are targeted. A necessary consequence of this cost minimization approach is that imaging functionality is kept to an absolute minimum.

3.2.2 Programmable imaging hardware

Having looked at some samples of dedicated imaging hardware, we are now ready to look at programmable imaging hardware. Many such devices have been built and sold commercially; we have picked an early sample (the TAAC-1 accelerator) and the more recent Texas Instruments MVP chip for discussion here.

TAAC-1 application accelerator The goal of the TAAC-1 application accelerator [6], introduced in 1988, was to speed up imaging, 3-D graphics, and volume visualization programs. These areas were, at that time, not the strong point of desktop RISC or CISC computers. Designed to be a programmable accelerator, the TAAC-1 was a set of two VME bus cards that plugged into the cabinet of deskside Sun workstations. The processor was based on a Very Long Instruction Word (VLIW) instruction set. Each instruction was 200 bits wide. Therefore, in one cycle, several operations could be carried out. The TAAC-1 block diagram is shown in Figure 3.6.

Because of its programmability, the TAAC-1 was able to perform a large variety of imaging and graphics functions, including convolution, Fast Fourier Transforms, and adaptive histogram equalization. In addition, the TAAC-1 accelerated various three-dimensional graphics and volume visualization operations.

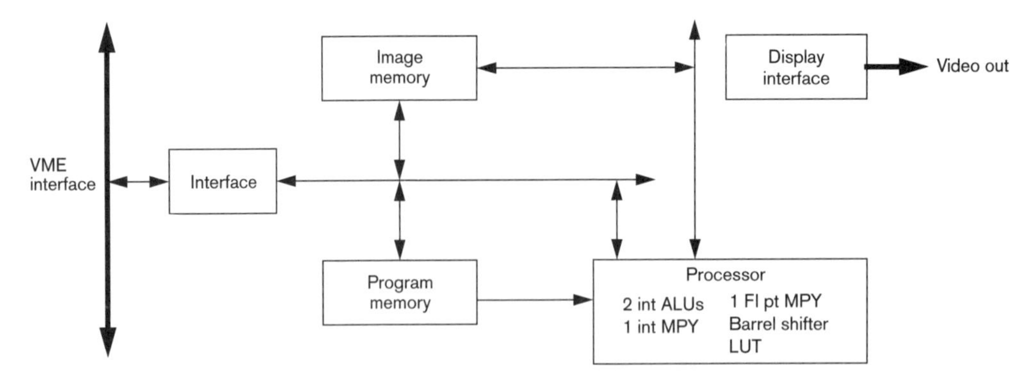

Figure 3.6 Block diagram of the TAAC-1 application accelerator

The user interacted with the TAAC-1 in one of two ways. The simpler method was to use the program *tademo*, an application program with a graphical user interface that enabled the user to exercise the TAAC-1's built-in image processing and graphics libraries.

For the more demanding user, it was also possible to write C code or TAAC-1 microcode to add new functions to this library, which could then be invoked from an application program. The program development environment for the TAAC-1 included a C compiler which generated the appropriate microcode.

The performance of the TAAC-1 for various imaging functions, on 512×512 images, is shown in Table 3.5 for 8-bit pixels.

Table 3.5 TAAC-1 performance

Function	Megapixels/sec
Copy	24.000
3×3 convolution	0.655
2D Fl pt FFT	0.256
Adaptive histogram eq.	0.058

The ability of the TAAC-1 in all areas of pixel rendering was very impressive for its time. The main disadvantage was the difficulty of programming; in particular, writing optimal TAAC-1 code required understanding of the datapaths of the TAAC-1 and filling the slots of each 200-bit VLIW instruction correctly.

Texas Instruments MVP chip The Multimedia Video Processor (MVP) chip from Texas Instruments [7] was designed to support multimedia processing needs on desktop computers. These needs include video compression and decompression, general image processing, document image processing, and graphics processing, as well as some generic floating-point functions. In addition, the MVP anticipates the integration of image analysis functions, such as object recognition, into the mainstream imaging marketplace.

Because of the diversity of these applications—and also because some of these applications are still emerging and may undergo changes—a programmable chip was called for. The sheer computation-demands (one example is MPEG 1 compression [Chapter 9], which requires about 1.2 billion operations per second) necessitated the use of parallel processors.

A block diagram of the MVP is shown in the Figure 3.7. The parallel processors perform integer operations on 32-bit integers. The master processor divides the work among the parallel processors and performs floating-point operations.

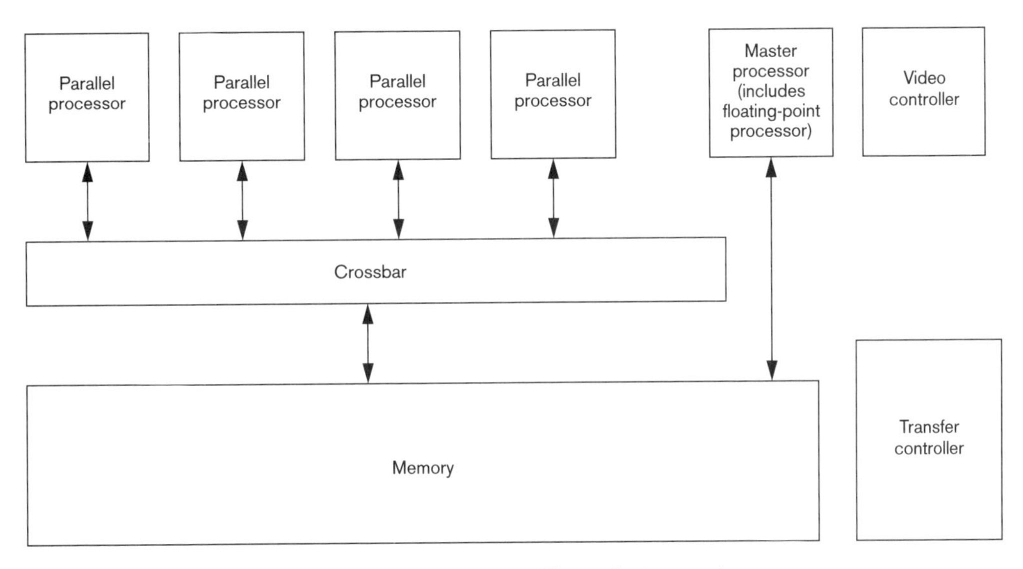

Figure 3.7 Block diagram of the MVP chip from Texas Instruments

The MVP chip is claimed to support 2 billion operations per second. A multimedia accelerator board based on the MVP, announced by Precision Digital Images [8], claims the performance numbers for 8-bit pixels shown in Table 3.6.

Table 3.6 MVP system (PDI) performance

Imaging function	Megapixels/sec
Lossless compression	26.20
Image registration	6.55
DCT	43.60
3 × 3 convolution	13.80
3 × 3 median	21.80
2nd order warp	13.10
Histogram	87.40

The main advantages of the MVP are its rich instruction set and the high performance that can be obtained by its inherent parallelism. The main disadvantage is that it is difficult to program. Despite a comprehensive program development environment, it takes much programming effort to optimize programs for maximum performance.

Before leaving programmable pixel processors, it should be noted that there are several other programmable processors on the market. For example, the wide variety of digital signal processing (DSP) chips fall into this category, as do some of the neural network chips. One processor of interest is the CNAPS processor, based on neural network principles, from Adaptive Solutions, Inc. The SX and the VIS, which are studied in detail in Sections 3.3 and 3.4, are also examples of programmable processors.

3.2.3 Imaging on the frame buffer

In Chapter 2, we examined the design of frame buffer displays, which are used to display images. In addition to their display capabilities, frame buffers—particularly those on UNIX computers—usually have some hardware acceleration for speeding up selected 2-D and 3-D graphics functions. This capability has been used for accelerating the performance of window systems as well as 3-D graphics applications such as texture mapping. Reflecting the growing importance of desktop imaging, frame buffer acceleration in recent years has been extended to include some imaging functions. In this section, we examine three frame buffers which are also useful as pixel processing engines for certain imaging functions.

Silicon Graphics Impact The Silicon Graphics Impact is a frame buffer accelerator designed for SGI workstations. It includes acceleration for various 2-D and 3-D graphics functions as well as some imaging functions. The Impact is a scalable device which comes in two flavors: High Impact and Maximum Impact. Maximum Impact is essentially twice as fast as High Impact.

The key to the Impact is accelerated texture mapping. This functionality also enables various image geometric operations discussed in Chapter 8. In addition, the Impact is also capable of performing 3×3 and 5×5 convolutions, as well as color lookup table functions.

Some performance figures for imaging functions reported for the High Impact frame buffer are shown in Table 3.7.

Table 3.7 High Impact performance (imaging, 8-bit pixels)

Imaging function	Megapixels/sec
3×3 convolution (general)	20
3×3 convolution (separable)	30
5×5 convolution (separable)	20
Color lookup	50

The main advantage of the Impact is its performance (particularly related to geometrical processing). The main disadvantage is a high price and the lack of programmability.

Hewlett Packard Image Accelerator The Hewlett Packard Image Accelerator is dedicated exclusively to imaging functions. Based on a PA-RISC core, this accelerator is a small card that plugs into a frame buffer card. This accelerator is capable of performing 3 × 3 convolution, pan and zoom, rotate and window and level (table lookup) functions. The interpolation scheme used in rotate and zoom is bicubic interpolation (see Chapter 8).

A key feature of this processor is the pipelining of all the above functionalities. Thus any combination of the primitive functions that it is able to accelerate runs at the speed of the slowest primitive function. The performance figures reported for this accelerator are shown in Table 3.8.

Table 3.8 HP Imaging Accelerator performance (on 715/80 workstation, 8-bit pixels)

Imaging function	Megapixels/sec
3 × 3 convolution (general)	13
Pan and zoom	42
Rotate	35
Table lookup (16 → 8)	15
All simultaneously	13

The main advantage of this accelerator is in the pipelining. Therefore, applications that require all of the above functions will benefit most from it. The disadvantage is that it is restricted to gray-scale images, and similar performance can be achieved by other microprocessors without add-on hardware.

Sun Microsystems Creator The Creator is a general-purpose frame buffer, which has been designed to accelerate graphics and imaging functions during the display of the image. A key innovation of the Creator is a special type of video RAM called 3DRAM [9], which integrates some processing capabilities within the memory itself. The functions accelerated include alpha blending, z-buffering, raster operations, and depth cuing.

A block diagram of the Creator is shown in Figure 3.8. When this diagram is compared with the generic frame buffer shown in Figure 2.12, we can see that much of the capabilities of the Update Processor is actually absorbed in the 3DRAM itself.

The imaging functions performed by the Creator hardware include alpha blending and fill. In addition, some of its 3-D graphics primitive functions can also be used for imaging. For example, the depth cuing function is used for performing a straight-line rescaling function.

The Creator plugs into the Ultra series of workstations, which are based on the UltraSPARC processor. This processor includes imaging acceleration using the Visual Instruction Set (VIS)

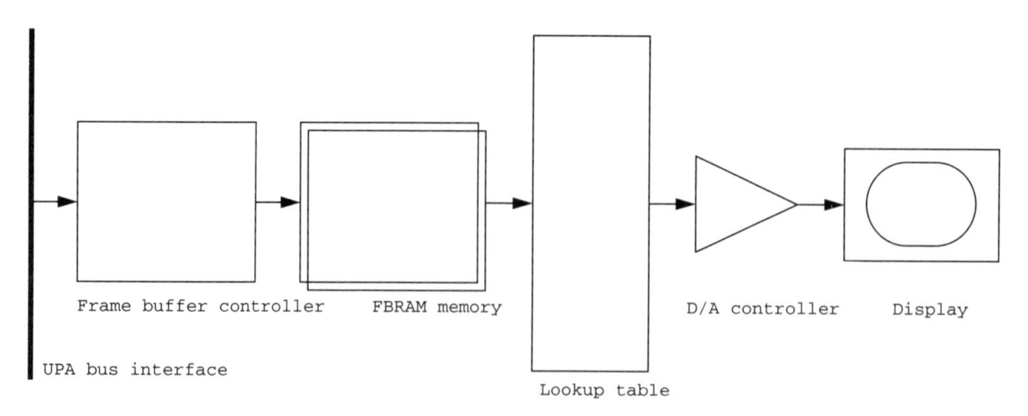

Figure 3.8 Block diagram of the Creator frame buffer

extensions, which are discussed in Section 3.4. Therefore, many common imaging functionalities, such as convolution and image zooming, which are accelerated by VIS, have not been duplicated on the Creator.

The performance of some imaging functions on the Creator, some of which also use the instruction set extensions of the system's microprocessor, are shown in Table 3.9. The main advantage of the Creator is the fast performance for imaging functions, some of which come for "free" from VIS. The main disadvantage is cost.

Table 3.9 Creator/VIS performance (8-bit pixels)

Imaging function	Megapixels/sec
3 × 3 convolution (general)	27
5 × 5 convolution (separable)	10
Copy	69
Scale (bilinear, 2×)	80
Table lookup (16 → 8)	50

Having examined several examples of specialized hardware for fast image processing, we are now ready to look at the general-purpose microprocessor as an image processing engine.

3.3 The microprocessor as a processing engine

For many imaging applications the pixel processing engine is the general-purpose microprocessor. The vast majority of these applications run on desktop computers; however, many complex imaging systems may also include on-board microprocessors for pixel processing. In this section we examine pixel computation using the microprocessor and look at some of the issues that are relevant for users and designers of these applications and their underlying systems.

Dedicated hardware was the only option for fast imaging well into the 1980s, but in recent years several improvements in microprocessors and the computer systems built around them have made fast imaging feasible on the desktop computer using the microprocessor as the pixel processing engine. These improvements include faster processors, better memory subsystems, better resolution of displays, sophisticated operating system features (such as virtual memory), and higher bus throughputs. As a result of these improvements, imaging on the desktop computer has gained acceptance along two distinct avenues.

First, many imaging products that used to depend on dedicated hardware imaging devices have gradually replaced some or all of the image processing hardware with general-purpose computers. Examples of these transitions come from medical imaging, color prepress, and other imaging application areas.

Second, new areas of desktop imaging applications have opened up with the advent of several imaging software packages that run on desktop computers. These packages include Adobe Photoshop, as well as a number of plug-in programs built around Photoshop. As a result, many imaging applications that needed a darkroom or an expensive stand-alone image processor can now execute on the desktop.

Whether the reader is considering the implementation of imaging functionality on a microprocessor, or simply using imaging software that executes on a microprocessor, several system features are of interest. We will discuss these features using a model of a general-purpose load/store microprocessor, simplified for our purposes and shown in Figure 3.9.

Our microprocessor consists of a load/store unit, a controller, and an Arithmetic Logical Unit (ALU). When a program executes, the controller decodes the instructions and issues commands to the load/store unit for loading the registers with data from the memory system. Then, the ALU performs the required operations in the register file. Finally, the load/store unit writes out the results from the register file into memory.

This is a rather simple description of a microprocessor's functions that illustrates some of the key features that need to be understood when considering a microprocessor as a pixel processing engine. These features are:

- *Clock frequency* The speed at which the ALU operates is directly dependent on the clock frequency of the processor. Therefore, the processor's clock frequency is important for high performance imaging.

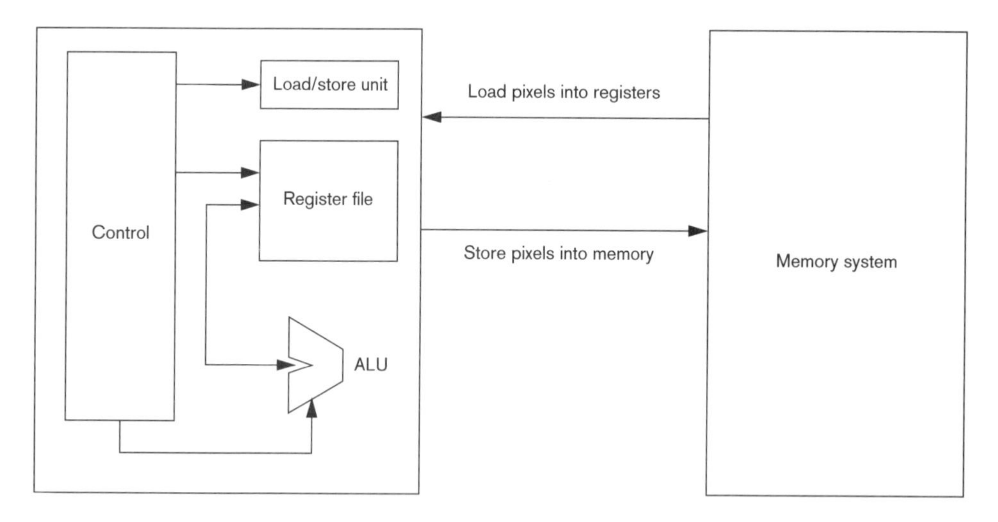

Figure 3.9 Simplified microprocessor model

- *Average memory access time* The time it takes for the load/store unit to read or write a pixel in memory is also important, since most images are large and require reading and writing a large number of pixels. However, this time may not be reflected by the clock frequency and is usually dependent on the design of the memory hierarchy of the computer (see Figure 2.13) as well as the access time of the memory chips being used.

  For example, a cache system designed and intended for a financial program will perform poorly in an imaging application, because the average image is much larger than the working data set of a financial program. Thus, many cache misses will result.

- *Instruction pipelining* Pipelining is the breaking down of execution of each instruction into a series of stages (usually four to seven), each performed in parallel. Today's processors also use *superpipelining* and *superscalar* techniques to improve performance. The former involves pipelines with many stages, whereas the latter involves the issuing of more than one instruction in one cycle.

All of the above features need to be taken into account when considering a general-purpose microprocessor for pixel computation.

Most of the currently available microprocessors can be divided into two classes: *reduced instruction set computers* (RISC) and *complex instruction set computers* (CISC). RISC microprocessors emphasize simplicity of the instruction set in order to attain high performance, while CISC microprocessors attempt to attain performance through a more complex instruction set. For example, in a RISC processor, all the operands of an arithmetic operation must be in the register file of the processor. However, in a CISC processor, the operands can be in registers as well as in memory locations.

Current examples of RISC processors include the Alpha family from Digital Equipment Corporation, the MIPS family from Mips Technology, Inc., the SPARC family from Sun Microsystems, Inc., and the PowerPC family from IBM/Motorola. The most prevalent example of CISC processors is the x86 family of microprocessors from Intel Corporation, of which the most recent member is the Pentium Pro or P6. Also available are the x86 clones from Advanced Micro Devices and Cyrix.

The relative merits of RISC processors compared to CISC processors have been the subject of a long debate [10]. In recent years, some of the distinctions between the two types have blurred. For example, before executing a complex CISC instruction, the Pentium Pro processor parses it into a sequence of *micro-operations*, which look like RISC instructions. The recent UltraSPARC-1 RISC processor includes a set of instruction set extensions, which is discussed in detail in Section 3.5. Arguably, these extensions stretch the boundaries of RISC towards CISC.

The performance of various microprocessors is a critical issue and large amounts of effort have been devoted into designing programs that benchmark this performance. One popular set of benchmark programs is the System Performance Evaluation Cooperative (SPEC). These programs have been chosen to reflect programs that users commonly execute on computers. Unfortunately, none of these programs include image processing functions, and they do not reflect the computational burdens placed on the microprocessor by imaging programs.

This deficiency has had two distinct impacts on the user who wants to do image processing on the desktop. First, the performance of the various microprocessors on the execution of imaging functions is still very uncalibrated. For instance, knowing the SPEC performance of a processor does not tell the user much about how fast the computer will be able to perform a 3×3 convolution on a large image. In addition, it does not tell the user how well the convolution will perform on a 512×512 image versus a 2048×2048 image. A systematic study of the imaging performance of one particular microprocessor can be found in reference 11.

Second, since the relative merits of microprocessors are measured largely by the results of these benchmarks, microprocessor and system designers have had little incentive to improve the design of their machines with respect to imaging functions.

The first problem still exists today in most formal benchmarks that measure computer performance. However, due to the gradual recognition that visual-based quasi-benchmarks such as Photoshop functions or MPEG decompression influence the user's perception of the performance of microprocessors, the second problem has been ameliorated somewhat on newer microprocessor models with instruction set extensions. We will take a detailed look at instruction set extensions later in this chapter.

From the programmer's point of view, a microprocessor is defined by its Instruction Set Architecture (ISA). The ISA is the programming model for the processor. It defines the instructions that the processor can perform, and it tells the programmer everything that is needed to program the processor correctly.

A great advantage of using general-purpose microprocessors for image processing, as opposed to using dedicated hardware for image processing, is the portability of the C and C++ programming languages. If an image processing function is written in C, then it will run on any microprocessor provided that a compiler is available to compile the C program into assembly language instructions for that processor. For programmers who are not concerned with performance, this virtually eliminates the need to understand the ISA. However, in order to achieve optimal performance, the programmer still needs to understand both the processor and the computer system based on it. While modern optimizing compilers have made the programmer's task easier by correctly scheduling processor instructions, the programmer still needs to be aware of potential speedups and bottlenecks associated with the the computer's design.

As we conclude this discussion of the general-purpose microprocessor for imaging applications, we should bear in mind that certain imaging areas still require significant hardware assistance. Some applications, including compression and decompression for video teleconferencing, and machine vision applications in industrial automation, have strict real-time requirements for a variety of complex image processing operations. Other applications requiring significant floating-point computations, such as image reconstruction in computed tomography, also require hardware assistance.

Having examined the general-purpose microprocessor as a pixel computation engine for imaging, we will now move on to two detailed examples of programmable imaging hardware: the SX and the VIS.

3.4 Detailed example 1: the SX accelerator

Our first example is the SX accelerator [12], which reflects the trend towards the integration of image processing acceleration on the computer itself. The SX, a part of the memory controller chip of a workstation, is an *integral* part of a desktop computer family. It is built into the workstation and provides "pixel acceleration" for imaging, video, and graphics. Once the workstation is paid for, the SX is free. The purpose of the SX is to provide acceleration for imaging as well as graphics at a level of low-end add-on accelerators without additional cost to the workstation user.

3.4.1 System architecture

The architecture of the workstation is shown in Figure 3.10. The workstation can have one or more CPUs; however, it can have only one SX. Instructions are sent to the SX by the CPU by "store" instructions. The "data" to be written are SX instructions and the storage location is the instruction queue of the SX.

The SX is part of the error-correcting memory controller of the system, located between the processor and the larger program memory (DRAM). Note the wider bus width between the SX and memory, which results in high data throughput. The machine is designed so that the SX does

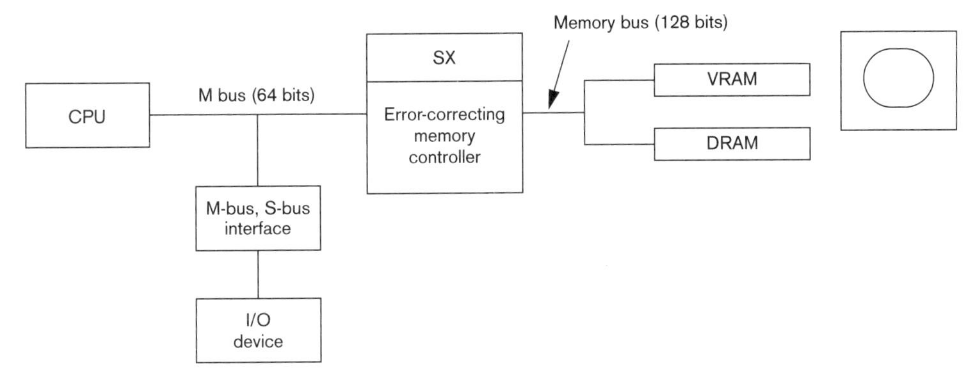

Figure 3.10 SX workstation architecture

not distinguish between DRAM (which is the normal program memory) and VRAM (the memory used for display).

3.4.2 SX architecture

The architecture of the SX processor is shown in Figure 3.11. The main parts of the SX are the instruction queue, the control unit, two ALUs, the I/O unit, and the register file.

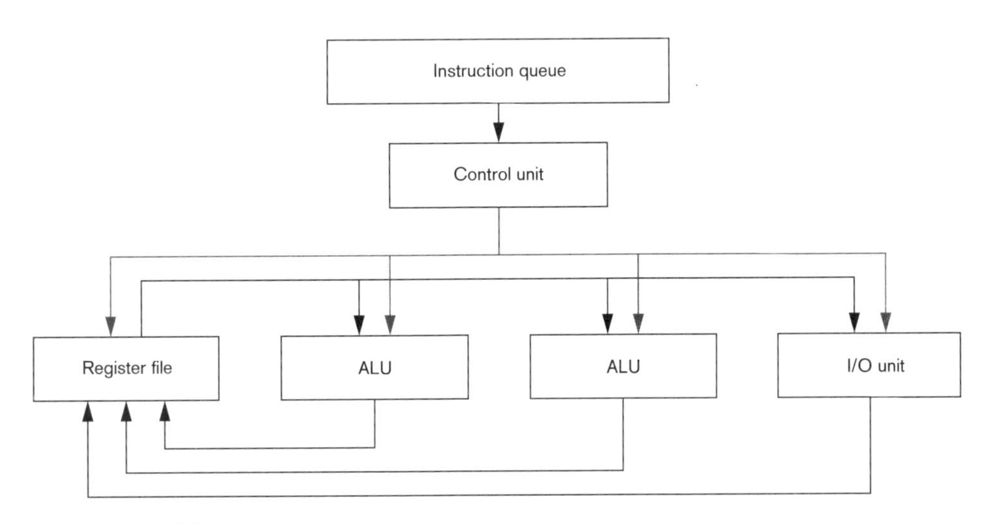

Figure 3.11 SX architecture

An SX program consists of instructions for the SX possibly interleaved with CPU instructions. SX instructions are written by the CPU into the instruction queue. The control unit takes these instructions and passes them to the appropriate processing unit. The two ALUs allow two pixels to be processed per clock cycle when the instructions are fully pipelined. The I/O unit reads data from memory into the register file. After the ALUs have processed the data, they are stored back into memory by the I/O unit.

The register file consists of 128 32-bit registers. The first ten registers of the register file are reserved for holding prespecified parameters used by SX programs. Therefore, 108 registers are available for the programmer to use in an SX program.

3.4.3 SX data types

The SX recognizes integer data only. This data, held in the SX register file, can be 8, 16, or 32 bits wide, signed or unsigned. These can be loaded to and stored from memory with a large amount of flexibility. In addition to monochrome 8-bit and 16-bit pixels, the SX is able to manipulate XBGR pixels in a flexible manner (e.g., treating the R channel as a separate individual image).

The usual fixed-point arithmetic (F8.8 or F8.16) is used to maintain precision in operations normally performed in floating point. Thus, for example, when convolving with a floating-point convolution kernel, the kernel values are scaled up by a large number (e.g., 65,536) and cast into integers. At the end of the convolution, the resultant pixel value is scaled down by the same scaling factor before being stored in memory. As we shall see, much of this scaling comes free of computational cost.

3.4.4 SX instruction set

There are two classes of SX instructions: *processor* instructions and *memory* instructions. We will momentarily examine each class of instruction in some detail, although due to space reasons we cannot offer exhaustive descriptions of the capability of each instruction. However, first we will examine the vector nature of SX instructions.

The SX is an *integer vector processor*, meaning all instructions operate on vectors—that is, spans—of pixels rather than individual pixels. This kind of processor is also known as a Single Instruction Multiple Data (SIMD) processor.

A demonstration of a vector operation, the addition of two vectors, is shown in Figure 3.12. The length of all the vectors is four. A vector of pixels, Source_1, is placed in registers 10–13; a second vector Source_2 is placed in registers 14–17. The result of the addition, the vector Dest, is placed in registers 18–21.

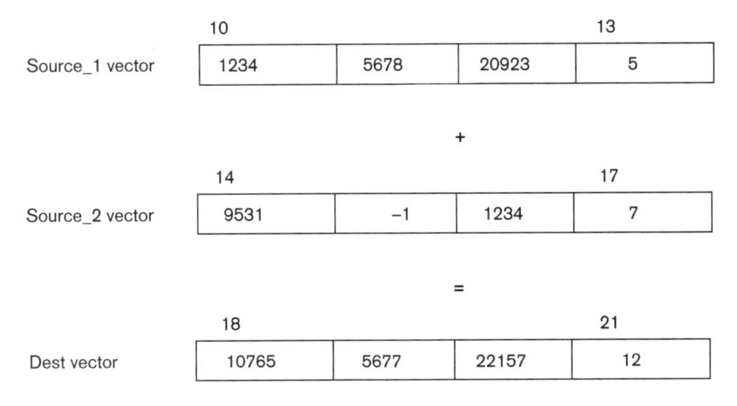

Figure 3.12 Adding two vectors of pixels

The SX instruction that executes the addition is:

```
add(10,14,18,4)
```

which tells the SX to add the two vectors starting at registers 10 and 14, and place the results starting at register 18. All vectors are of length 4.

Memory instructions The basic memory instructions of SX are *load* and *store*. Load reads data from memory into the SX registers. Store writes data from the SX registers to memory.

The memory locations accessed by the load/store instructions can be addressed in either *direct* mode or *array* mode. In direct mode, consecutive pixels in the vector correspond to consecutive words in physical memory. In array mode, a table of offsets is used to calculate the addresses in physical memory corresponding to consecutive pixels in the vector.

Bytes can be accessed in four formats. These are *normal, quad, channel*, and *packed*. Normal access means that each byte in memory is associated with a corresponding SX register. Quad bytes enable the processing of each channel of a four-channel (e.g., XBGR) image separately. Channel bytes allow accessing of only the bytes of one particular channel of a four-channel image. Packed bytes allow treating four bytes in one 32-bit register, thus speeding up data movements and logical functions on pixels. These different load functions are shown in Figure 3.13.

Several operations can be performed free of cost during memory instruction execution. During loading, sign extension (to 32 bits) and predefined left shifts (of 0, 8, 16, and 24 bits) can be done. During storing, *mask*, *plane*, and *clamping* controls can be enabled. Mask control enables masking of elements within a vector. Plane control masks particular bit planes of each pixel, and clamping control clips the values of the pixel to be written within the largest and smallest values for that data type.

An example of storing eight 16-bit unsigned pixels with mask and plane controls is shown in Figure 3.14. Since the plane control is set to fff0, the least significant four bytes of each word

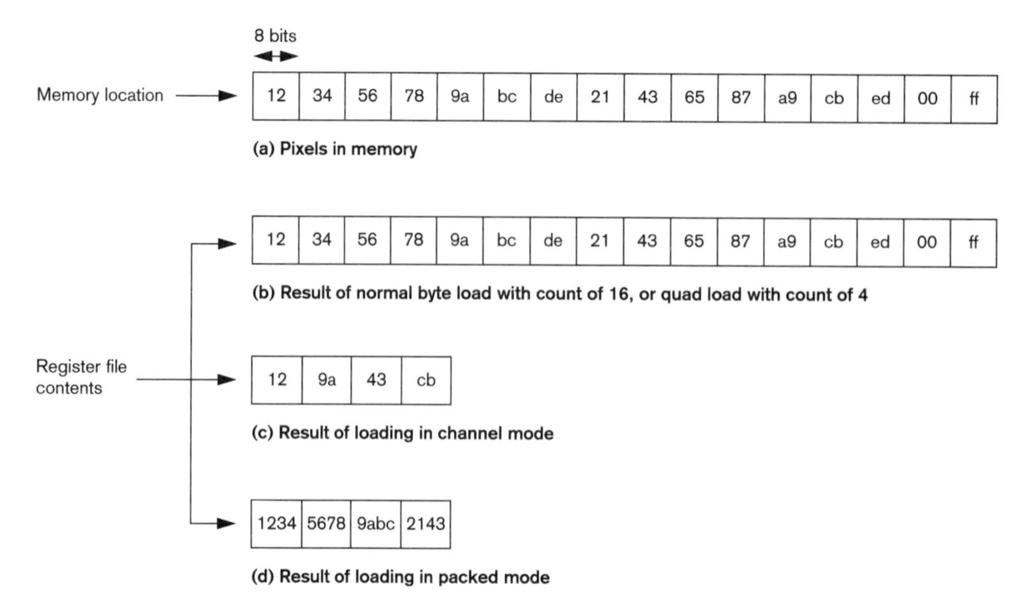

Figure 3.13 Loading of bytes into SX registers

in the destination memory are left untouched (marked x). Since the mask control is set to *ef*, corresponding to a binary 11101111, the word corresponding to 0 is left completely untouched.

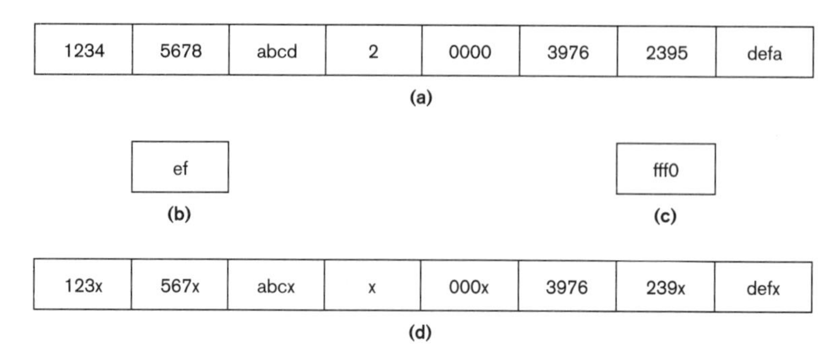

Figure 3.14 Storing unsigned shorts using masking and planemasking control: (a) quantities in registers, (b) mask register, (c) plane register, (d) quantities written to memory. The store was issued with a count of 8.

Processor instructions The second class of SX instructions is called processor instructions. The operands for these operations can be *vectors* or *scalars*. As before a count is always included in the instruction. The important classes of processor instructions are described below.

- *ALU monadic and dyadic instructions* Monadic operations are arithmetic operations between a pixel and a constant. Dyadic operations are arithmetic operations between two pixels. These instructions `add`, `sub`(tract), `or`, `and`, and `xor` between two values. For monadic variants (for example, add a constant to a pixel), one source vector and an immediate constant must be specified. For dyadic instructions, two source vectors must be specified.

- *Multiply instructions* There are three flavors of multiply instructions. These are `mul`, `dot`, and `saxp`. Multiplications can be performed in 16×16 or 32×16 precisions, and the result can be shifted right 0, 8, or 16 bits—all at no additional cost. The `mul` instruction multiplies two source vectors, element by element. The `dot` instruction computes the dot product of two vectors. The `saxp` instruction combines two vectors by multiplying one vector with a constant and adding the result to the second vector. The constant must be placed in a special register reserved for this purpose.

- *Shift instructions* These perform *arithmetic*, *logical*, and *funnel* shifts. Arithmetic and logical shifts move each element within a source vector either by an immediate operand or by the elements of a destination vector. Funnel shift treats the entire source vector as one concatenated string of bits.

- *Compare* This instruction compares each element of a vector to those of another vector or a scalar. Comparison functions are *equal to*, *greater than*, *greater than or equal to*, *less than*, and *less than or equal to*. The bits in the mask register are set depending on the result of comparison of each pair of elements. The destination vector is replaced with the elements (from either source or destination vector) for which the comparison was true.

- *Scatter and gather* The *scatter* instruction copies consecutive elements from the source vector to nonconsecutive but equally spaced elements in the destination vector. The spacing is specified in the instruction. *Gather* is the inverse operation to scatter.

- *Miscellaneous instructions* There are several other instructions used by the SX. The *rop* instruction performs a ternary Boolean operation between two source vectors and a reserved `ROP` register into a destination vector. The `select` instruction selects elements from two vectors or two scalars to place into the destination vector. The selection criterion is defined by bits set in a reserved mask register. The `plot` and `delt` instructions are used to draw shaded lines. `delt` cumulatively adds four delta values to four successive elements in the source vector for each four values that are output to the destination vector. Plot instruction performs Bresenham interpolation [13], creating a vector of offsets from major and minor offsets and error deltas.

We now proceed to show some programming examples on the SX.

3.4.5 Programming examples

Programming the SX to perform image processing functions requires some preparation. Since the SX lies between the CPU and DRAM memory, any image memory that the SX works with must be flushed from the CPU cache and marked *uncacheable*. Also, in addition to the normal virtual memory address associated with the memory, a second address to be used by the SX must be obtained by the operating system.

Here, we show two examples of SX programming: zooming an image and convolving an image with a 3×3 kernel.

Image zooming The first example zooms an image by a factor of 2 using nearest neighbor interpolation [14]. Image zooming using different interpolation methods has been covered in detail in Chapter 8. The $2\times$ zoom requires that each source pixel is responsible for the creation of four destination pixels, which are copies of the source pixel.

The algorithm is vectorized before implementation on the SX. That is, instead of one pixel, vectors of pixels are processed. Only the inner loop of the program is shown in the following code segment. The source pixels are assumed to be at the location `src_addr`, and the destination pixels are to be written to `dest_addr`. The register file of the SX is assigned as follows: The first 32 registers are assigned for the source pixels. The next 64 are assigned for the destination pixels. `IN` and `OUT` correspond to the beginning of these two vectors. The variable `dlb` is the number of bytes between successive rows of the destination image.

The corresponding SX code inside the inner loop is as follows:

```
load(src_addr, Rsrc, N);
scat(Rsrc,2,Rdest,N);
scat(Rsrc,2,Rdest+1,N);
store(dest_addr,Rdest,2N);
store(dest_addr+dlb,Rdest,2N);
```

The execution of this code segment is shown graphically in Figure 3.15.

Before we leave image zooming, it should be noted that the SX has also been programmed to perform interpolated zooms. In particular, both bilinear and bicubic interpolation (see Chapter 8 for details) are supported.

Convolution The second example of SX programming is the inner loop of a convolution of the image by a 3×3 convolution kernel. Convolution is covered in detail in Chapter 9. Briefly, the operation moves the kernel over the image, and for every position, the pixels underneath the kernel are multiplied by the kernel elements. These products are then added to yield the destination element for that position. For a 3×3 convolution, nine multiply-accumulation operations are needed.

The following algorithm does enough work to yield a vector in the destination N pixels wide. Three register banks, `Rrow0`, `Rrow1`, and `Rrow2` are assigned to hold N pixels of the source rows each. Another register bank, `Raccum`, is used to accumulate the resulting N pixels.

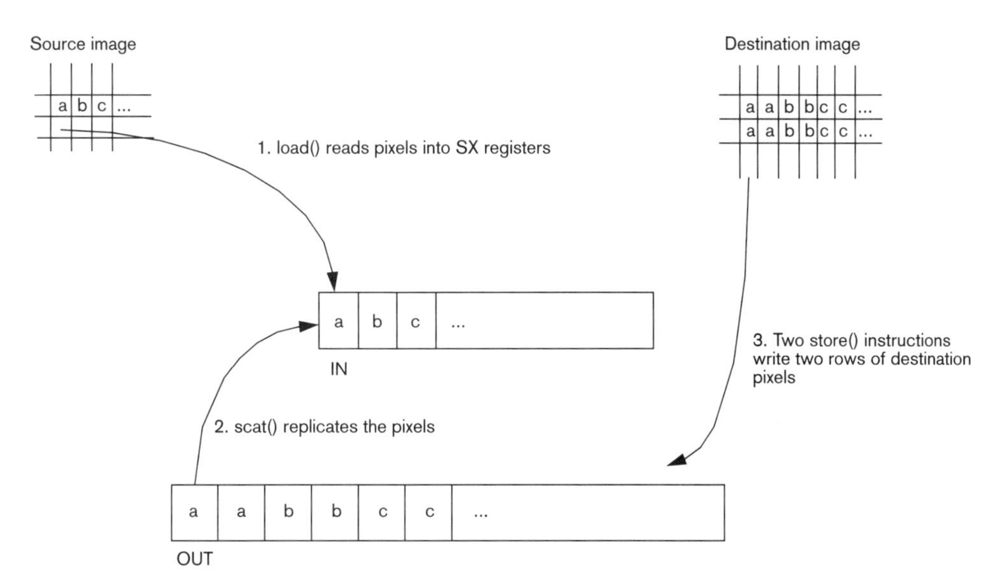

Figure 3.15 Performing a 2× nearest neighbor zoom using the SX

```
/* read the pixels */
load(src_addr, Rrow0, N+2);
load(src_addr+slb, Rrow1, N+2);
load(src_addr+2*slb, Rrow2, N+2);
/* convolve the first row */
write(Scam, kernel[0][0]);
saxp(Rrow0, 0, Raccum, N);
write (Scam, kernel[0][1]);
saxp(Rrow0+1, Raccum, Raccum, N);
write (Scam, kernel[0][2]);
saxp(Rrow0+2, Raccum, Raccum, N);
/* convolve the second row */
write(Scam, kernel[1][0]);
saxp(Rrow1, 0, Raccum, N);
write (Scam, kernel[1][1]);
saxp(Rrow1+1, Raccum, Raccum, N);
write (Scam, kernel[1][2]);
saxp(Rrow1+2, Raccum, Raccum, N);
/* convolve the second row */
write(Scam, kernel[2][0]);
saxp(Rrow2, 0, Raccum, N);
write (Scam, kernel[2][1]);
saxp(Rrow2+1, Raccum, Raccum, N);
write (Scam, kernel[2][2]);
saxp(Rrow2+2, Raccum, Raccum, N);
/* store scaled */
store(dest_addr, Raccum, N);
```

A similar algorithm can be used for convolution with 5×5 and 7×7 kernels. Separable convolutions can also be performed with kernels as large as 9×9. However, since the SX has 128 registers, larger kernels cannot be accommodated.

3.4.6 Performance improvements

As mentioned earlier, the performance goals of the SX were to provide imaging and graphics performance at about the same level as low-end add-on hardware accelerators. Typically 2 to 4× the speed of the CPU is achieved for most accelerated functions.

There are primarily two ways the SX is able to achieve this performance. First is the faster access to memory. As we shall see in Chapter 4, memory access can be a bottleneck for many imaging operations, and fast memory read/write cycles improve the performance of most imaging functions. Second is the presence of two ALUs in the chip. This means that for most instructions, after an initial overhead is paid, two pixels can be processed per clock cycle.

The clock rate for some SX instructions is shown in Table 3.10. The performance of some typical imaging operations is shown in Table 3.11.

Table 3.10 Performance of SX instructions in clocks

Processor instruction	Cost/vector element	Overhead/instruction
Most operations	0.5	0.5
Compares	0.5	1.5
mul,saxp,dot	16×16:0.5, 32×16:1.0	0.5
sum, funnel shifts, scatter	1.0	0
delt	1.0	1.0
Memory instruction		
Store direct	0.5–1.0	1 or 4
Store array	2 or 7	0
Load direct	0.75–1.0	1 or 9
Load array	3 or 8	5

The SX performs according to the numbers in Table 3.11, provided that the image is already uncached and a second virtual address to the memory has been obtained. The SX provides between 200 and 500 percent speedups over CPU-only execution of the same operations. This speedup—available "free" once the computer is purchased—is the main advantage of the SX.

Table 3.11 Performance of imaging algorithms on the SX

Operation	Performance (megapixels/sec)	Bits/pixel
Copy	26.0	32
Monadic operations	29.0	16
Dyadic operations	20.0	16
Alpha blend	3.0	32
Convolve (3×3 full)	5.3	8
Convolve (5×5 separable)	2.5	8
Lookup $8 \rightarrow 8$	11.0	8
Transpose	13.0	32
Zoom 2× (nearest neighbor)	23.0	32
JPEG decompression	25–29 frames/sec	24

A significant disadvantage of the SX is its requirement for operating on uncached memory. If a sequence of imaging operations is being executed, and one of them cannot be executed by the SX (and must be executed by the CPU), then there is a serious performance penalty associated with moving the image data into cache (for the CPU to execute this operation) and moving it out of cache (for the SX to resume operation).

Many of the lessons learned from the SX were applied to the next device in the evolution of programmable imaging devices at Sun Microsystems. The architect of the SX was also one of the architects of the Visual Instruction Set, which we discuss next.

3.5 Detailed example 2: Visual Instruction Set (VIS)

The SX accelerator described in the previous section is representative of a trend towards integration of imaging on desktop computers. In this section we will discuss the integration of imaging acceleration into the microprocessor using *instruction set extensions*. After discussing the evolution of these extensions, we will examine one particular implementation—the Visual Instruction Set (VIS) extensions of the UltraSPARC-1 processor—in some detail.

Originating in the Intel i860 microprocessor, instruction set extensions are special instructions that can be executed on a processor in addition to the general-purpose RISC or CISC instructions that the processor normally executes. These instructions usually perform *partitioned arithmetic*, allowing the acceleration of certain integer operations that are normally performed on pixels for imaging, video, and graphics applications.

The idea behind partitioned arithmetic is as follows. The internal registers of a microprocessor often have much higher precision (usually 32 or 64 bits) compared to the precision needed to represent image pixels (usually 8- or 16-bit integers). Therefore, several pixels can, in principle, be processed in parallel in each register, yielding better performance.

For example, given a microprocessor with 64-bit internal registers, we could load four 16-bit pixels into a register. If we added two registers loaded in this manner, we would effectively add four pixels simultaneously, and the resultant 64-bit register would have four 16-bit values, each corresponding to the sum of the corresponding 16-bit source pixels. The result would be almost correct—except that any overflow at the most significant bit of a pixel would be carried over to the least significant bit of the next pixel. Partitioned arithmetic hardware performs this operation correctly by disabling the carry operation.

In addition to the i860, partitioned arithmetic was also used in the 7100LC processor from Hewlett Packard [15], which included five special instructions for accelerating multimedia operations. In particular, one target was MPEG decompression (discussed in detail in Chapter 9), which is an extremely expensive computational process requiring intense usage of addition, subtraction, and multiplication.

These 7100LC extension instructions are tabulated in Table 3.12. They all operate on 32-bit registers. The `HADD` and `HSUB` instructions treat 32 bits as two 16-bit quantities. The `HAVE` instruction averages two halfwords. The `SHnADD` instruction is useful for multiplying a 32-bit register by a (small) constant.

Table 3.12 H-P 7100LC instruction set extensions

Instruction	Function
HADD	Add two halfwords at a time
HSUB	Subtract two halfwords at a time
HAVE	Average of two halfwords
SHLADD	Shifts 32-bit value left by 1, 2, or 3 bits and adds second 32-bit value
SHRADD	Shifts 32-bit value right by 1, 2, or 3 bits and adds second 32-bit value

Performance improvements achieved by the 7100 LC due to these instructions have included a real-time MPEG-1 system decoder, as reported in reference 16.

More recent examples of instruction set extensions have come from Sun Microsystems and Intel Corporation. The Visual Instruction Set (VIS) from Sun Microsystems, Inc. [17] is intended as a general-purpose multimedia instruction set, and is implemented on the UltraSPARC-1 microprocessor. Following closely on this model, Intel has announced the MMX Instruction Set [18], which is very similar to the Visual Instruction Set. MMX, which is also intended for multimedia applications, will be available in certain future versions of the x86 processors. The main differences between MMX and VIS are that MMX contains multiply-accumulate instructions and saturation arithmetic. VIS, on the other hand, has the pdist instruction for motion estimation, as well as a larger register file.

In the following sections, we take a close look at details of the VIS, including the processor supporting them, the actual instructions, the program development environment, and some programming examples. Further information on VIS can be obtained from the VIS User's Guide [19].

3.5.1 UltraSPARC-1: the processor for VIS

The UltraSPARC-1 microprocessor is a 64-bit microprocessor based on the SPARC Version 9 architecture [20]. It is a *four-way superscalar* processor, meaning it can sustain the execution of up to four instructions per clock cycle. A simplified block diagram of the UltraSPARC-1 is shown in Figure 3.16.

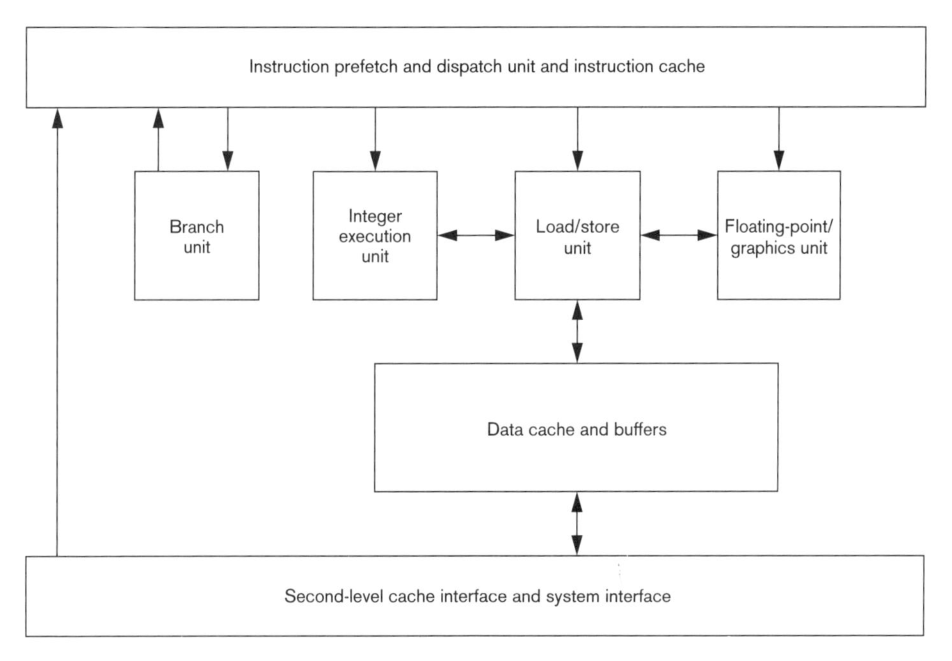

Figure 3.16 Block diagram of the UltraSPARC processor

The UltraSPARC-1 consists of a floating-point/graphics unit (FGU), an integer execution unit (IEU), a branch unit (BU), a load/store unit (LSU), and an instruction prefetch/dispatch unit (PDU).

The FGU contains both floating-point processors and arithmetic processors. The former are used for execution of floating-point instructions. The latter are used for the execution the VIS instructions described in Section 3.5.4.

The IEU performs all arithmetic/logical operations for integer (non-VIS) data. The BU is responsible for dynamic branch prediction during program execution. The LSU executes all instructions that transfer data between memory and the registers of the UltraSPARC-1. The PDU works in conjunction with the BU to prefetch instructions from the instruction cache and prepare them for execution.

The superscalar nature of UltraSPARC-1 allows the execution of a maximum of two FGU instructions, one load/store instruction, and another IEU or branch instruction per clock cycle. Thus, each cycle can be pictured as having four slots, into which certain groups of instructions can fit. These groups are governed by the so-called *grouping rules* of the processor [21]. Another way to look at this is that the instructions must be *scheduled* correctly in order to attain superscalar performance. As we shall see later, this superscalar behavior contributes much to the performance enhancements obtained by VIS.

Now that we have seen the basic structure of the UltraSPARC-1 processor, we can proceed to look at the VIS instructions in detail, starting with the data types.

3.5.2 VIS data types

The main VIS data types are shown in Figure 3.17. VIS operates on single-precision (32-bit) and double-precision (64-bit) registers. For the purposes of VIS, the single-precision registers can contain four 8-bit quantities or two 16-bit quantities, and the double-precision registers can contain four 16-bit or two 32-bit quantities. These quantities may represent pixels or constant values (for example, convolution kernel values).

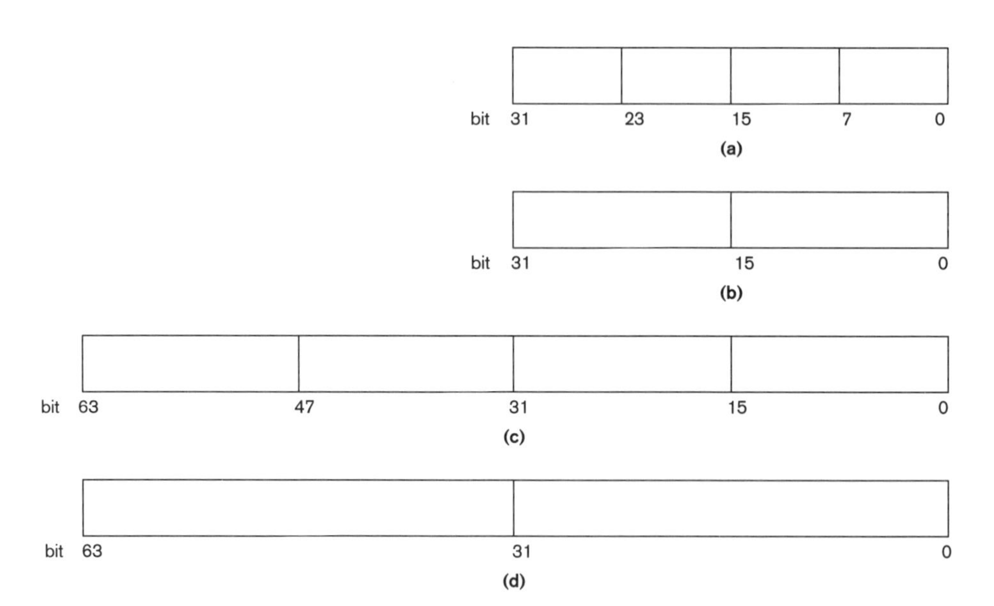

Figure 3.17 VIS data formats: (a) 4 × 8-bit pixels, (b) 2 × 16-bit pixels; (c) 4 × 16-bit pixels, (d) 2 × 32-bit pixels

3.5.3 VIS instructions

In this section we look at the individual VIS instructions. These can be classified into the following types: *data formatting* instructions, *address manipulation* instructions, *arithmetic/logic (ALU)* instructions, *memory access* instructions, and *special* instructions.

Several instructions operate under control of a special register of the UltraSPARC called the Graphics Status Register (GSR). The GSR has two fields relevant to VIS. These are *scale_factor* (bits 3 to 6) and *align_offset* (bits 0 to 2), as shown in Figure 3.18. The relevance of these fields will be explained when we encounter the GSR in the following instructions.

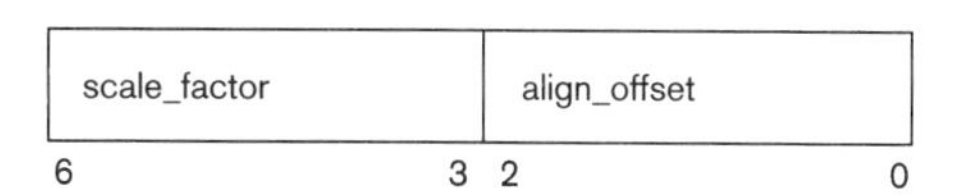

Figure 3.18 Fields of the GSR

Data formatting instructions These instructions convert between various data formats supported by VIS. The instructions include the following:

- `FEXPAND` Expands 8-bit pixels to 16-bit pixels, four at a time. This is needed so that 8-bit pixels can be converted to 16-bit pixels for maintaining sufficient intermediate precision during arithmetic operations. The source register is 32 bits (single precision) and the destination register is 64 bits (double precision). Each 8-bit value is left shifted by four bits and zero extended. `FEXPAND` is shown in Figure 3.19.

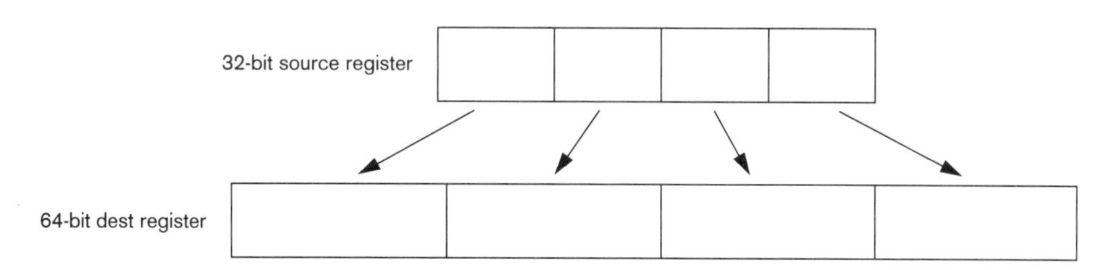

Figure 3.19 The FEXPAND instruction

- `FPACK16` This is the inverse instruction of `FEXPAND` in the sense that four 16-bit pixels in a 64-bit register are converted into four 8-bit pixels in a 32-bit register. One can also left shift the data using this instruction. Each 16-bit component is first shifted left by the amount represented by the bits in `GSR.scale_factor`. Then, it is clipped (between 0 and 255) and truncated to an 8-bit component starting at bit 7. `FPACK16` is shown in Figure 3.20.

- `FPACK32` This instruction merges parts of two 64-bit registers into a 64-bit destination register. The two 32-bit components of the first source register are left shifted by 8 bits and copied into the top 24 bits of each 32-bit component of the destination register. The bottom 8-bit components of the partitions of the destination register are created from the two 32-bit

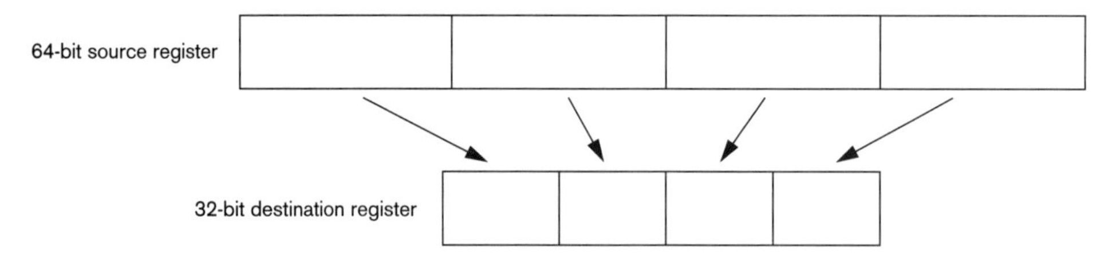

Figure 3.20 The FPACK16 instruction

components of the second source register by left shifting these components by `GSR.scale_factor` and then clipping and truncating at bit 23. This is shown in Figure 3.21.

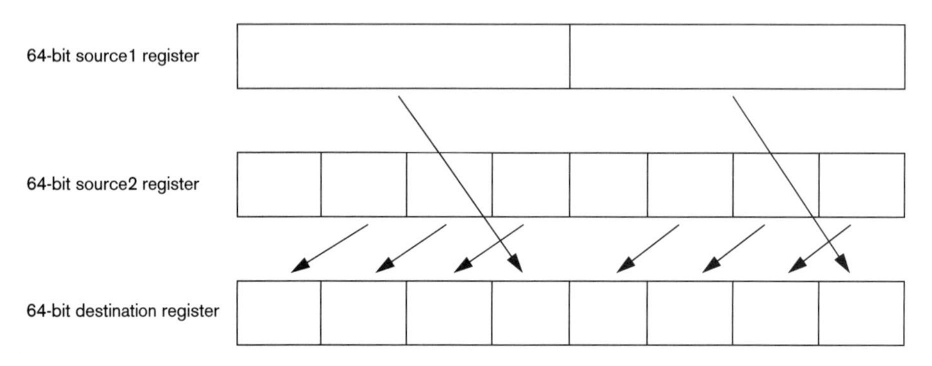

Figure 3.21 The FPACK32 instruction

- `FPACKFIX` This instruction packs two 32-bit pixels of a 64-bit source register into two 16-bit pixels of a 32-bit destination register. The source components are left shifted by `GSR.scale_factor`, and clipped and truncated after bit 16 before being packed into the destination register. This instruction is shown in Figure 3.22.

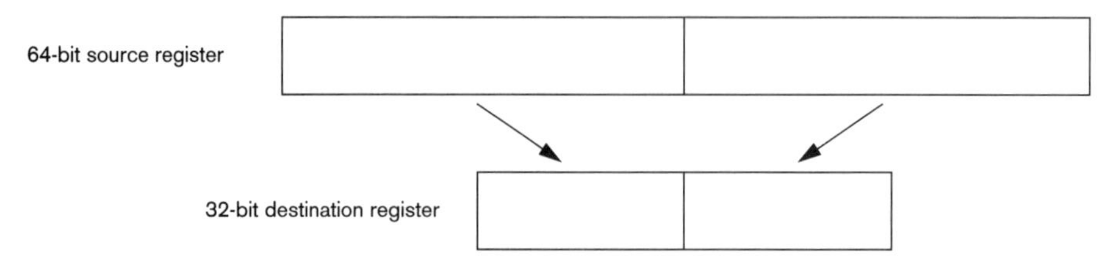

Figure 3.22 The FPACKFIX instruction

- `FPMERGE` This instruction interleaves the eight bytes of two 32-bit source registers into a 64-bit destination register. This is shown in Figure 3.23.

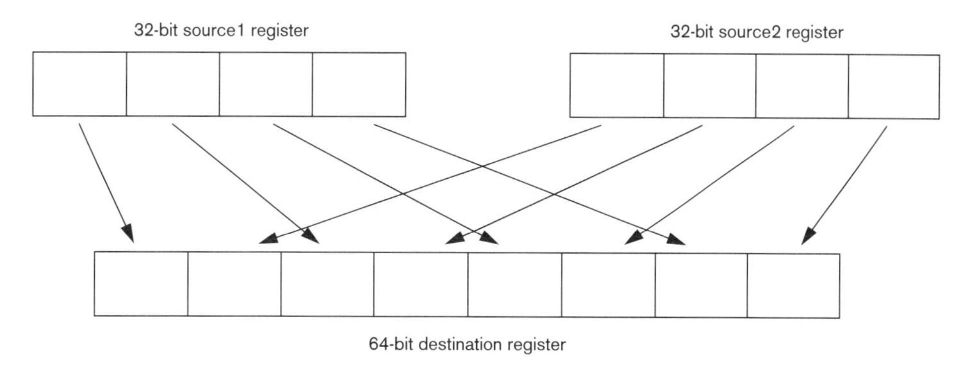

Figure 3.23 The FPMERGE instruction

Address handling instructions These instructions are needed because several VIS instructions require that the memory address be aligned to 64 bits (that is, the last 3 bits of the addresses must be zero). They facilitate the handling of these constraints.

- `ALIGNADDR` This adds two 64-bit source registers, `rs1` and `rs2`, and stores the result in the destination result with the least significant 3 bits set to zero. The least significant 3 bits of the result are stored in the `GSR.align_offset` field. In a typical example, `rs1` may be a base address and `rs2` an offset. Then the destination register will contain the double word–aligned address corresponding to (`rs1 + rs2`), and the offset will be stored in `GSR.align_offset`, which is used by `FALIGNDATA`.

- `FALIGNDATA` Concatenates two 64-bit registers and extracts eight bytes out of this using the offset specified in `GSR.align_offset`.

- `EDGE` This instruction is used for generating a mask for use in the partial store instructions, so that handling the end of scan lines is easier and special code need not be written. Bits 0 to 3 of the GSR are used as the mask, and the partial store instructions use this mask to selectively write out pixels into memory.

Arithmetic/logical instructions These instructions perform partitioned arithmetic operations on either 16-bit or 32-bit components. These include instructions for addition/subtraction, logical operations, multiplication, and comparison.

The instructions for addition are `FPADD16`, `FPADD16S`, `FPADD32`, and `FPADD32S`. The suffix *S* indicates that the operands are single-precision registers (32 bits.) These are tabulated in Table 3.13. The subtraction instructions are similar.

Table 3.13 ADD instructions

Instruction	Operation
FPADD16	Four 16-bit additions
FPADD16S	Two 16-bit additions
FPADD32	Two 32-bit additions
FPADD32S	One 32-bit addition

VIS also has instructions for a set of logical operations. The double-precision versions of these are shown in Table 3.14. As before, the single-precision versions have a suffix *S*.

Table 3.14 VIS instructions for logical operations (double precision)

Instruction	Operation
FZERO	fill with zeros
FONE	fill with ones
FSRC1	copy SRC1
FSRC2	copy SRC2
FNOT1	Negate (1's complement) src1
FNOT2	Negate (1's complement) src2
FOR	Logical OR
FNOR	Logical NOR
FAND	Logical AND
FNAND	Logical NAND
FXOR	Logical XOR
FXNOR	Logical XNOR
FORNOT1	Negated src1 OR src2
FORNOT2	Src1 OR negated src2
FANDNOT1	Negated src1 AND src2
FANDNOT2	Src1 AND negated src2

There are seven flavors of the partitioned multiplication instructions, all of which perform 8×16 multiplication.

- `FMUL8x16` Operates on two source registers (`rs1` is 32 bits and `rs2` is 64 bits) to create a 64-bit destination. It multiplies each 8-bit component of `rs1` with the corresponding 16-bit component of `rs2` and places the upper 16 bits of each result in the destination. This is shown in Figure 3.24.

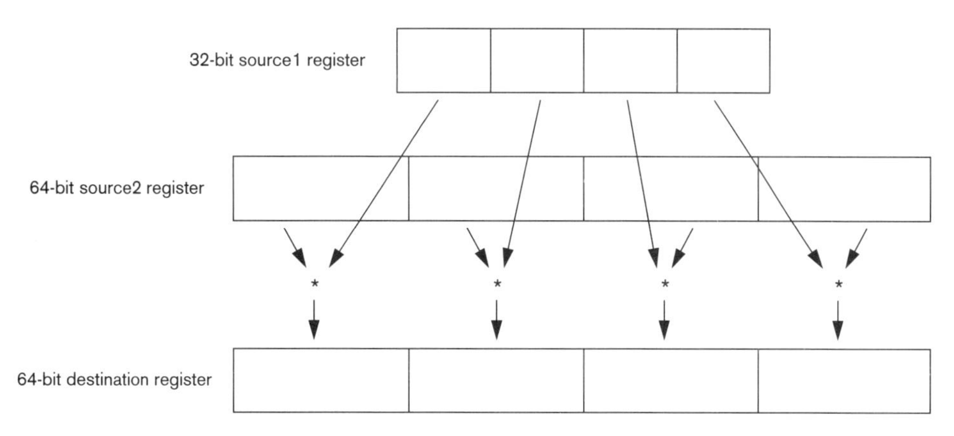

Figure 3.24 The FMUL8x16 instruction

- `FMUL8x16AU` Operates on two 32-bit source operands to create a 64-bit destination. It multiplies each 8-bit component of `rs1` with the most significant 16 bits of `rs2`, and the most significant 16 bits of the result are stored. The instruction is shown in Figure 3.25.

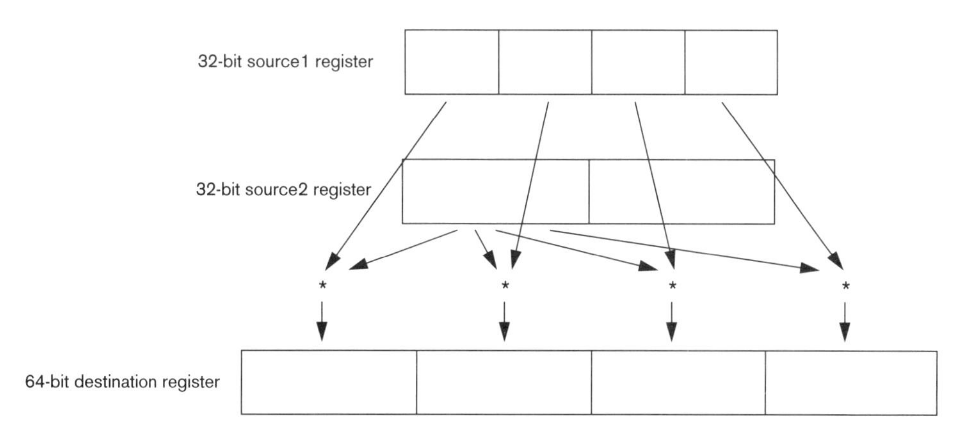

Figure 3.25 The FMUL8x16AU instruction

- `FMUL8x16AL` Identical to `FMUL8x16AU`, except that the least significant 16 bits of `rs2` are used as the scalar.
- `FMUL8SUx16` Takes two 64-bit source operands to create a 64-bit destination operand. The upper 8 bits of each 16-bit component of `rs1` are multiplied by the 16-bit components of `rs2`, and the upper 16 bits of the result are rounded and stored in the corresponding places in the destination register. This is shown in Figure 3.26.

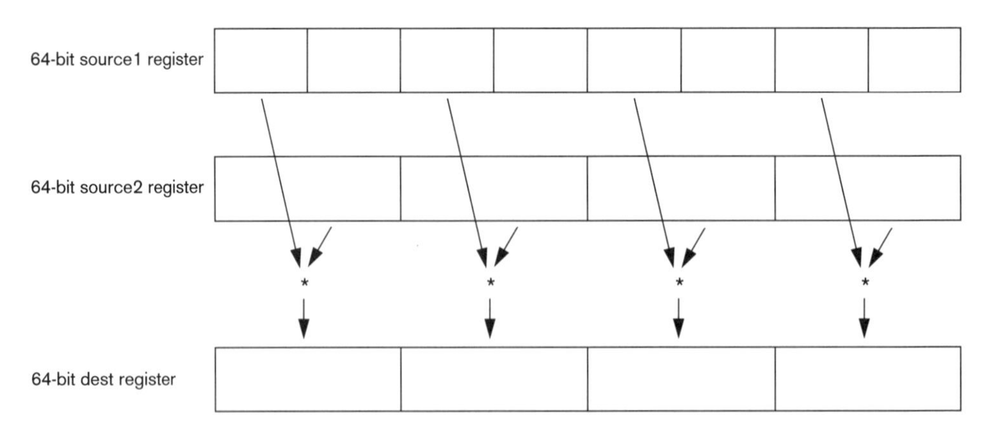

Figure 3.26 The FMUL8SUX16 instruction

- `FMUL8ULx16` Similar to `FMUL8SUx16` except that the lower 8 bits of `rs1` are used for the multiplication. The 24-bit components are sign extended to 32 bits and the upper 16 bits of these components are stored in the destination register.

  Note that `FMUL8SUx16` and `FMUL8ULx16` can be used together to perform a 16×16 multiplication of all the four components of the registers. This is shown in a programming example in Section 3.5.5.

- `FMULD8SUx16` Operates on two 32-bit registers, `rs1` and `rs2`, to create a 64-bit destination register. The upper 8 bits of each 16-bit component of rs1 are multiplied by each 16-bit component of `rs2`, and the two 24-bit results are left shifted by 8 bits and stored in a 64-bit destination. This is shown in Figure 3.27.

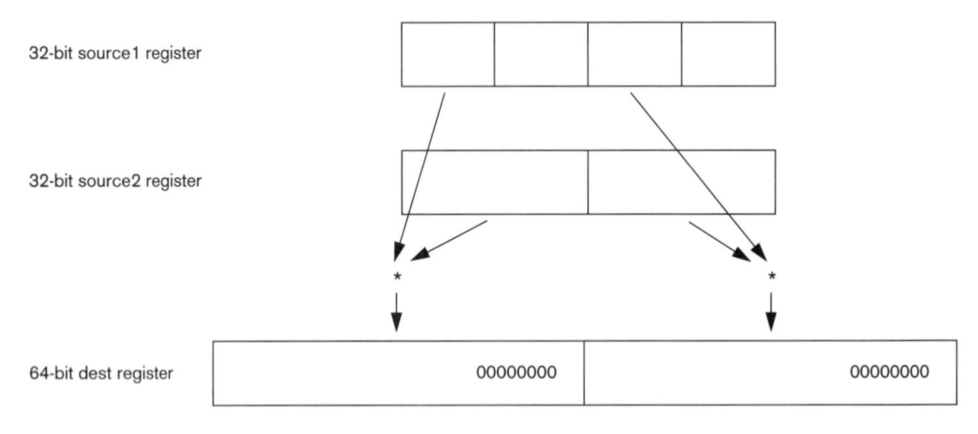

Figure 3.27 The FMULD8SUX16 instruction

- `FMULD8ULx16` Identical to `FMULD8SUx16` except that the lower 8 bits of each component of the `rs1` register are used for the computation.

VIS also has a set of instructions to compare values of pixels stored in two source registers. These instructions perform four 16-bit comparisons or two 32-bit comparisons in parallel. The results of the comparisons are indicated by setting the least significant 4 or 2 bits in the 32-bit destination register. These instructions are shown in Table 3.15.

Table 3.15 VIS compare instructions

Instruction	Operations
FCMPGT16	Four 16-bit compares > (set bit if src1 > src2)
FCMPGT32	Two 32-bit compares > (sset bit if src1 > src2)
FCMPLE16	Four 16-bit <= compares (set bit if src1 <= src2)
FCMPLE32	Two 32-bit <= compares (set bit if src1 <= src2)
FCMPNE16	Four 16-bit != compares (set bit if src1 != src2)
FCMPNE32	Two 32-bit != compares (set bit if src1 != src2)
FCMPEQ16	Four 16-bit = compares (set bit if src1 == src2)
FCMPEQ32	Two 32-bit = compares (set bit if src1 == src2)

Memory access instructions These instructions are for loading and storing data to and from the register set of the processor. There are three classes of load and store instructions: partial, short, and block.

Partial store instructions store to a 64-bit aligned address and use a mask register to decide which pixels of the destination register to write into the destination address. The masking enables a read-modify-write just like the SX. The use of these instructions is shown in the programming example "Adding a constant to an image" in Section 3.5.5. In addition, there are short load and store instructions, which move one or two bytes between memory and 32-bit registers. Finally, there are block load and store instructions, which transfer data between 64 bytes of memory and a group of eight 64-bit registers.

Miscellaneous instructions In addition to the above classes of instructions, VIS has several other instructions that are useful for particular operations.

The PDIST instruction is useful for motion estimation in image data compression (see Chapter 9). It computes the absolute difference between the 8-bit components of two double-precision registers, in the form of the following expression:

$$d = \sum_{i=0}^{7} |a_i - b_i| \tag{3.1}$$

This instruction is estimated to save about 45 assembly instructions. Its goal is to perform compression in real time (for example, for videoteleconferencing applications) without added hardware.

The ARRAY instruction converts a three-dimensional address (x,y,z) into a linear address. It is useful for applications that deal with three-dimensional data, such as volume visualization.

Having looked at the individual VIS instructions, we are ready to examine how a VIS program is put together. The VIS program development environment is intended to simplify this.

3.5.4 VIS program development environment

As we saw in Section 3.5.1, the superscalar nature of the UltraSPARC-1 necessitates careful instruction scheduling according to the grouping rules of the processor. This is certainly possible by writing the programs in SPARC assembly language and hand-tuning them; however, the code developed with such an approach rapidly becomes illegible and unmaintainable. Therefore, a more friendly development environment is needed.

This environment is provided by a set of VIS program development tools. These tools include a set of C-callable inline macros to invoke the VIS instructions, a compiler that knows how to optimize and schedule for UltraSPARC-1 (SPARC Compiler 4.0), a set of utility inlines that allow manipulation of data in registers, and an instruction cycle accurate simulator for the processor that tells us the instruction slots being filled during each cycle.

The inline macros allow the programmer to call the VIS instructions directly from a C program. The syntax of these inlines is as follows:

```
dest_register = VIS_instruction(source_registers);
```

As an example, consider the following inline macro:

```
!double vis_fpadd16(double /*frs1*/, double /*frs2*/);
        .inline vis_fpadd16,4
        std %o0, [%sp+0x40]
        ldd [%sp+0x40],%f4
        std %o2,[%sp+0x48]
        ldd [%sp+0x48], %f10
        fpadd16, %f4, %f10, %f0
        .end
```

If this C instruction sequence is written:

```
        double A01, B01, C01;
        C01 = vis_fpadd16(A01, B01);
```

then the compiler, using the above inlines and some clever optimization techniques, generates the following assembly language code from this sequence:

```
        fpadd16 A01, B01, C01
```

In addition to the inlines, there are several utility inlines that help the C programmer to manipulate registers. Examples of these inlines, which are used in the coding examples below, are:

```
vis_read_hi (double)—returns top 32-bit portion of a double
vis_read_lo (double)—returns bottom 32-bit portion of a double
vis_write_gsr()—writes the GSR
vis_write_hi(double)—writes top 32-bit portion of a double
vis_write_lo(double)—writes bottom 32-bit portion of a double
```

Once the code is generated, the compiler must schedule the code in such a way that the maximum number of instructions are issued during each clock cycle according to the instruction grouping rules.

The execution of the program can be observed using the simulator. During each clock cycle, the grouping of the instructions is displayed. This can be used to track down performance bottlenecks.

3.5.5 Example programs

We now look at some example VIS programs.

Data alignment While VIS is able to execute on chunks of data 64-bits wide, reading and writing these chunks from memory require that the memory addresses be aligned to 64 bits. The following example shows the use of `ALIGNADDR` and `FALIGNDATA`, which are used together to read a group of eight bytes from an arbitrarily aligned address `addr`:

```
void *addr, *addr_aligned;
double data_hi, data_lo, data;

addr_aligned = vis_alignaddr(addr,0);
data_hi = addr_aligned[0];
data_lo = addr_aligned[1];
data = vis_faligndata(data_hi, data_lo);
```

Note that when we need to access a stream of data, it is not necessary to execute both `ALIGNADDR` and `FALIGNDATA` every time. Rather, the address is aligned once using `ALIGNADDR`. Thereafter, `FALIGNDATA` is invoked, usually within a loop, to read the data.

16 × 16 multiplication The following code performs a 16 × 16 multiplication of two multiplicands, `d1` and `d2`, each of which contains four 16-bit pixels.

```
double d1, d2, result_hi, result_lo, result;

result_hi = vis_fmul8sux16(d1,d2);
result_lo = vis_fmul8ulx16(d1,d2);
result = vis_fpadd16(resultu, resultl);
```

Changing pixel format The following example shows the use of the FPMERGE instruction to change the format of an image from band-sequential to band-interleaved. The source image is composed of four separate bands for red, green, blue, and alpha. These bands are interleaved in the destination image. Note that this example assumes that all data are aligned to 64 bits and the row widths are multiples of eight pixels.

```
double *red, *green, *blue, *alpha, *abgr;
double r, g, b, a, ag, br;
int times;

/* ... compute times based on length of pixel span .... */

for (i=0; i<times; i++) {
    /* read 8 consecutive pixels of each band into a double */
    /* for example, load the pixels r0r1r2r3r4r5r6r7 into r*/
    /* load the pixels g0g1g2g3g4g5g6g7 into g, and so on*/
    r = red[i];
    g = green[i];
    b = blue[i];
    a = alpha[i];

    /* do the first four pixels */
    /* begin merging... ag will contain a0g0a1g1a2g2a3g3a4g4 */
    /* br will contain b0r0b1r2b2r2b3r3 */
    ag = vis_fpmerge(vis_read_hi(a), vis_read_hi(g));
    br = vis_fpmerge(vis_read_hi(b), vis_read_hi(r));

    /* continue merging to get final result */
    /* abgr[4*i] will contain a0b0g0r0a1b1g1r1 and so on */
    abgr[4*i] = vis_fpmerge(vis_read_hi(ag), vis_read_hi(br));
    abgr[4*i+1]=vis_fpmerge(vis_read_lo(ag), vis_read_lo(br));

    /* now do the last four pixels in a similar manner */
    ag = vis_fpmerge(vis_read_lo(a), vis_read_lo(g));
    br = vis_fpmerge(vis_read_lo(b), vis_read_lo(r));

    abgr[4*i+2] = vis_fpmerge(vis_read_hi(ag), vis_read_hi(br));
    abgr[4*i+3] = vis_fpmerge(vis_read_lo(ag), vis_read_lo(br));
}
```

Adding a constant to an image Our last programming example [22] shows a complete subroutine that adds a constant to a pixel. It makes no assumptions about the alignment of the pixels or the width of the image. The ALIGNADDR and FALIGNDATA instructions are used for getting correctly aligned data. In addition, the EDGE instruction is used to make sure that only the appropriate pixels are written out to the destination.

```
/* da and sa are pointers to the destination and source images */
/* dlb and slb are linebytes for destination and source */
/* w and h are the width and height of the image; k is the constant */
```

```
void addk(char *da, int dlb, int w, int h,
    char *sa, int slb, unsigned short k[4])

{
        char *d, *de;
        int mask, off;
        double *s, k01, s01, s23, i01, r01, r23, d01, d23;

        vis_write_gsr(3<<3);             /* scale bits in GSR */

        k01 = vis_to_double((k[0]<<(16+4))|
                        (k[1]<<4);
                        k[2] << (16+4)) |
                        k[3] << 4));

        while (--h >= 0) {
             d = da & ~0x7;
             de = da + w - 1;
             mask = vis_edge8(da,de);
             off = d-da;
             s = (double *) vis_alignaddr(sa,off);
             s01 = s[0];
             do {
                  s23 = s[1];
                  i01 = vis_faligndata(s01,s23);
                  r01 = vis_fexpand(vis_read_hi(i01));
                  r23 = vis_fexpand(vis_read_lo(i01));
                  d01 = vis_fpadd16(r01,k01);
                  d23 = vis_fpadd16(r23, k01);
                  r01 = vis_write_hi(r01, vis_fpack16(d01));
                  r01 = vis_write_lo(r01,vis_fpack16(d23));
                  vis_stdfa_ASI_PST8P(r01, d, mask); /* partial store */
                  d += 8;
                  s++;
                  s01 = s23;
                  s23 = s[1];
             } while ((mask = vis_edge8cc(d,de)) >= 0);
             sa += slb;
             da += dlb;
        }
}
```

3.5.6 *Performance of VIS*

Performance enhancements obtained by using VIS to perform imaging functions can be substantial. Up to 4× speedup over a non-VIS implementation of a computational operation appears to be the limit; however, with various other optimizations, such as block loads and stores, this speedup can become significantly higher. The key to obtaining the performance is scheduling the instructions. The UltraSPARC is a four-way superscalar microprocessor. That is, it is able to issue four instructions every clock cycle. However, there are restrictions on what these four instructions can be. To gain benefit from the superscalar nature of the processor, the instructions need to be scheduled correctly before being executed.

In addition, as with the SX, the algorithm must be vectorized before being used by VIS. Also, since moving data between the floating-point registers and the integer register file is expensive, the entire algorithm must be cast in terms of VIS instructions.

Table 3.16 summarizes some of the performance enhancements that were obtained on an Ultra-1 workstation using VIS. Note that many of the C code examples could probably be significantly speeded up using the algorithms and techniques that are described in Part 3 of this book.

Table 3.16 Performance improvements due to VIS

Function	VIS code Mpixels/sec	C code Mpixels/sec	Speedup
Zoom (2.5×, bicubic)	21.35	0.53	4000%
Zoom (2×, bilinear)	90.00	38.30	235%
Convolution (3 × 3)	21.00	2.48	847%
Table lookup (8 → 8)	50.00	35.70	140%
Table lookup (8 → 32)	22.50	1.60	1400%
Color conversion	1.67	0.48	350%

In addition to the above performance gains, VIS has also been shown to be effective in improving the performance of video algorithms [23]

In summary, the advantage of VIS is that it allows high-speed imaging without the need for add-on hardware. A disadvantage is that this capability is available only to those who can afford a computer with the UltraSPARC processor. The VIS program development environment substantially reduces the difficulty of writing VIS code; however, it is still more complicated than writing C code.

3.6 Conclusion and further reading

In this chapter we have examined several classes of pixel computation devices. These include specialized image processing machines, as well as general-purpose microprocessors with built-in image processing capabilities. We have evaluated the benefits and disadvantages of these devices and provided performance figures wherever available.

There are several avenues that can be explored further. For a discussion of various issues related to computer architecture, see reference 10. The Web sites for the various companies (Sun, Intel, Adaptive Solutions) offer detailed information on pixel computation products. Detailed information about the other products may be obtained from the manufacturers.

3.7 References

1 *Image Processing System Architectures*, J. K. and M. J. B. Duff, eds., Letchworth, Hertfordshire: Research Studies Press; New York: Wiley, 1985.

2 J. Adams, E. C. Driscoll, Jr., and C. Reader, "Image Processing Systems," *Image Processing Techniques*, M. Ekstrom, ed., Orlando, FL: Academic Press, 1984.

3 W. K. Pratt, "Intelligent Image Processing Display Terminal," *Proceedings of the SPIE,* Vol. 199, San Diego, CA, August 1979, pp. 189–194.

4 Data Translation Model 2868 Image Processor product brochure, Data Translation and Imaging Technology, Inc.

5 Imaging Technology Series 150/4 product brochure, Imaging Technology, Inc.

6 TAAC-1 Application Accelerator technical notes, Sun Microsystems, Inc., 1988.

7 K. Guttag, R. J. Gove, and J. R. Van Aken, "A Single Chip Multiprocessor for Multimedia: The MVP," *IEEE Computer*, November 1992, pp. 52–64.

8 "Workstation Houses TI's MVP," *EE Times*, March 28, 1994.

9 M. Deering, S. Schlapp, and M. Lavelle, "FBRAM: A New Form of Memory Optimized for 3D Graphics," *Proceedings of SIGGRAPH*, 1995.

10 J. Hennessy and D. Patterson, *Computer Architecture: A Quantitative Approach*, San Mateo, CA: Morgan Kaufmann, 1988.

11 D. Rice, "High-Performance Image Processing Using Special-Purpose CPU Instructions: The UltraSPARC Visual Instruction Set," *MS Report*, UC Berkeley, 1996.

12 W. Donovan, P. Sabella, I. Kabir, and M. Hsieh, "Pixel Processing in a Memory Controller," *IEEE Computer Graphics and Applications*, January 1995.

13 J. Bresenham, "Algorithm for Computer Control of Digital Plotter," *IBM System Journal*, Vol. 4, No. 1, 1965, pp. 25–30.

14 I. Kabir, M. Hsieh, A. Jabbi, B. Radke, and W. Donovan, "Image Processing in a Memory Controller," *Proceedings of the IEEE International Conference on Image Processing*, 1994.

15 "New PA-RISC Processor Decodes MPEG Video," *Microprocessor Report*, January 24, 1994.

16 R. Lee, "Realtime MPEG Decode via Software Decompression on a PA-RISC Processor," *Proceedings of COMPCON*, 1995.

17 L. Kohn, G. Maturana, M. Tremblay, A. Prabhu, and G. Zyner, "The Visual Instruction Set in UltraSPARC," *Proceedings of COMPCON*, 1995.

18 *MMX Technology Overview*, Intel Corporation, March 1996. Available from *http://www.intel.com.*

19 *Visual Instruction Set User's Guide*, Mountain View, CA: SPARC Technology Business, 1995.

20 *The SPARC Architecture Manual*, Englewood Cliffs, NJ: Prentice Hall, 1994.

21 *Spitfire Programmer's Manual*, Mountain View, CA: Sun Microsystems, Inc., 1995.

22 C. Zhou, D. Rice, I. Kabir, A. Jabbi, and X. Hu, "MPEG Decoding on the UltraSPARC," *Proceedings of COMPCON*, 1995.

23 S. Howell, personal communication.

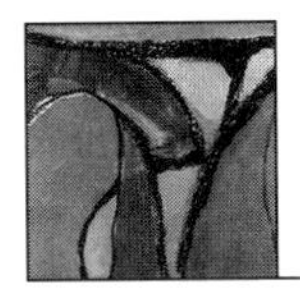

chapter 4

Imaging software design

4.1 Introduction

This chapter is about the design of imaging software. It describes a sequence of steps that the implementor can take to ensure that the imaging product being implemented is reliable and of high quality. The underlying premise is the following: The more upfront effort there is in the design, the more likely it is that the program will able to perform as required. The less effort spent on design, the more likely it is that the product will be buggy and will ultimately prove expensive to the manufacturer.

The design and implementation of reliable, high-quality software is a well-investigated subject. In this chapter we focus on the design issues peculiar to imaging software with which the implementor has to deal. As imaging software has grown in sophistication over the years, increasing demands have meant an increasing necessity to pay attention to design details.

In the beginning, imaging software existed only in college and research laboratories where the challenge was to invent algorithms. This software executed on mainframe and minicomputers, and met the demands made on it fairly easily.

As dedicated imaging hardware (e.g., Comtal, IIS, Vicom) became more common in the late 1970s and early 1980s, it was accompanied by software that was essentially dedicated to the hardware. The user interface was usually a command-line interface. The input and the output images were displayed on a dedicated monitor. The main function of the imaging software was to parse the requested command (with its parameters) and set the appropriate hardware registers. If the requested operation was not accelerated by hardware, then the software trapped to a CPU (host) subroutine that executed the operation. Since the individual operations were relatively simple and cleanly defined, it was not necessary to exert a great deal of upfront design effort into the software development process. Once the algorithm was debugged, everyone was happy.

Towards the middle of the 1980s, some interactive programs became available on hardware image processors, including those with zoom, scroll, and paint-like functions. In addition, programmable hardware started appearing. Both these developments increased the complexity of imaging software.

Today's imaging products, such as Adobe Photoshop, must deal with additional complexities, such as interactive functionality, window systems, and multiprocessor systems, as well as peripherals such as scanners, printers, and video input/output devices. In addition, there is frequently the need to provide support for add-on accelerators, which may provide acceleration for a select few functions. If the software needs to run on more than one hardware platform, there is a need to separate device-dependent portions from the device-independent portions so that it can be quickly and cleanly ported across different platforms. These developments in the evolution of imaging software are summarized in Table 4.1.

Today's demands on imaging software require that a good deal of thought be invested in design before a software imaging product is implemented. The rest of this chapter presents one way to focus these thoughts systematically.

Table 4.1 Examples in the evolution of imaging software

Time frame	Imaging system example	Target hardware	Functionality of software	Interactivity of software
Mid-1960s to late 1970s	Research software	Mainframes and minicomputers	Algorithms under investigation	None
Late 1970s to mid-1980s	DAISY, UC Davis	IIS hardware and Vax minicomputer	Basic imaging algorithms	Command-line interpreter (CLI)
Early 1980s to mid-1980s	Vicom s/w from Vicom Systems, Inc.	Vicom h/w and 68000 microprocessor	Wide range; mix of h/w and CPU functions	(CLI) + interactive zoom, scroll, paint functions
Mid-1980s	Data Translation Software	Data Translation accelerator on the PC bus	Limited to h/w capabilities	None
Late 1980s	TAACLib from Sun	TAAC Accelerator card on Sun	Programmable; spatial and Fourier functions	Basic GUI (menus, buttons, sliders, etc.)
Mid-1980s to now	Photoshop from Adobe	Macintosh, PCs, workstations	Wide range of spatial functions	Completely interactive (arbitrary regions of interest)
Early 1990s	XIL from Sun	SPARC, add-on accelerators, x86 processors	Most common imaging functions	Allows interactivity through window systems

The sections of this chapter follow the natural progression in the life of an imaging software package. First, a software *requirement* must be generated, which sets forth the intended market and the criteria that the software must meet to succeed in that market. Then, the software needs to be *specified*—that is, its parts, functionality, performance, and verification methodology must be defined. Next, it must be *designed* and the design *reviewed* for possible flaws. *Implementation* of the product follows the design. This consists of writing, debugging, and performance tuning the code. Once the product has been completed, it must be *maintained*, in order to provide customer satisfaction, for several years before it is superseded by another product or it is withdrawn from the market. Finally, if there is any plan for porting the software to other platforms, then a *porting guide* must be produced.

After a brief look at the hierarchy of imaging software in Section 4.2, we look at a broad overview of the software development process in Section 4.3. Requirements for imaging software are examined in Section 4.4. The specification of the software is discussed in Section 4.5. In Section 4.6 we present methodology to do a detailed design of the software. Section 4.7 is about implementation of the software. Maintenance of the product is discussed in Section 4.8. In Section 4.9, we look at the porting guide.

Since there are numerous styles of writing specifications, we do not propose any particular format for the design documents. Instead, questions and issues that the specifications should address are discussed in an informal manner with examples.

Project management issues of the software development effort, such as scheduling and manpower allocation are *not* covered in this chapter. There are numerous excellent references on software project management, and there is nothing unusual about the management of imaging software development to warrant coverage in this book.

While the novice implementor may be tempted to "jump in and start coding," experience has shown that a systematic design effort will, in the long run, end up saving him or her a great deal of frustration and save the company a good deal of money. Following the methodology presented in this chapter is one way to ensure that the design effort is systematic and comprehensive.

4.2 Imaging software hierarchy

In this section we define the types of imaging software with which the rest of this chapter is concerned.

When the software is designed for a single hardware platform—which is the case for many imaging applications such as turn-key imaging products or medical imaging products—the typical software hierarchy is as shown in Figure 4.1.

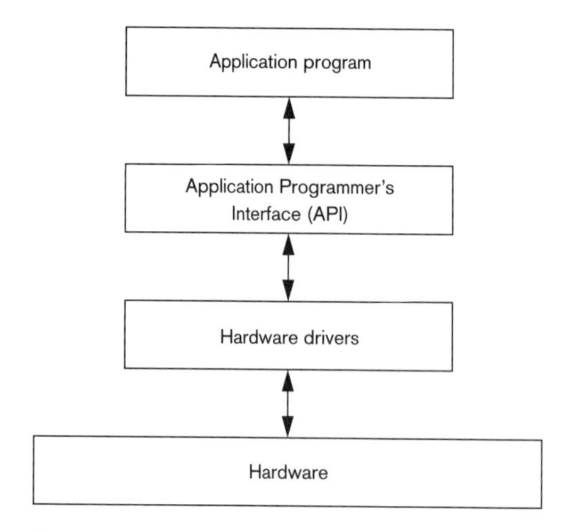

Figure 4.1 Software hierarchy for a single-platform imaging application

In this example, the application program accepts commands from the user. The user interface can be the buttons and dials on a medical ultrasound machine, or a command-line interpreter on a hardware image processor, or the Graphical User Interface (GUI) of a program executing under a window system. The application program reads the user input and sets up the correct parameters for the requested operation.

The program then makes the appropriate calls to a software library of imaging functions called an Application Programmer's Interface (API). The purpose of the API is to decouple the image processing functions from the rest of the application program. When the image processing functions are concentrated into the API, they can be used by other application programs as well.

The API, in turn, invokes driver software to execute that command. The driver software is separated from the API level to decouple hardware commands from the image processing functions. If the user requests a function that the hardware does not accelerate, then the API executes the command in software. The communication process is usually two-way, because values computed by the hardware can be transmitted back to the user.

A more complex hierarchy for imaging software is shown in Figure 4.2. The main difference from the single-platform software shown in Figure 4.1 is that this software executes on multiple platforms. On the topmost level is the application program, with which the user interacts. The application program accepts all user inputs and makes the appropriate calls to the API.

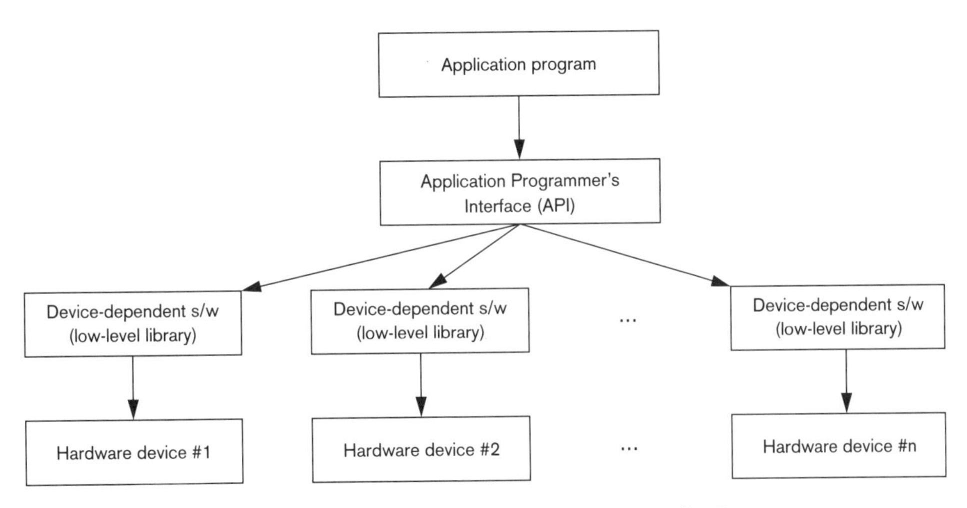

Figure 4.2 Software hierarchy for a multiplatform imaging application program

Distinct from the previous example, the API in this architecture may not contain the code for all the image processing functions. It is possible, instead, that the API controls the device being used, validates all the parameters being called, and performs general utility functions.

The API, in turn, detects the device being used—usually through a configuration file or an environment variable, or by directly checking the hardware—and calls the appropriate low-level library, called the device-dependent layer. This layer does the actual computation and display of images.

Since multiple hardware devices are supported by this architecture, one instance of a device-dependent module exists for each device.

4.3 Imaging software development process

The process for design and development of imaging software follows a set of clearly defined steps. The process is shown in Figure 4.3.

Initially, a requirement must be generated that addresses the reason the software is needed. Following this, functional and internal design are completed and reviewed. The software is then coded and debugged. After this, a verification method is employed to ensure that the software works correctly for a large number of cases. Once the software is known to work correctly, its performance is estimated and tuned. After the software begins to ship, it must be maintained by releasing new versions containing new features and bug fixes.

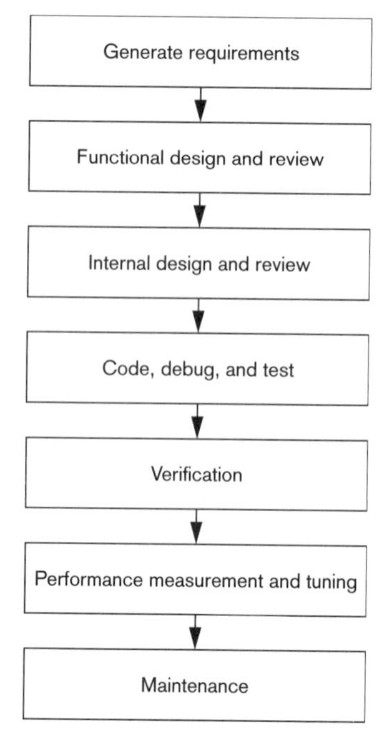

Figure 4.3 Imaging software design and development process

4.4 Imaging software requirement

The requirement for an imaging software defines the need for the software and the criteria that must be satisfied to meet that need. It is an encapsulation of the product's vision.

The requirement is usually based on input from the marketing department, where the target market for the product, as well as the functionality required for success, is decided. The requirement also defines the attributes that distinguish the product from its competitors. In addition, the requirement describes how the product interacts with other products from the same company, and its position in the product line.

Some questions that the requirement should answer are:

- *Who will buy this product?* This is the target market of the product. Everyone involved with the product must agree on the target market. The target market for the application program is an end-user who buys and uses the program. For APIs, the immediate target is the manufacturer of the application program (also known as Independent Software Vendor, ISV). For low-level drivers, the immediate target is the manufacturer of the API.
- *Why will they buy it?* Having determined who the product is meant for, the answer to this question lies in the functionality, price, and performance of the product.

- *What is the functionality of the product?* The functionality of the product must be specified in broad terms here. This is also a good place to specify several key attributes of the software, such as the pixel types supported, various image attributes, and how the software interacts with other pieces of software such as the window system.
- *What is the minimum acceptable performance?* This defines the lowest common denominator for the product. If the product has direct competition, then it must perform at least as well, or better, in order to gain customer acceptance.
- *What are some competitors?* A brief analysis of the main competition is needed here.
- *What features will enable it to beat the competitors?* The features that distinguish the product from the competition need to be mentioned here.
- *How does it fit in with the company product line?* If the company manufactures other products, then the position of this product relative to other similar products must be understood. A consistency of vision is needed, one that sends a clear message to the customer about the commitment of the company to this particular application area. The manner in which this product interacts with the other products must also be examined.

Once the requirements for the product are defined and agreed upon, the specification can proceed.

4.5 Specification of imaging software

The engineering design of the software begins with the specification. Before the software can be implemented, the following questions must be answered:

- What does it do?
- What pieces is it made of?
- How does the user use it?
- How well does it do its job?
- How do we know it works?

The software specification addresses these and other questions.

The functional specification, described in Section 4.5.1, answers the first three questions. In Section 4.5.2, we address the specification of performance, which answers the fourth question. The verification methodology tells us how to make sure that the product works (and keeps on working as it undergoes changes to fix bugs and improve performance). It is described in Section 4.5.3.

4.5.1 *Functional specification*

The functional specification describes the parts making up the software and the behavior of the software to its user. It is necessary to define this in detail because unless the programmer knows what the program is supposed to do, he or she cannot implement it. There are two special things that a functional specification must specify: the *architecture* of the product and its *external interface*.

Architectural specification The architectural specification is the *big picture* of the imaging software being designed. It describes the different parts of the software and how they interact with each other. Optionally, this specification also describes how the software interacts with other software that may be relevant (for example, how an imaging API interacts with a graphics API from the same manufacturer). The architectural specification may also describe the manner in which external hardware devices interact with the software. In general, this specification is short and to the point, concentrating only on the larger issues relating to the software.

The criterion used to divide the software into its parts is commonality of functions. As shown in the software hierarchies of Figure 4.2 and Figure 4.3, pieces of the software that perform similar operations are usually kept together.

The example in Figure 4.4 shows the architecture of the XIL imaging API library. This library provides a device-independent interface to a rich set of commonly used image processing operations. It is designed to execute on more than one platform. The library is divided into a device-independent core and a set of device-dependent *loadable* drivers. The XIL core is the portion of the library that parses the function call, checks the parameters for inconsistencies, and opens the device-dependent drivers needed to execute that function.

The device-dependent drivers are divided into three classes: *storage*, *compute*, and *input/output*. In general, one instance of each driver must be written for each device that is supported.

The storage driver is responsible for image memory management. An instance of the storage driver, dedicated to a particular hardware, allocates image memory as needed on that hardware. It also converts from one type of image memory to another. For example, it converts from CPU images (allocated on main memory) to SX images (allocated on the integrated hardware accelerator SX, described in Chapter 2).

The compute driver is responsible for the actual pixel processing. It is responsible for handling all the specified image types and their attributes, and for performing all imaging functions. If the underlying hardware is unable to perform the requested operation—because, for example, the image format is not supported—then the compute driver returns a "failure to perform" message to the core, which then tries the next available device.

The input/output driver is responsible for all input and output devices. Since the images are displayed on the monitor through a window system, the I/O driver is also responsible for interacting with the window system.

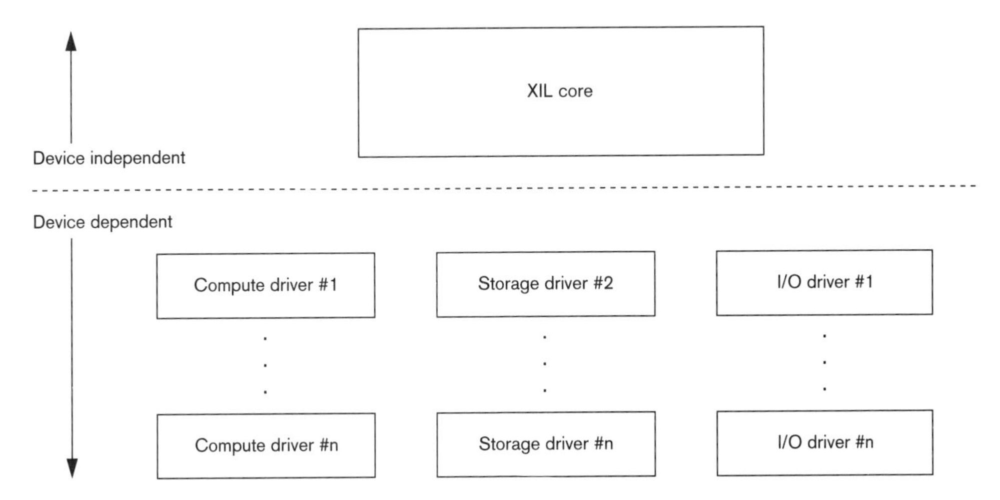

Figure 4.4 XIL architecture

In addition to these parts, XIL also interacts in various ways with the window system to display images. XIL is also able to interoperate with the graphics library XGL, and accepts data structures generated by various X Windows calls.

The architectural specification for this imaging software product needs to present all the above information. For the sake of brevity, some other parts of XIL (such as deferred execution) have been omitted here. The real specification would also need to discuss those parts.

External interface The external interface defines, in precise terms, what the product does for its user. The user can treat the software like a black box, and expect it to behave in the way that the external interface specifies. A technical writer should be able to take an external specification and write a user's guide from it without additional information.

External interface for an application program Since most application programs interact with the user via a GUI, this specification is a list of all the items that make up the GUI, and the function that each item performs. Each of these functions must be precisely defined, including the equation being executed, and the precision of the operation.

The external interface for an application program defines very clearly what the behavior of the program is to the user. For an interactive program, it must provide the following details:

- Any initialization that the user must do before execution of the program, and the effect of the setup parameters
- How images are loaded and saved
- More generally, how the user creates and destroys objects, such as images, regions of interest, convolution kernels, lookup tables, and so forth, and what the default values are

- List of all the menu options
- Precise definition of the effect of choosing each menu option
- Image file formats that are supported
- How external input/output devices are invoked
- How imaging accelerators are invoked
- What constitutes an error on the user's part
- How the program behaves when the user errs
- Any special functionality such as "undo"

This specification must also define the result of errors, both on the part of the user and the environment in which the program is executing.

External interface for an Application Programmer's Interface (API) For an API library, the external interface consists of a definition of all the functions in the library. It must also define what the parameters to these functions are. The following questions must be clearly answered in this specification:

- *What is the definition of the operation?* A mathematical equation describing the operation is most preferable, but if that is not possible, a clear description of the operation should be given.
- *What is the definition of the parameters?* All the parameters needed by the operation must be defined. For example, for a lookup table function, the components of the lookup table must be defined.
- *What types of data are handled?* The types of image pixels that are handled by the operations must be specified.
- *How do different image attributes, such as region of interest (ROI), affect the operation?* Effect of image ROIs (discussed in Chapter 1) on the operation must be specified. Usually, source image ROIs are treated as read masks, and destination image ROIs are treated as write masks. For example, in XIL, if the operation does a geometric transformation on the image, then the same transformation is applied to the source ROI, and the transformed ROI is intersected with the destination ROI to find the area in the destination where pixels can be written. If the image has origins, the effect of an origin on the operation must be defined.
- *What is a precise definition of corner or edge conditions where there is room for ambiguity?* For neighborhood operations, such as convolution or interpolated geometric functions, a method must be defined that specifies what happens at the edge of the

image, when the source pixels needed to perform the operation lie outside the image (and may not exist).

- *What is the level of acceptable tolerance from different hardware devices on which the program executes?* For a program that is intended to execute on several hardware platforms, some indication should be given for acceptable differences in precision.
- *What are the error conditions?* Error conditions and the behavior of the program under those conditions need to be specified. Of course, all possible user errors cannot be enumerated; however, a consistent error handling strategy should be used.
- *How does the user import and export images?* A method is needed to allow the user to bring his or her image into the program, or take out an image from the control of the program. This may be trivial for programs that do not hide the data, but difficult for programs that hide the pixels from the user.

External interface for a device-dependent library For a low-level device-dependent library the external specification is similar to the API specification. However, the objects involved can be a good deal less complex. For example, while the API works on images only, it is desirable to write the low-level library so that it accepts pointers to the first pixels of the relevant images in order to reduce overhead.

Another important difference occurs in error reporting. In order to reduce overhead, the error processing in the low-level routines is kept to a minimum. For example, it is frequently assumed that consistency and type checking performed by the API function prior to calling this library is good enough, and a bare minimum "assertion" type of checking is done by the low-level routine. The physical characteristics of the image, including its layout, are important in the low-level routine.

4.5.2 Performance specification

After correct functional behavior, performance is perhaps the most important characteristic of any imaging product. It is important because most imaging algorithms involve reading, processing, and writing large sets of data.

In many cases, performance, as perceived by the customer, can make or break a new imaging product. An example is Live Picture. This product is able to subsample and manipulate very large image files (100 MB or more) in a fraction of the time it takes for other imaging software, and this has enabled it to be successful.

When specifying the imaging software, performance specification (or goals) must be delineated as precisely as possible. For most general-purpose computers, this is an extremely difficult task.

In theory, estimating performance is simple. It consists of counting the number of clock cycles (C) required to perform the imaging operation for one pixel. Then, if the clock speed is M MHz, the performance, in megapixels per second, is computed as:

$$P = M/C \text{ megapixels/sec} \quad (4.1)$$

As discussed in Chapter 3, imaging operations executing on many of today's computers follow the three steps shown in Figure 4.5.

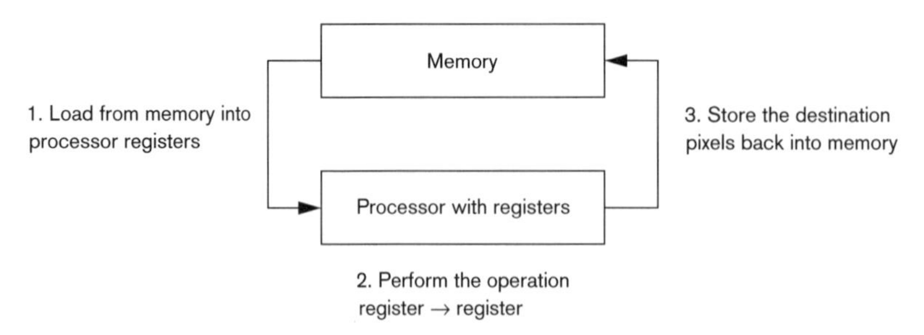

Figure 4.5 Sequence of instructions in a load/store microprocessor

Since a variety of sophisticated techniques (e.g., caching at multiple levels, write buffers, and data prefetch) are used in the load and store steps, it is extremely difficult to estimate the number of cycles it takes for the memory access part of the operation. In addition, in virtual memory systems, if the imaging operation accesses parts of an image that have not been paged in, then a page fault can occur, further complicating such estimation.

Therefore, in practice, a combination of methods, such as studying the performance of similar algorithms on the machine and older machines, and counting the number of operations, are all used to arrive at a ballpark figure for the estimate.

For software executing on dedicated hardware devices, arriving at these goals is somewhat simpler. Knowing the characteristics of the device and the proposed algorithm, the time for completing the operation can be estimated. Once this is done, a derating factor can be applied to account for software overhead and other system latencies.

An example of performance estimation is shown in the following example of a 2× pixel–replicated zoom on the SX accelerator (the SX accelerator is described in detail Chapter 2).

Example: performance estimation for 2× zoom on the SX The SX has two types of instructions: memory instructions and ALU instructions. The interface to memory is relatively clean, so that the number of cycles required for load/store operations is well understood. The ALU instruction timings can also be predicted reliably. These timings have been discussed in Chapter 3, where we have also discussed zooming an image using the SX.

The actual code for an optimal implementation of the 2× zoom is shown below. This is somewhat different from the example code shown in Section 3.4.6, which was written for clarity. In reality, we want to load as many pixels as possible at a time.

```
load(sa,INREG,32);                  /* read as much as possible */
scat(INREG,2,OUTREG,16);            /* replicate first 16 2x ...*/
scat(INREG,2,OUTREG+1,16);          /* ...in the hor. direction */
scat(INREG+16,OUTREG+32,16);        /* replicate next 16 2x ... */
scat(INREG+16,OUTREG+33,16);        /* .. in the hor. direction */
store(da,OUTREG,32);                /* store result 64 pixels */
store(da+32,OUTREG+32,32);          /* 32 at a time (max) */
store(da+lb,OUTREG,32);             /* on two lines */
store(da+lb+32,OUTREG,32);          /* to get 2x vertically */
```

The operation is shown in Figure 3.15. To minimize load and store overhead, 32 pixels are read in. Using the scatter instructions, these pixels are replicated 2× in the horizontal direction to create a row of 64 pixels. Four scatters are needed because scatter, being an ALU instruction, can process a maximum span size of 16 pixels. Finally, the result is stored on two consecutive rows in the destination.

The number of cycles required for processing 128 destination pixels is 17 (for load) + 64 (for scatters) + 48 (for stores) = 129 cycles. For the SX processor running at 40 MHz, this is 39–40 million destination pixels/sec. While this is a relatively simple example for which performance estimation is straightforward, performance of more complex imaging functions on the SX can also be estimated to a reasonable degree of accuracy primarily due to the clean memory interface.

4.5.3 *Verification methodology*

One of the most difficult tasks in any software endeavor is verification. Verification is an attempt to answer several questions, such as: how do we know that the software is working correctly for all possible inputs, and how do we know that the software keeps on working correctly as new features are added, bugs are fixed, and performance is tuned?

Any software of reasonable complexity can have countless execution paths and options. The problem becomes even more acute in imaging because there are millions upon millions of pixels that must be verified for each operation that the software performs.

It is, therefore, critical to decide on a verification strategy at the outset. Since it is impractical to verify correctness of every pixel by hand, verification methods usually rely on comparing results with some known reference which is assumed to be correct.

The standard way to test software is to use a *test suite*, a group of tests designed to be orthogonal and test as many possible cases as possible. The programs in the test suite exercise the software and compare the results against the reference.

There are two different ways of verifying against a reference: using a *reference software package*, or using a set of *reference images*. In the former, the result of each operation is compared to the result produced by the same operation using a reference package, and any difference outside a predefined tolerance is flagged as an error. In the second method, the resultant image is stored after each operation and compared to a corresponding reference image. These methods are shown in Figure 4.6.

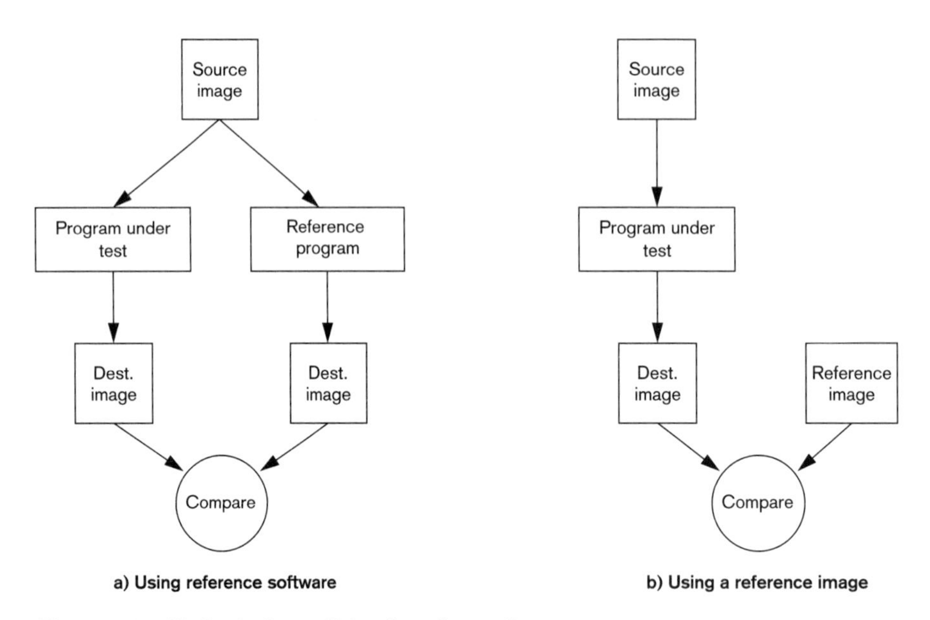

Figure 4.6 Methods for verifying imaging software

The comparative advantages of each method are tabulated in Table 4.2.

Table 4.2 Comparison of verification methods

Verification method	Advantages	Disadvantages
Reference s/w package	Uses less disk space	Need reference package known to work
	No need to generate images	Package may be buggy
		All functions may not be matched exactly
Reference images	No need for reference package	Needs lots of disk space
	Can be done in a "blind" manner	Susceptible to corruption of the reference images
		Need to generate images
		For a large number of tests, images need to be in "sync" with the tests

Choice of reference software The choice of reference software is often difficult. Essentially, this is the standard against which the correctness of the software under development will be judged. Occasionally, older versions of the software, which have been in use for some time and are known to be bug-free, can be used. At other times, an initial version of the current software, painstakingly examined to be correct, is used. A third option is to use an external software package that offers a similar set of functionality. An imaging standard, such as the Programmer's Imaging Kernel System (PIKS), is a good candidate for this.

An example of verification using reference software is the Xilch verification suite used to verify the XIL API library. Xilch is described in more detail later in this chapter.

Choice of reference images If reference images are to be used, then they must be generated somehow, which again requires software known to work correctly. But this software need not be very elaborate. Once the set of reference images is defined, the programs needed to generate them can be written. This scenario is useful particularly when special-purpose hardware is involved, since this kind of hardware usually performs a limited set of functions. An example is the verification process used during the development of Sun Microsystems' Genesis medical image processing product. A set of images were generated using programs executing the same algorithms that the hardware executed. The hardware/software combination was tested against these reference images. Of course, any of the above-mentioned options for reference software can be used to generate the images.

The choice of image is largely a function of the application. In general, it is desirable to test as many orthogonal combinations of attributes—such as diversity of pixel values, image size, image subsets, and regions of interest—as possible.

The test matrix The test matrix is the list of all input images that need to be tested. As mentioned above, orthogonality in image attributes is desirable. For a function that requires the specification of other parameters (such as convolution kernel for convolve), the test matrix also includes several samples of these parameters.

Accounting for hardware differences If the software being tested is expected to run on different hardware platforms, then additional complexities arise because of precision differences between platforms. Thus, a platform using fixed-point arithmetic is likely to produce different results than one using floating-point arithmetic. These differences are more acute for 16-bit pixels than for 8-bit pixels.

As long as the imaging function being tested is one where the algorithm modifies only the pixel value, but not the pixel location—and this includes almost all imaging operations except geometric modifications of the image—a *tolerance* can be used in the test program to account for hardware differences, should they arise. Thus, for example, a tolerance of one should account for differences observed on 8-bit images in different hardware. Use of tolerance in testing against different hardware is shown in Figure 4.7.

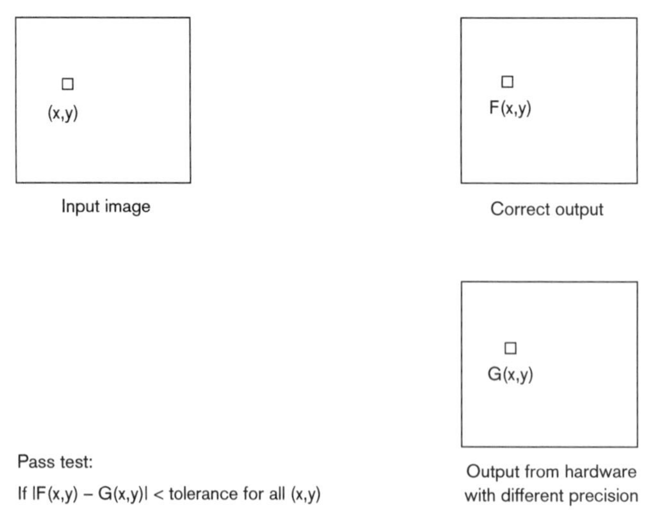

Figure 4.7 Verifying against tolerance

For operations that modify the geometry of the image, using, for example, complicated pixel address computations, it is difficult and impractical to check pixel locations within a tolerance. The problem is illustrated in Figure 4.8. Therefore, reference images should always be used when verifying these operations.

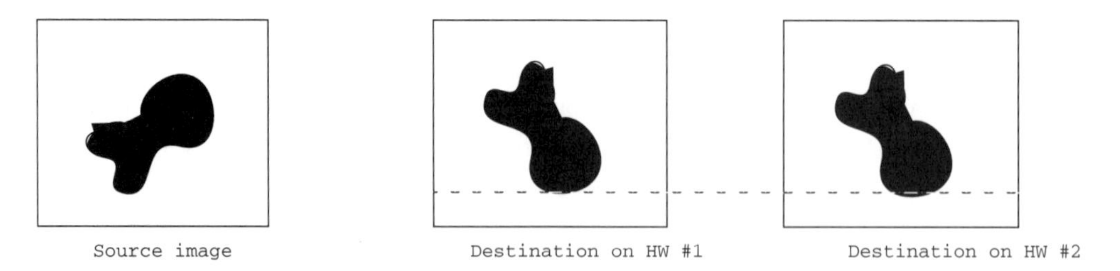

Figure 4.8 Why tolerances may not work for geometric operations

Example: The XILCH test suite The Xilch test suite was designed to test the XIL API library from Sun Microsystems. Xilch has undergone several modifications since its inception; in this section we restrict ourselves to Xilch accompanying XIL Version 1.1

Xilch's goal is to verify correctness of different releases of the software across different platforms. Verification is based on reference software, which is a known-to-be-correct, "reference" version of XIL executing on a known host. Xilch control flow is shown in Figure 4.9.

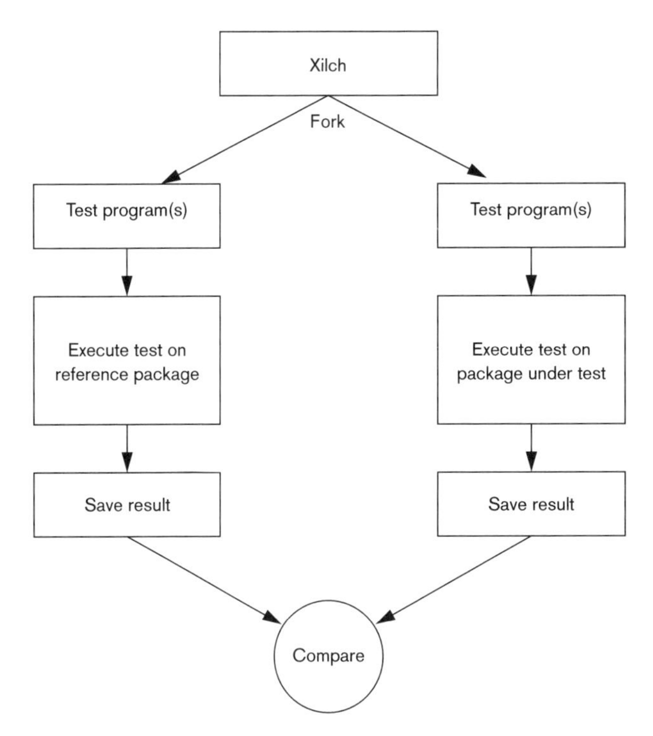

Figure 4.9 Flow of Xilch testing

Before starting the test, the user needs to set up two directories with the two versions of the XIL libraries: the reference and the one under test. The environment variables `XILCHREFHOME` and `XILHOME` are set to point to these directories.

The program `Xilch` is now executed. It uses the UNIX `fork()` command to create two processes, which are told the test programs to execute. One process uses the `XILCHREFHOME` library; the other uses the `XILHOME` library. Each process then executes the test programs. At the end of each program, the resultant images from the two versions are compared. An error is flagged if a difference is found.

In addition to verification using this methodology, `Xilch` also offers several other options. Since both the list of test programs and the list of images are specified in text files, new test programs and image matrices can be added provided that they are written in the `Xilch` format. Results can be saved in a log file. A different machine can be used as the reference host. Two different modes of verbosity (i.e., how much of the results are printed out) are allowed. `Xilch` also allows a tolerance to be used during testing.

4.6 Detailed design of imaging software

After functionality of the imaging software has been specified completely, a detailed internal design can proceed. The goal of this design is to plan an implementation that meets the functional specification and the requirements, and to scrutinize it for possible flaws.

The detailed design needs to address the implementation from several perspectives. It is useful for the internal design to define, in detail, the objects necessary for the software to function. Therefore, object-oriented design techniques are helpful. In addition, the algorithms that will be used need to be decided. This design is also an opportunity for tracing the control flow of the program through its various execution paths. Since performance is important, the inner loop of the program, the portion that does the actual computation over all the pixels, must be examined closely for possible suboptimal coding.

4.6.1 Object-oriented design in imaging

Object-oriented (OO) software design [1], an approach that enables rapid production of reliable and maintainable code, revolves around classes of objects. While software implementation in several languages lend themselves to OO design, C++ is the one used most commonly. In the following text, we discuss the advantages that can be gained from using OO design techniques in imaging using examples in C++. It is assumed that the reader is familiar with the fundamentals of OO design, which is covered very well in a number of texts.

The principles of OO design that are particularly useful for imaging software are *data abstraction and encapsulation*, *function overloading*, and *inheritance and derived classes*.

There are several benefits to be derived from object-oriented techniques when designing imaging software. Since many imaging operations are similar in nature, and require the use of the same objects (images) or similar objects (convolution kernels and structuring elements), code can be developed faster using the object-oriented techniques. The frequent dependence of imaging software on multiple hardware devices necessitates the separation of the device-independent portion of the code from the device-dependent part. This can be accomplished simply using object-oriented techniques.

Data abstraction and encapsulation Data abstraction is the definition of new data types that are well-suited to the application being programmed. When a new data type, or a *class*, is created, C++ enables the programmer to separate the *specification* of the object from its *implementation*.

As an example, consider the specification of a class for region of interest (ROI) in an API library. An ROI in an image is an area of the image over which the image processing operation is done. There are two general ways in which an ROI can be represented: as a list of rectangles or as a bit mask.

The *specification* of the ROI is the manner in which it is represented. It is contained in the file `Roi.h`:

```
struct Rect {int x, int y, int width, int height};
struct BitMask{ int x, int y, int lb, int width, int height};

class Roi {
private:
     Rect        *RectList; /* linked list of rects */
public:
     RectList    *getRectList();/* returns a rectlist */
     BitMask     *getBitMask();/* returns a bitmask */
}
```

Here, the variable `RectList` is called a *member variable* and the functions `getRectList()` and `getBitMask()` are called *member functions*. The implementation of the functions `getRectList()` and `getBitMask()` are in a file `Roi.cc`, which is compiled into `Roi.o`.

Now, suppose an application program called ImageEdit is written on this API. The header file `Roi.h` is included both in the file `ImageEdit.cc`, as well as in `Roi.cc`. This is shown in Figure 4.10.

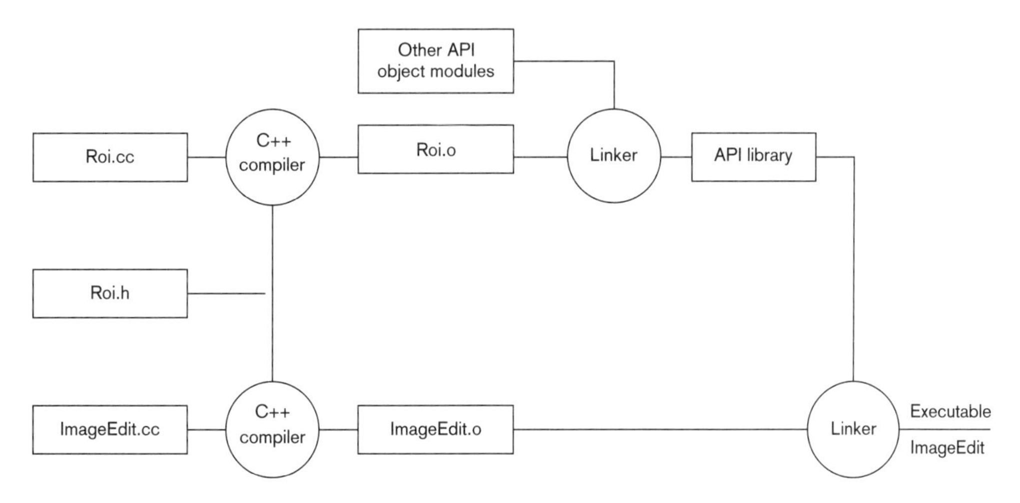

Figure 4.10 Data encapsulation using C++

Whenever ImageEdit needs an ROI in a RectList format, it calls `getRectList()`. If it needs an ROI in a BitMask format, it calls `getBitMask()`.

But—and here is a crucial advantage—ImageEdit cannot touch the internal representation of the ROI (in this case the *private* linked list of rectangles). The data for the ROI have been *encapsulated* in the ROI class.

The most important advantage of this encapsulation is that the API can, in the future, change the internal representation of the ROI (e.g., changing the coordinates to be floating point instead of integer, or representing it as a bit mask instead of a rectlist) without breaking the application ImageEdit. All that it needs to do is ensure that the functions `getBitMask()` and `getRectList()` behave in the same manner.

Another advantage is that during debugging, one knows exactly which function touches the internal representation of the ROI, reducing the number of candidates to search for bugs.

Inheritance and derived classes Inheritance is another useful feature of OO programming that can be used to the software designer's advantage when designing imaging software. It enables the programmer to express the commonality among different objects. First, a *base class* is defined with all the common functions. Then, objects are *derived* from this class, taking on its properties.

To design a base class, we look for functions that will be common to several types of objects. These functions can then be abstracted in the base class.

Consider, for example, the design of an imaging API library. Some objects used in this library are images, ROIs, lookup tables, convolution kernels, and structuring elements. There are some common functions associated with all of them. For example, they can be created and destroyed. There is a unique version number and type associated with each object, which can be read or written. Using this information, a class hierarchy can be constructed, as shown in Figure 4.11.

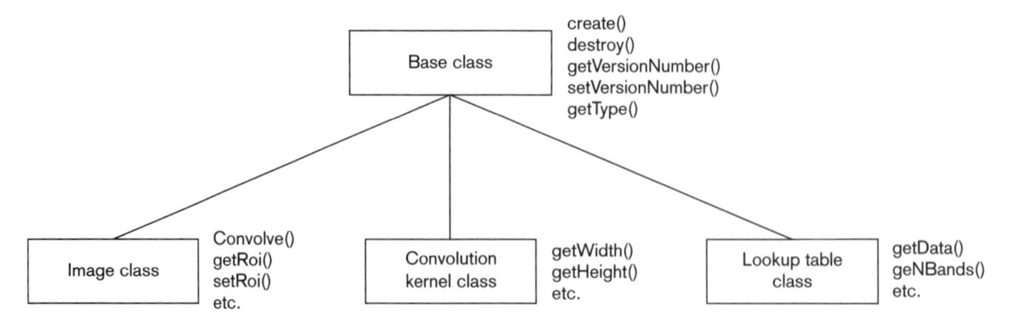

Figure 4.11 Derived classes for an imaging API

The base class is shown with its member functions `create()`, `destroy()`, `getVersionNumber()`, `setVersionNumber()`, and `getType()`. In the derived classes, these function names are overloaded with functions that do the appropriate operation for that class.

Three examples of classes derived from this base class—image, convolution kernel, and lookup table—are also shown in this figure. Next to the classes are shown member functions that are unique to that class. Thus, Image has many member classes that are unique to it, including all the image processing operations such as `convolve()`. Since there is an ROI associated with each image, this class also includes `getRoi()` and `setRoi()`. The convolution kernel and lookup table classes also have their unique functions. The lookup member function `getData()` is another example of data encapsulation, whereby the internal representation of the table is hidden from the user, and `getData()` is guaranteed to return a pointer to the table data in a prespecified format.

Function overloading Function overloading enables the same function name to be used for the different objects. For example, to create an object, a member function called `create` can be used. Then, `create()` can be used to create images, ROIs, convolution kernels, lookup tables, and so on.

This increases the readability of the code. If several programmers are working on the project, it makes it easier for them to remember the name of the function.

Having summarized some of the useful features of OO design, we can now examine the internal design specification in more detail.

4.6.2 Internal design specification

The internal design specification should contain details of how the imaging software is implemented internally. It should contain information about the objects being used and the general control flow of the program. It should also contain details of the inner loop of every image processing operation in the package.

Internal design considerations: useful objects The most important object is the image. In addition, other important objects include regions of interest, convolution kernels, lookup tables, structuring elements, and geometric transformation matrices.

The following set of questions serves as a guideline for the information that should be present in the document. A common set of questions for all objects is followed by more specific questions for individual objects.

Common set of questions for all objects:

- *How and when is the object created and destroyed?* For an application program, these questions are answered relatively simply, However, for an API or lower-level library, this can be quite tricky. For example, XIL follows an on-demand model where images can be created any time, but they are allocated only when there is actual demand for the image. Destruction of objects is straightforward, but deciding when to destroy them requires examination of the trade-offs between possible memory leaks and the expense of having to create and destroy objects frequently.

- *What parts are private and what parts are public? Why?* Again, this is not so much of an issue for an application program as it is for an API library. In general, the rule is to make public only those parts that are needed by the user.

- *How are device-independent and device-dependent parts of the object separated?* For a software that executes on multiple platforms, the separation of device-dependent and device-independent parts must be explicit. For example, an image object can have a public member function that returns a description of the device-dependent portion of the image.
- *What is the precision of the data in the object? (For example, for images, are the pixels 1/8/16/32 bits? Are convolution kernel values in floating point?)* Unless it is designed for a special market, the imaging software needs to handle 1/8/16/32-bit pixels. Floating-point pixels are helpful but not necessary for the general imaging market, but they are necessary for technical applications. The precision requirements are driven by the precision of current image acquisition devices, as well as the intermediate precisions required by the image processing functions.
- *Is there a version number to indicate when the data is changed?* It is useful to assign a unique version number to the object. The version number changes if the data are changed. This is useful for caching the data.

Additional questions for the image object:

- *What is the precision of the pixels?* This question has been addressed above. The precision of pixels supported is a key feature of the software.
- *What image attributes are there (origin, region of interest)?* Some common image attributes have been discussed in Chapter 1. One consideration is that the addition of an attribute causes substantial increase in the software complexity. This calls for a conservative approach. Only those attributes that are absolutely necessary for the target user need be supported. For example, if the target application of an API library performs geometric rotations of the image, then it makes sense to support image origins, since images are usually rotated around an origin.
- *Are subsets of images allowed? What are their attributes?* If the user wants to use only a portion of the image, then subsets of images may be used. This is distinct from the ROI, which usually represents a read mask for the source image(s) and a write mask for the destination image(s). If image subsets are allowed, methods must be presented for their creation and modification. The implementation should be capable of supporting the subset as if it were an independent image.
- *How are the bands of the image structured?* Two common formats of the bands of an image are band-sequential and band-interleaved (discussed in Chapter 1). Hardware memory access modes need to be considered before choosing one (or both).
- *What is the method moving the image from one device to another?* If images can reside on multiple devices, then the software needs to provide some mechanism for moving them.

- *What is the method for making the pixels of the image available to the user?* This question is for API libraries. Since the gamut of image processing functions is so large and diverse, no API library can supply all the possible functions (and special cases of functions) that an application program may need. A method is needed to enable the user to access the pixels. The implementation needs to support this method correctly and efficiently.

Additional questions for the ROI object:

- *How is the ROI represented internally?* Two possible internal representations are a bit mask or a list of rectangles.

- *What is the precision of the internal representation? Is it sufficient for all applications?* For most imaging functions, integer precision for the coordinates is sufficient. The exception is in geometric manipulation of images, where floating-point precisions may be necessary.

- *What is the method of getting a list of rectangles from the ROI? What is the method of getting a bit mask from the ROI?* Functions must be specified for the user to get the ROI in a given format.

- *How are the ROIs manipulated (e.g., intersection, rotate, scale)?* ROIs need to be intersected for most imaging functions. Geometric operations also need ROIs to be rotated, scaled, and so forth. The implementation of these operations must be addressed.

Additional questions for the lookup table (LUT) object:

- *How is the LUT's data represented internally?* LUTs can be single-banded or multi-banded. Data layout is easy for single-banded LUTs but requires some thought for multi-banded LUTs.

- *If clipping is allowed, how it is implemented?* Clipping of lookup functions is addressed in Chapter 5. If clipping is supported, then the implementation needs to address that.

Other useful objects are convolution kernels, geometric transformation matrices, and morphological structuring elements.

Internal design considerations: choice of algorithms If high performance is the goal of the software, then the choice of algorithms can be thought of as the "strategy" required to meet that goal. That is, the algorithms represent the longer-term planning that will make the product successful. (Local optimization and code-tuning are the "tactics," as we will see in Section 4.5.5.)

Once the functionality is chosen, the algorithm used to implement the functionality must be decided. The decision of the chosen algorithm depends to a large extent on the architecture of the machine for which the software is being designed.

Some questions to consider when choosing the algorithm are:

- *What internal precision is required for execution of the algorithm? Does it match the capabilities of the machine?* Many algorithms require floating-point precision of the intermediate values. For example, some fast algorithms to implement the Discrete Cosine Transform require floating-point precision. If the computer does not have a dedicated floating-point processor, or a microprocessor capable of floating-point instructions, this would be an unfortunate choice. On the other hand, if a floating-point processor is available, then it can be exploited for faster algorithms.

- *What types of instructions does the algorithm execute most of the time? Do they match the hardware?* If a multiplication is done using a software trap on the processor, then the algorithm should be such that the number of multiplications is minimized. An example of this is the implementation of convolution using a lookup table.

- *What is the order of data access of the algorithm (i.e., the order in which pixels are read and written)? Does it match the hardware?* Many "fast" algorithms, which reduce the number of operations required for a particular function, do so at the expense of reading and writing pixels in a non-scan-line fashion. Therefore, care must be taken to accurately assess the cost of lost locality of reference and the associated cache misses.

- *Does the algorithm have the potential to exploit the cache structure of the machine? Can important data structures be cached to reduce read/write overhead?* For example, if the algorithm relies on a table that can be cached, it is better off. An example of this is the convolution function using lookup tables.

- *For special-purpose hardware, is the algorithm matched optimally to the primitive operations of which the hardware is capable?* Frequently, we come across hardware that was optimized to do a small set of imaging operations. In this case, casting the algorithm in terms of the hardware primitives yields big gains. An example is the use of a convolver and lookup table hardware to do morphological image processing (discussed in more detail in Chapter 6).

- *For parallel processors, is the algorithm parallelized to the fullest extent possible? For multiprocessor systems, is the algorithm amenable to splitting up the work between different processors?* In general, the more the parallelization, the better the speedup. Considerable work must often be expended to parallelize algorithms.

Internal design considerations: control flow The control flow of the device-dependent portions of most imaging software is relatively simple. The driver gets the parameters from the device-dependent part, then crunches all the pixels according to the algorithm.

Some questions that the control flow needs to answer are:

- *What are the execution paths for each function?* This can be illustrated using a flowchart of pseudocode.

- *How are errors handled?* Error handling needs to be graceful, particularly if it occurs at the application level. On the other hand, low-level drivers cannot afford to perform elaborate error checking for performance reasons.
- *What is the execution path of the interaction between device-dependent and device-independent parts of the software?* The internal interactions between device-dependent and device-independent parts of the software need to be examined in detail. If the user is capable of transmitting device-specific hints (e.g., "use fast memory for this operation") through the device-independent layer, the implementation must support this.

Internal design considerations: the inner loop The inner loop of the program processes all the pixels by looping over the entire image. It should be designed so that it implements the chosen imaging algorithm in a manner that handles all possible image formats correctly and efficiently.

Image layout plays a large role in the format of the inner loop. A simple case occurs when the rows of the image are contiguous (case A in Figure 4.12). We illustrate the inner loop for this case by showing a lookup table operation:

```
/* initialize input image and output image pointers */
...
/* loop over the entire image */
count = rows*cols + 1; /* one more because of the pre-decrement */
while (--count) {
     *out++ = *(lut + *in++);
}
```

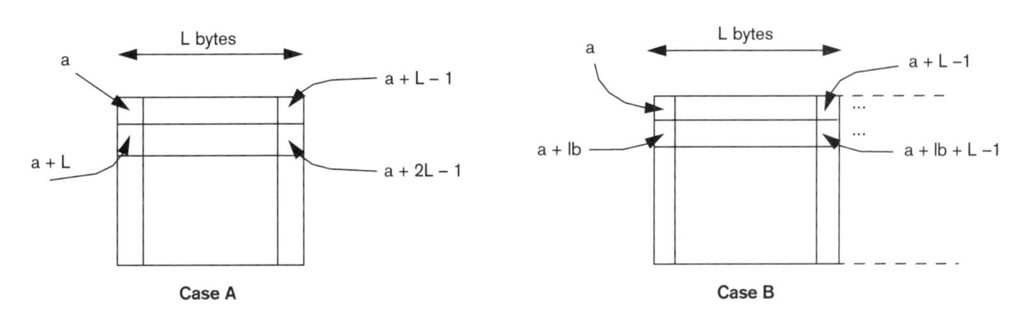

Figure 4.12 Image layout: contiguous and noncontiguous rows of an image

A more general image layout introduces a new variable called *linebytes* (*lb* in the above diagram). This is done to allow, for example, many-banded images. In this case, the inner loop actually becomes two loops, and our lookup table example becomes:

```
/* initialize input and output pointers */
/* save # of cols, initial input and output pointers */
...
/* loop over the entire image */
while (--rows) {
     while (--cols) {
          *out++ = *(lut + *in++);
     }
     in = (in_saved+=linebytes);
     out = (out_saved+=linebytes);
}
```

Note that both the above examples assume that the image is single-banded. For multi-banded images, another nested loop at the innermost level is required to loop over the bands of every pixel.

From the performance standpoint, the design of the inner loop is critical. This is because any suboptimality on the inner loop is amplified by the number of times it is executed, which is very large. We shall study this program in more detail in Section 4.6.4 to see how to performance-tune the inner loop.

4.6.3 Design and code reviews

In any engineering project, reviews can serve a useful purpose. By exposing the design (before it is implemented) to the scrutiny of others competent enough to examine it critically, flaws can be found and the soundness of the design ideas can be tested.

For imaging software, design and code reviews can be used to great advantage to produce reliable and efficient software in a reasonable amount of time. The hierarchy of the reviewing process is shown in Figure 4.13.

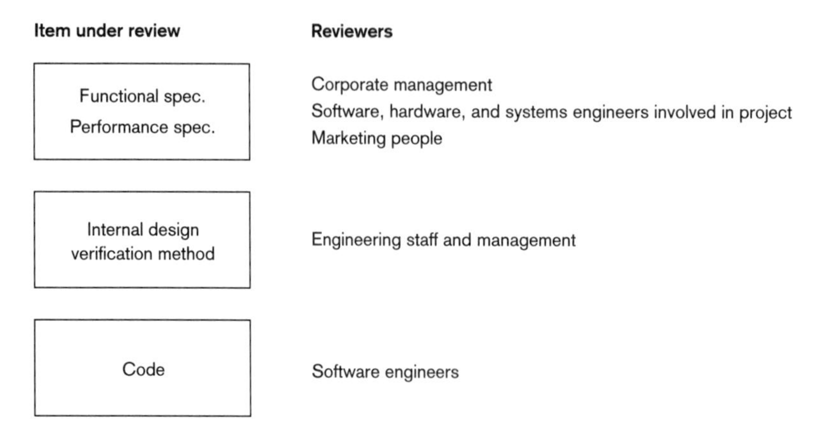

Figure 4.13 Review process for a software imaging product

The functional specification of the product should be reviewed by people with different tasks. From the corporate viewpoint, it should be examined to see how it fits in with the goals of the company. From the marketing perspective, it should meet the requirements of the product. Engineers should examine it to make sure that it is possible to accomplish with the resources at hand. It is very important, at this stage, to detect deficiencies in the external interface of the product. The performance specifications also need to be reviewed here, but it makes more sense for the verification specification to be reviewed with the internal design.

The internal design is to be reviewed primarily by engineers to ensure a sound design and detect any flaws in it. This should include examination of the appropriateness of the objects and algorithms, the flow of control, and the logic of the inner loop.

As implementation proceeds, samples of the code should be examined by peer engineers to catch possible bugs and suboptimalities. Code reviews are useful particularly for programs where suboptimalities are suspected.

4.7 Implementation of imaging software

After the design of the software is reviewed and approved, it can be implemented. Several choices must be made before implementation can proceed, however. These include the choice of algorithms and components of the development environment. These choices have to do with the development environment—for example, the tools used and the way code is managed. After the code is complete and correct, we need to examine the performance and "tweak" as needed.

4.7.1 Development environment and tools

The software environment where the product is developed plays an important role in the quality and speed of the software development. Some components of the development environment are the operating system, compilers, debuggers, performance tools, and code management systems. It is necessary for the development environment to be friendly to the software development effort.

Operating system The duty of the operating system (OS) is to manage the resources of the computer. Imaging software development poses various demands on the operating system. These range from the need for multitasking (for debugging) to virtual memory (for large images) to the ability to fork (for verification purposes). The OS also needs to support an accurate timer function for performance studies.

In most situations, once the target platform has been decided, there is little the programmer can do to influence the choice of the development OS, which ends up being the same as the target's.

Compiler While most modern compilers are of good quality, the compiler being used for the development effort must be scrutinized for its optimizing capabilities. In particular, it must understand the strengths and weaknesses of the target processor and be able to optimize the inner loop accordingly.

Debuggers Perhaps the simplest debugging tool for imaging software is displaying of the result of the imaging operation, where blatant errors in addressing and precision are detected easily. However, for more subtle errors, a source code–level debugger (also known as a symbolic debugger), such as the SPARCworks Debugger, is required. The debugger must be able to do address calculations easily, since debugging imaging software often requires examining the values of many pixels. It must be able to print out the contents of memory in whatever format is convenient for the programmer. If C++ is being used, the debugger must be able to understand C++ classes, particularly derived classes.

A more advanced class of debuggers tracks the memory addresses that are accessed by the program. These debuggers are useful for detecting memory leaks, which are a type of insidious bug caused by the failure to free up allocated memory. When a program is kept running for some time, it starts to slow down because the available memory decreases. Advanced debuggers, such as Purify, Sentinel, or Centerline, detect these errors.

If the target environment is different from the development environment, then of course the debugger needs to execute on the target environment.

Debugging utilities While debugging imaging software, it is often helpful to write additional debugging utilities. These are short programs that print out the pixels in a given format. An example is the function `pci_printblock`.

```
pci_printblock(char *sa,int xoff,int yoff, int w, int h, char *s)
where    sa = address of first pixel in the block
         xoff, yoff = x and y offset of area to be printed
         w,h = width and height of area to be printed
         s = character string to be printed as title
```

Code management tools If several people are working on a development team, a code management system, which keeps track of the history of the software under development, such as the Source Code Control System (SCCS) or Teamware, is necessary.

SCCS allows one master directory of the source files. Individual programmers can "check out" files, edit them, and "check in" the files after changes are made. One person can edit one file at a time.

Teamware defines *workspaces* which are hierarchies of SCCS directories. Programmers may have their own workspaces, which they "bring over" or "put back" into the master workspace. This enables more than one person to work on one file.

Performance profiling During performance-tuning, the question most frequently asked is, "Into what portion of the software is the time going?" To answer this question, a performance profiler, such as `prof`, `gprof`, or the SPARCworks pair Collector/Analyzer, is helpful. All of these tools collect statistics on how many times subroutines are being called and what time is spent inside them, which is vital for performance analysis.

Simulators Simulators are useful when software must be developed in conjunction with hardware, and debugging must take place before the actual hardware is available. The simulator needs to model the hardware machine accurately.

Other tools Many other software tools can be useful in imaging software development. These include make for compiling and linking programs, m4 macros for minimizing the source code needed for similar functions, lex and yacc for generating parsers, and various clever editors. However, all tools have a time and a place for use, and the utility of a tool is largely a function of the soundness of the programmer's judgment in deciding to use it.

4.7.2 Utility software

Most imaging software requires some support software in addition to the main software. This software is usually written by the programmers writing the imaging software. For standard operations such as file conversion, off-the-shelf software may be available for some platforms.

Utility software includes software that reads, writes, and converts between the dozens of different types of image file formats. If compressed file formats, such as .jpg or .mpg, are to be supported, then the software must also be able to perform JPEG and MPEG compression/decompression. Other kinds of support software include those that parse commands, maintain a linked list, sort and search, handle generic region of interest operations (such as adding a rectangle), and in some instances, provide deferred execution support. For unit testing, utility programs that generate test images are useful.

4.7.3 Coding, compiling, and unit testing

Once the design and the development environment are in place, coding can proceed. After each piece is coded, it needs to be compiled and linked. Then a short test program can be written that exercises that functionality of the code. The programmer can verify the correctness of the output by executing the program with one or more known source images.

4.7.4 *Common errors and debugging hints*

As in any other software development process, the software engineer encounters many different types of errors when developing imaging software. However, the principal kinds of error can be broadly classified into three categories: addressing errors, precision errors, and memory leaks. These errors can be corrected relatively simply by using the proper tools.

Addressing errors Addressing errors occur when pointers are not updated correctly. They are most common in the inner loop. They are easy to detect if the output image can be displayed, because the problem becomes visible. Then, running a debugger and stopping at the coordinates where the error first occurs should resolve this problem. A trickier version of this error occurs if a pointer incorrectly tries to access outside the program's address space, causing a Bus Error or Segmentation Fault. Again, the bug can be readily detected by running a debugger on the program.

A more serious error occurs when the code accesses outside a data array, which may cause intermittent crashes. These errors can be caught with a symbolic debugger, as well as other software debugging tools (such as Purify and Sentinel).

Computation errors These errors are caused by incorrect computation of the pixels. They can be simple or tricky depending on the gravity of the error; in general, the more blatant the error is, the easier it is to find and fix.

Frequently, computation errors are caused by the use of fixed-point arithmetic to emulate floating-point precision in the hardware. When debugging this kind of error, it is helpful to use a calculator to simulate the fixed-point arithmetic by hand to make sure that the desired result is indeed obtained by the method used. If the hand calculation yields incorrect results, then the method, or algorithm, is flawed, and a new method of doing the computation must be designed. If the correct result can be arrived at, then there is a bug in the code, which can be located by carefully examining the intermediate values generated by the code.

Memory leaks Memory leaks are caused when dynamically allocated memory is not freed up after its use. In the early days of imaging, when programs were executed on a "one-shot" basis, memory leaks were not important because when the program exited after execution, any dynamically allocated memory automatically freed up. However, modern imaging programs run under window systems and do not exit after executing just one function. Every time the leaky code is executed, it takes up more of the swapspace, until the swapspace becomes very small and the program performance becomes sluggish.

Memory leaks are potentially treacherous because they cannot be detected by the usual testing methods, which check for correct functionality of every function in the program. They show up when the program is being demonstrated or being used by a customer on a regular basis.

Since memory leaks are difficult—extremely difficult when source code is not available—to detect, commercially available tools should be used to track them down. Two of these tools are Purify and Sentinel.

4.7.5 *Measuring and tuning performance*

Performance measurement and tuning is an integral part of imaging software development. It is done only after the software becomes reasonably stable and bug-free, because functional errors almost always take precedence over suboptimal performance.

Measuring performance Once the program has been written, performance measurement can proceed. An initial estimate of the performance can be done using a timer function, such as the UNIX command `time`. Executing the program test under time:

```
time test
```

prints out the time it took to execute `test`. This kind of timing is very rough and can be actually misleading because there may be a large portion of setup time and system time.

A more involved effort requires the use of the UNIX subroutine `gethrtime()`. A code fragment illustrating this is as follows:

```
#include <sys/time.h>

main()
{
        hrtime_t start, end, delta;
        float seconds;
        int iter = 10;

        start = gethrtime();
        for (i=0; i<iterations; i++)
             /* call the imaging function here */
        end = gethrtime();
        delta = end - start;
        seconds = ((double) (delta))/(1000000000.*iterations);
        printf ("iterations: %d. Time per call %f sec\n",iterations, sec-
onds);
        printf ("Mpixels/sec: %f\n",(float)((HEIGHT*WIDTH)/(1000000.*sec-
onds)));
}
```

Performance-tuning Performance-tuning is an iterative process where the software is tweaked until the desired performance is achieved. In theory, the software should run correctly and optimally the first time it is executed. In practice, however, the correctly executing program rarely runs optimally the first time.

The debugging and tuning of an imaging program typically follows this sequence:

1 Write program.

2 Verify correct execution.

3 Measure performance.

4 If performance is unacceptable…

5 Analyze the program and find bottleneck.

6 Fix bottleneck.

7 Go to 3.

Now we illustrate this methodology with a simple example of performing a lookup table operation. A naive implementation of the lookup table program is shown in the following code segment.

```
#include <stdlib.h>
#include "pci.h"

/*
        Function : pci_lookup

        Operation: Perform an 8-bit table lookup operation on an image
                   Naive, suboptimal implementation

        Input:     src is the source image
                   lut is the lookup table

        Output:    dst is the destination image
*/

void
pci_lookup(pci_image *src, pci_image *dst, unsigned char *lut)
{
        unsigned char *spix, *dpix;
        int dlb, slb, w,h;
        int i,j;

        /* get parameters from the images */
        spix = src->data;
        slb = src->linebytes;
        dpix = dst->data;
        dlb = dst->linebytes;
        w = dst->width;
        h = dst->height;

        /* pass the image through the table */
        for (i=0; i<w; i++)
             for (j=0; j<h; j++)
                  *(dpix+dlb*j+i) = *(lut + *(spix+slb*j+i));
```

There are several things wrong with this program. The image is being traversed along rows rather than columns, resulting in cache misses. Also, there are two unnecessary multiplications inside the inner loop. The following program is used to measure the performance of this routine.

```
main(int argc, char **argv)
{
        pci_image src, dest;
        unsigned char table[256];
        hrtime_t start, end, delta;
        int i, iterations = 10;
        double seconds;

        /* allocate the images */
        pci_alloc_image(&src,HEIGHT, WIDTH, 1);
        pci_alloc_image(&dest, HEIGHT, WIDTH, 1);

        /* initialize the lookup table */
        pci_init_lookup_8(&table[0]);

        /* load the image */
        ...

        start = gethrtime();
        for (i=0; i<iterations; i++)
             /* perform the lookup */
             pci_lookup(&src, &dest, &table[0]);
        end = gethrtime();
        delta = end - start;
        seconds = ((double) (delta))/(1000000000.*iterations);
        printf ("iterations: %d. Time per call %f sec\n",iterations, sec-
onds);
        printf ("Mpixels/sec: %f\n",(float)((HEIGHT*WIDTH)/(1000000.*sec-
onds)));
}
```

On a SPARCstation 10SX, it runs at a little over 2 Mpixels/sec. What can be done to improve this performance? A second attempt at rewriting the inner loop results in the following code.

```
/*
        Function : pci_lookup2

        Operation: Same as pci_lookup
                  Somewhat more optimal implementation

        Input:     src is the source image
                  lut is the lookup table

        Output:    dst is the destination image
*/


void
pci_lookup2(pci_image *src, pci_image *dst, unsigned char *lut)
{
        unsigned char *spix, *dpix, *dpixptr, *spixptr;
        int dlb, slb, w,h;
```

```
        int i,j;
        unsigned char *lutptr = lut;

        /* get parameters from the images */
        spixptr = spix = src->data;
        slb = src->linebytes;
        dpixptr = dpix = dst->data;
        dlb = dst->linebytes;
        w = dst->width;
        h = dst->height;

        /* pass the image through the table */
        /* move through one row at a time to minimize cache misses */
        for (i=0; i<h; i++) {
             for (j=0; j<w; j++) {
                  *dpixptr++ = *(lutptr + *spixptr++);
             }
             dpixptr = (dpix += dlb);
             spixptr = (spix += slb);
        }
}
```

This program, when compiled with the optimization flag turned on and executed, processes over 8 Mpixels/sec.

For RISC machines, a further improvement can be made to this program by "unrolling" the inner loop. This helps because the autoincrement does not come for free on RISC machines; it is yet another instruction. This would result in code that looks like this:

```
        /* pass image through the lookup table */
        while (count -= 8) {
             dst[0] = *(lutptr+src[0]);
             ...
             dst[7] = *(lutptr+src[7]);
             dst += 8;
             src += 8;
        }
```

Another source of optimization is to load 32-bit quantities into the registers, and then strip off one byte at a time and perform the lookup on these bytes, one at a time. This method works for the cases where loading and storing are expensive operations. By carefully studying the assembler code generated by the compiler, many such time-saving tricks can be employed to execute the code faster.

Use of a profiler Most real-life programs are not as simple as `lut.c`. When a program does not perform as expected, it is seldom easy to locate where the time has gone. A profiler, such as `prof` or `gprof` in UNIX or the Collector/Analyzer tool available under the SPARC-works debugger, prints out statistics from execution of the program, such as where the most time was spent, how many routines were called, and so forth.

When the Collector and Analyzer are used to analyze the performance of the initial version of the lut program, statistics on the program are first gathered by executing the program under the debugger with the collector turned on. After the program finishes, the Analyzer is executed for the gathered data. The information supplied by the Analyzer will point to the fact that the multiplication inside the inner loop is taking up the lion's share of the time.

4.8 Maintenance of imaging software

After an imaging software product is introduced and shipped to customers, it must be maintained. This means that bugs must be fixed, features added, and the product improved over time. Here we discuss some of the software engineering techniques that can be used to maintain an imaging software with the least expense.

4.8.1 Tracking and fixing bugs

As bugs begin to be reported from the field, it is important to keep track of them in a systematic manner. A bugtracker is a database for maintaining a list of bugs.

Each record of the database corresponds to a unique bug. The record has fields for the software version which has the bug, summary of the bug, detailed description of the bug (including a method to replicate it), its severity, the priority for fixing, the person who reported the bug, and the responsible manager, as well as assorted information about the software and hardware release under which the bug was seen.

In addition, to track the progress of the bug, a number of states are associated with each bug. One possible set of such states is shown Figure 4.14. After the bug is reported, it is examined and replicated by the relevant engineer. If it is really a bug (as opposed to user error), it is accepted. Then the cause of the bug is investigated and a fix is found. This fix is applied and the program tested to ensure that the bug is indeed fixed. After this, the bug fix is integrated into the software (for example, by checking in the file into SCCS, or doing a putback under Teamware). The software is then built at a neutral area and the bug fix is verified, after which the bug can be closed. During these stages, the state of the bug database is always kept updated so that anyone interested in the bug can find out its status.

4.8.2 Adding functions and features

Users of the software imaging product frequently ask for new functions and features. This happens because the number of imaging algorithms that have been developed over the years is very

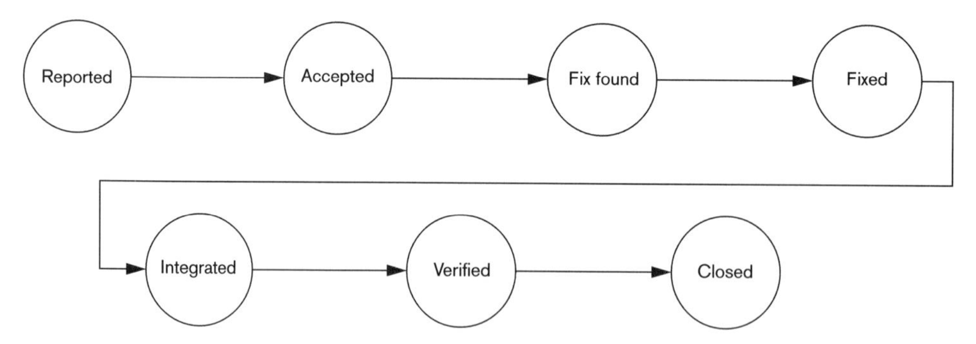

Figure 4.14 States of a bugtracker program

large, and it is not possible for one software to support all of them. Unfortunately, it is usually not possible to accommodate all the requests and they must, therefore, be prioritized. Some questions to ask before adding a new feature or a new function are:

- Why is it important to add this feature (more sales, aesthetically pleasing, an important customer wants it, etc.)?
- Will it be used by many people?
- How much engineering effort is required to add this feature?
- Will the addition of this feature break anything else?
- What is the cost of not adding this feature (e.g., fall behind competition)?

Once these questions are answered favorably, the new feature can be added for a future release.

4.8.3 Controlling code complexity

Rarely does the same software engineer stay on to maintain the imaging program he or she developed. Therefore, it is important to control code complexity. This means that the next programmer who is called in to fix a bug should be able to read and understand the code and decide why things were done the way they were.

For a target machine which is a general-purpose computer, this is taken care of relatively easily by putting appropriate comments in the code, which explain *why* rather than *how*. For special-purpose hardware it is more difficult to maintain the code and OO techniques are very useful.

4.9 The porting guide

The porting guide is intended to help software engineers port the imaging software to different hardware devices. It should contain all the information needed by the engineer about the interfaces that are available to interact with the core portion of the software.

The ideal porting guide consists of a set of *interface* definitions which the ported device driver must use to communicate with the core software. This interface consists of methods to allocate and access memory, methods to add device-specific functions to the software, and methods to access input/output devices. These methods should be clarified with examples.

For example, for the XIL API library, whose architecture is shown in Figure 4.2, porting to a new hardware device consists of writing a new set of hardware devices for the storage, input/output, and compute drivers. Therefore, the porting guide should consist of all the interfaces that are available to the software engineer to write a set of three drivers for a particular device. As an example, consider a function that allocates image memory. There is a prescribed way to allocate the memory (or to propagate it between different devices if it has already been allocated).

4.10 Internet programming using Java

In this section we will discuss software based on the Java programming language and its implications for imaging software. Java—along with the Internet—represents a new model for computer usage. We will describe this model, briefly discuss the Java language, and examine the implications for imaging software design.

4.10.1 The Java programming model

The Java programming model, shown in Figure 4.15, enables the transmission of *executable content* over a network.

The model consists of two phases: program compilation and program execution. During compilation, a Java program is compiled by the Java compiler. However, whereas most "normal" compilers compile a high-level program into assembly language for a particular microprocessor, the Java compiler compiles the Java program into *Java bytecode*. The bytecode represents the assembly language for the *Java virtual machine*—that is, a microprocessor that may be implemented in software or hardware. The Java bytecode therefore represents an executable program for the Java virtual machine.

Once the bytecode is constructed, it can be sent over the network for execution on a Java virtual machine in a computer possibly different from the one where it was compiled. Once it

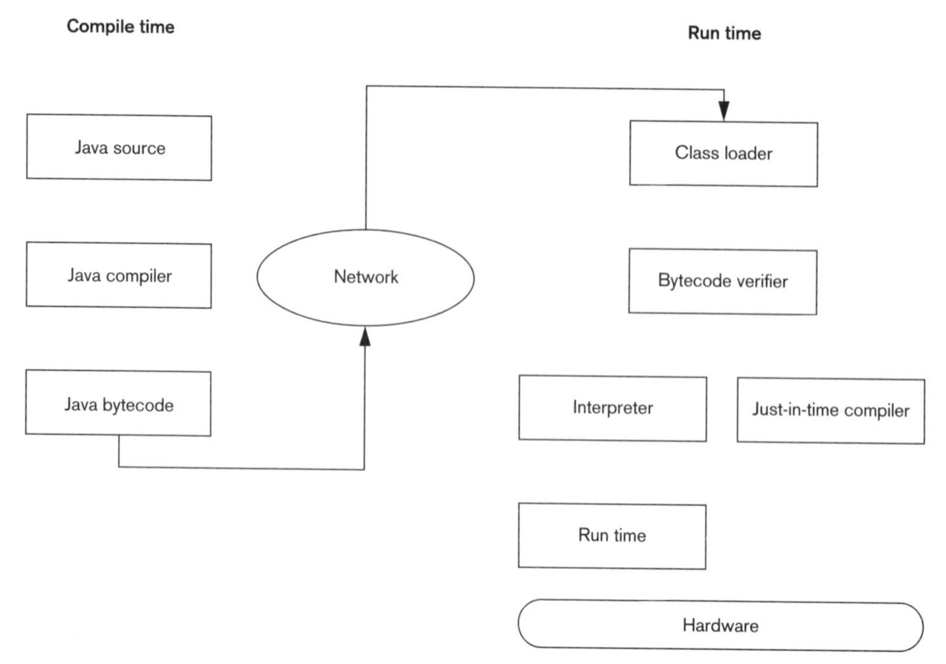

Figure 4.15 The Java programming model

arrives at the new computer, a bytecode verification is performed in order to verify that the executable code is not doing inappropriate things.

During program execution, the bytecode can be executed in one of two modes: interpreted or compiled. It can be interpreted using the Java interpreter; or it can be compiled into the underlying hardware's assembly language and executed as a compiled program.

A critical advantage of the Java model is that the Java virtual machine can be constructed on any computer platform. It has already been written for several popular computers including Windows, Windows NT, Sun, and Apple Macintosh. Therefore, the Java bytecode represents an executable that can be sent over the network and executed on any machine without the need to recompile.

4.10.2 *The Java language*

In this section we will briefly touch upon some of the key features of Java. Several texts have appeared on Java programming; the reader is advised to refer to them for detailed information.

Java is an object-oriented language that looks very much like C and C++. However, some key differences between Java and C/C++ are:

- *No pointer arithmetic* Java has the concept of arrays but not pointers. There are two reasons for this. First, since studies have shown that pointers are a major source of bugs in C and C++, their elimination reduces the number of potential bugs. Second, pointers can be used to access areas of memory that are out of bounds for the program. By performing strict bounds checking on its arrays, this security risk is eliminated by Java.
- *No typedefs, #defines, or preprocessor* These are no longer needed since all this information can be encapsulated in Java class libraries.
- *No need to free memory* Java has an automatic garbage collector, which frees up memory not being used. This eliminates memory leaks in programs.

In addition to the above changes, some other C/C++ features that have been removed include structs, unions, multiple inheritance, and operator overloading. Some of the new features available in Java include built-in support for multithreading and for the TCP/IP protocol for the Internet.

4.10.3 Implications for imaging software

There are several implications for imaging software design in the context of the Java programming model. This model enables portability and a level of dynamic programming that was not possible before. There are also certain performance implications.

The portability afforded by Java through its virtual machine concept is obvious. What is perhaps less obvious is the dynamic nature of the programming model. To illustrate this, consider the example of a Java application program, an Internet browser, contrasted with a more traditional Internet browser such as Mosaic. Examples of Java browsers include the programs HotJava and Netscape Navigator.

During execution, the Internet browser can go into many Internet sites and read data to be displayed on the user's computer. Some of this data can be image or video files. Since there are numerous image and video file formats, such as JPEG, GIF, MPEG, and so forth, the traditional browser such as Mosaic needs to have built-in support for all these file formats in order to properly display them. As a result, this browser can never be quite complete, and as features are added it becomes more unwieldy.

However, the Java-based browser does not need to have all this built-in support. Instead, the Internet site that provides the data also provides the Java bytecode to decode the image files. If the browser encounters an image file it does not know how to read, it simply asks the server for a program to decode the file.

This process can be generalized in terms of a client program (such as the browser program) and a server program (such as that which supports the Internet site), as shown in Figure 4.16. This feature can enable a new type of imaging application on the Internet, one that enables a higher degree of interactivity.

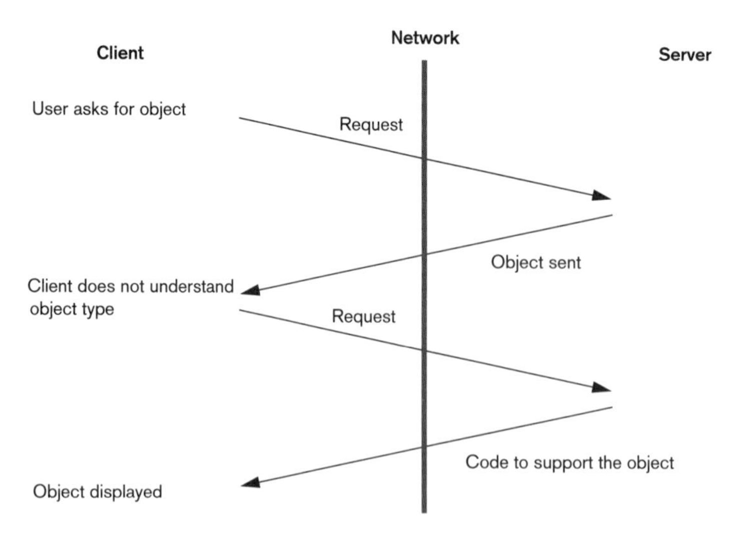

Figure 4.16 Dynamic nature of Java

Another important consideration for imaging programs is performance. Since Java is normally an interpreted language, it runs quite a bit slower than a corresponding C program. This, however, can be ameliorated is two ways. First, a just-in-time compiler can be used to compile the Java bytecode into native assembly code for performance-critical applications. Second, Java allows the invocation of processor-specific code and libraries using an interface called native methods. Using this, for example, an imaging library that has been written for a particular processor can be invoked by a Java application program.

Finally, before leaving the topic of Java altogether, we note that the Java program development environment comes with a rich set of class libraries, which will eventually include some imaging and multimedia capabilities.

4.11 Conclusion and further reading

Good imaging software is never an accident. It is the result of a thorough and meticulous design and implementation process. In this chapter we have delineated all the steps necessary for producing imaging software of high quality. These steps are generating the requirements, completing the architectural, functional, and detailed design; implementation; performance tuning; and maintenance. We have described each step in detail and tried to list the issues that need to be considered. It is hoped this chapter will encourage the implementor to develop his or her own design philosophy that can be useful in producing imaging software of high quality.

There is not much published literature on software design for imaging. However, there are several books available for object-oriented design. Meyer's book (reference 1) is considered the classic. Java is the subject of a spate of new books, such as reference 2. For an overall understanding, see reference 3.

4.12 References

1 B. Meyer, *Object-Oriented Software Construction*, Englewood Cliffs, NJ: Prentice Hall, 1988.

2 A. van Hoff, S. Shaio, and O. Starbuck, *Hooked on Java*, Reading, MA: Addison-Wesley, 1996.

3 B. Simpson, et al., Making Sense of Java, Greenwich, CT: Manning, 1996.

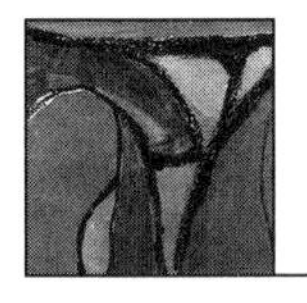

chapter 5

Image point operations

5.1 Introduction

Point operations are paradoxically some of the simplest, yet most powerful functions in image processing. A diverse group of operations can be grouped together in this category. Their common characteristic is this: The value of the destination pixel at a particular location depends on the value (or values) of the source pixel (or pixels) at that location. This distinguishes point operations from other imaging operations such as neighborhood operations where a neighborhood of input pixels can contribute to a single output pixel, or transform operations where an entire block of source pixels contributes to the computation of each destination pixel.

Point operations are useful for a variety of reasons. The simplest point operation is to copy an image from one memory location to another. Copying is used for moving images between physical memory and the frame buffer, moving images in and out of accelerators, and for updating image displays (for example, for a cine-loop display of a sequence of images to show motion). While copying appears to be a trivially simple operation, virtually all imaging products must implement copy in an efficient and robust manner. We will examine some subtleties involved in the implementation of the copy operation.

Another class of point operations is arithmetic and logical operations, which allows us to combine the pixels of one image with those of another image (*dyadic* functions) or with a constant (*monadic* functions) using arithmetic and logical functions. ALU operations make up the backbone of many imaging libraries. They are used in a wide variety of applications, from image enhancement to image analysis and medical diagnosis.

A third important class of point operations is table lookup, where the pixels of an input image are mapped to an output image using a mapping function loaded into a lookup table. Lookup is a versatile operation. It is difficult to think of an area of imaging where table lookup is not useful. While we have encountered the use of lookup in several areas of imaging, in this chapter we are concerned with a general table lookup function that can be used to perform a wide range of mapping functions.

Other point operations include histogram-based operations, image compositing using alpha blending, and band combination. The histogram gives us insights into the distribution of pixels in an image, which can then be used to enhance the image. Histogram stretching and equalization use the histogram to build a mapping function that can be used in a table lookup to increase the contrast of low-contrast images. Compositing of images is useful for blending different partial images to create a whole image, as well as for merging together more than one image in a controlled manner. The band combination function is another versatile function that allows interband combination of bands of an image, enabling, among other things, color space conversions.

This chapter covers these point operations commonly used in imaging. In Section 5.2 we examine image copying. Section 5.3 is about image ALU operations. The ubiquitous table lookup is considered in Section 5.4. The image histogram and operations based on it are considered in Section 5.5. Image compositing using alpha blending and the band combination of images are the subject of Section 5.5.

5.2 Image copying

A common, and often overlooked, operation in imaging is *copying* an image from one memory location to another. The memory location can be either physical memory, a frame buffer, or image memory in an accelerator. A robust copy function must be able to efficiently copy between all these devices. This is easiest when all of these memory locations are memory-mapped to a virtual address space.

While copying is an apparently trivial function, complications arise when we want to support copying into a display. We can no longer assume that the source and destination images are nonoverlapping, since images may be displayed on windows that overlap one another. Consider the examples in Figure 5.1, which show variations arising due to overlap between source and destination images.

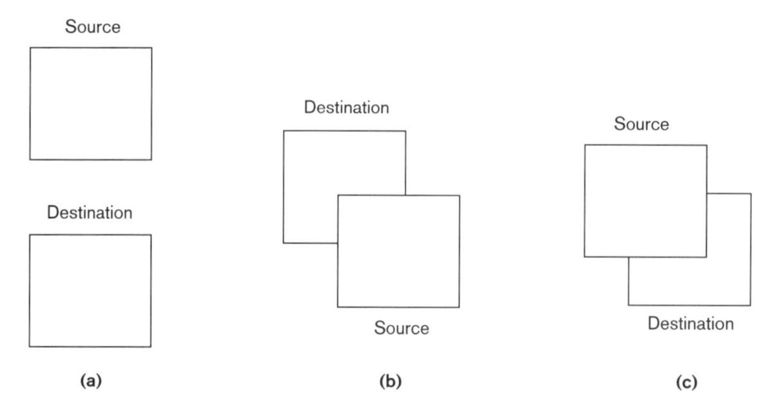

Figure 5.1 Variations of copying an image

During overlap, the goal is to correctly copy the entire source image into the destination image. In examples (a) and (b) in Figure 5.1, copying row by row can commence from the top row of the source and continue until the last row. However, in example (c), copying the top row of the source into the destination overwrites the source pixels occupying the same location as the top of the destination before they are used.

It can be shown that all the overlapping cases can be handled by two flavors of the copy function. These are:

- *copy()* This function copies source raster to destination raster, row by row, starting at the top row and moving down.
- *coyrl()* This function copies source raster to destination raster, row by row, starting at the bottom row and moving up.

In Figure 5.1, (c) is done by `copyrl()`, (b) by `copy()`, and (a) by either of the two.

5.2.1 Implementation of copy

The key to the efficient implementation of copy is to use available hardware features to the fullest extent. For example, graphics hardware often has built-in functions for performing "bitblt," an abbreviation for bit-block-transfer. These functions should be invoked whenever possible.

In general-purpose microprocessors, copy is a "load from memory" followed by a "store to memory." Often, there is an optimal size of this data that can be moved by these microprocessors. For example, in many microprocessors, it is optimal to move data in units of "double words," which are 64-bit quantities. A more general case is a "block" move instruction, which loads and stores blocks of data, in chunks of 16 bytes or larger. The block load and store instructions of the Visual Instruction Set discussed in Chapter 3 are examples of this kind of instruction.

Example: Duff's device A straightforward implementation of a copy routine is as follows:

```
copy (char *src, char *dest, int count)
{
        do
             *dest++ = *src++;
        while (--count > 0)
}
```

On a CISC processor, the inner loop translates into a move and a subtract-and-branch type of instruction. As we have seen in Chapter 4, one way to optimize the inner loop is to perform loop unrolling, which, in this case, decreases the number of subtract-and-branch operations executed. If this is done, we are left with a remainder portion that must also be executed.

Tom Duff [1] invented a clever way of handling this remainder loop, as shown:

```
copy(char *dest, char *src, int count)
{
        int n = (count+7)/8;
        switch (count%8) {
        case 0: do {      *dest++ = *src++;
        case 7:           *dest++ = *src++;
        case 6:           *dest++ = *src++;
        case 5:           *dest++ = *src++;
        case 4:           *dest++ = *src++;
        case 3:           *dest++ = *src++;
        case 2:           *dest++ = *src++;
        case 1:           *dest++ = *src++;
               } while (--count > 0);
        }
}
```

Experimental results on Sun workstations indicate that Duff's device offers about a 20 percent speedup over the brute-force approach shown in the earlier example. The speed of Duff's device is about the same as that of a loop-unrolling optimization.

Having looked at image copying, we now proceed to image point operations based on arithmetic and logical operations.

5.3 Image ALU operations

Image ALU operations are fundamental operations needed in almost any imaging product for a variety of purposes. In this section, we will refer to operations between an image and a constant as *monadic* operations, and operations between two images as *dyadic* operations.

5.3.1 Monadic image operations

Monadic image operations are ALU operations between an image and a constant. These operations are shown in Table 5.1—$s(x,y)$ and $d(x,y)$ are the source and destination pixel values at location (x,y), and K is the constant.

Table 5.1 Monadic image operations

Function	Operation
Add constant	d(x,y) = s(x,y) + K
Subtract constant	d(x,y) = K– s(x,y)
Multiply constant	d(x,y) = Ks(x,y)
Divide into constant	d(x,y) = K / s(x,y)
Divide by constant	d(x,y) = s(x,y) / K
Or constant	d(x,y) = K or s(x,y)
And constant	d(x,y) = K and s(x,y)
Xor constant	d(x,y) = K xor s(x,y)
Absolute value	d(x,y) = abs (s(x,y))

Monadic operations are useful in a variety of situations. For example, they can be used to add or subtract a bias value to make a picture brighter or darker. Multiplication and division are useful for simulating noise that can occur during image data acquisition.

Implementation of monadic operations is straightforward. However, in order to avoid possible confusion, one must address the issues listed in Section 5.3.3 before proceeding with an implementation. An example of a monadic operation on an image (Figure 5.2) is shown in Figure 5.3.

Figure 5.2 Snog Rock original

Figure 5.3 The number 0x66 xor'ed with the Snog Rock image

5.3.2 Dyadic image operations

Dyadic image operations are arithmetic and logical functions between the pixels of two source images producing a destination image. These functions are shown below in Table 5.2—$s1(x,y)$ and $s2(x,y)$ are the two source images that are used to create the destination image $d(x,y)$.

Table 5.2 Dyadic image operations

Function	Operation
Add	d(x,y) = s1(x,y) + s2(x,y)
Subtract	d(x,y) = s1(x,y) – s2(x,y)
Multiply	d(x,y) = s1(x,y)s2(x,y)
Divide	d(x,y) = s1(x,y) / s2(x,y)
Min	d(x,y) = min(s1(x,y), s2(x,y))

Table 5.2 Dyadic image operations (continued)

Function	Operation
Max	d(x,y) = max(s1(x,y), s2(x,y))
Or	d(x,y) = s1(x,y) or s2(x,y)
And	d(x,y) = s1(x,y) and s2(x,y)
Xor	d(x,y) = s1(x,y) xor s2(x,y)

Dyadic operations have many uses in imaging. For example, the subtraction of one image from another is useful for studying the flow of blood in digital subtraction angiography. Addition of images is a useful step in many complex imaging algorithms. An example of a dyadic operation is shown in Figures 5.4 and 5.5 (two originals) and Figure 5.6 (result of adding the originals).

Figure 5.4 Museum original

Figure 5.5 Bus original

Figure 5.6 Dyadic operation: Museum + Bus

5.3.3 Implementation issues in ALU operations

While ALU imaging functions appear to be trivially simple, there are several details that must be addressed before implementation. These are listed below.

- *What is the precision of arithmetic operations?* The desired precision of the arithmetic operation must be defined. For maximum precision, all the pixels and constants are floating-point numbers and the computations are performed in floating point. However, floating point is computationally expensive and acceleration hardware often performs fixed point arithmetic. Therefore, the acceptable tolerance must be understood and specified for arithmetic operations. (For example, if the operation is performed in integer mode, then adding the quantity 0.65 to a pixel may yield a result that is different from one obtained using floating-point arithmetic.)
- *What is the result of multiplication?* When an 8-bit pixel is multiplied with another 8-bit pixel, the result is a 16-bit number. While it is reasonable to assume that the destination pixel value will be the most significant 8 bits of these 16 bits, this must be clearly spelled out.
- *What happens when the pixel overflows or underflows?* In cases where pixels are represented in fixed point—for example, as bits, bytes, or shorts—the result of an ALU operation may be larger than the largest number (or smaller than the smallest number) that can be placed in one pixel. For example, for a pixel type of unsigned byte, the result of adding two pixels may be larger than 255. In this case, the most reasonable option is to *clip* the value of the destination pixel to remain between 0 and 255. The result of overflow and underflow must be specified for arithmetic operations.
- *How are constants specified for monadic operations?* For arithmetic operations, the specification of constants is straightforward—they are either signed integer or signed

floating-point quantities. However, for logical operations, there is a need to precisely specify how the constant is used in the operation. For example, what does it mean for an image of unsigned byte pixels to be or'ed with –5? Should the "Or constant" function allow only unsigned bytes as its input?

- *How are multibanded images handled?* In the case of multibanded images, we need to specify how ALU operations are performed. For monadic operations, the scalar constant is replaced by a vector of constants. For dyadic operations, the individual bands of each pixel are operated on separately.

5.4 Table lookup operations

Table lookup is one of the most versatile operations in imaging. In this operation, the value of the source pixel is used as an index into a table from which the destination pixel is looked up. Given a source pixel $S(x,y)$, a destination pixel $D(x,y)$, and a table $LUT[]$, the operation is defined as:

$$D(x, y) = LUT[S(x, y)] \tag{5.1}$$

In other words, table lookup operations are a point-to-point *mapping*, which maps the source pixels to destination pixels using a mapping function.

5.4.1 Uses of table lookup

Table lookup is useful in many areas of imaging. These areas include image enhancement, technical image processing, geometric image modifications, medical imaging, pixel data conversion, and so forth. Many implementation techniques described in this book use lookup tables. In this section, however, we are mainly interested in the use of table lookup in image enhancement, and in a general lookup table function that can be used for a variety of purposes.

Image enhancement The goal of image enhancement is to make an image more suitable for display or machine analysis. During the process of acquisition, the digitized image may suffer quality degradations due to poor scanning or due to limits in the acquisition process. A large arsenal of image enhancement techniques is available for improving the quality of these images [2].

Table lookups are useful in several classes of image enhancement operations. These include contrast enhancement including the so-called window-leveling function, other point mapping functions, and histogram-based functions. Here, we cover the first two classes of operations; the histogram-based functions are covered in detail in Section 5.3.5.

In contrast enhancement, the goal is to increase the contrast of a possibly low-contrast image. If the display dynamic range is larger than the dynamic range of the image, then the image contrast can be stretched using a table lookup. This is shown in example (a) in Figure 5.7, where the dynamic range of a source image is limited between $s0$ and $s1$, while the display dynamic range is between $d0$ and $d1$. By programming the lookup table with the straight line function shown, the dynamic range of the image is increased significantly. The function $LUT[x]$ is generated using:

$$LUT[x] = \frac{(d1 - d0)}{(s1 - s0)}(x - s1) + d1 \tag{5.2}$$

A variation of this method is used in an interactive method called window-leveling, widely used in medical imaging while viewing an image. The function loaded into the lookup table is shown in example (b) in Figure 5.7. The slope and intercept of the line can be adjusted interactively. It is desirable for the lookup table operation to update the image at interactive speeds as the line is updated.

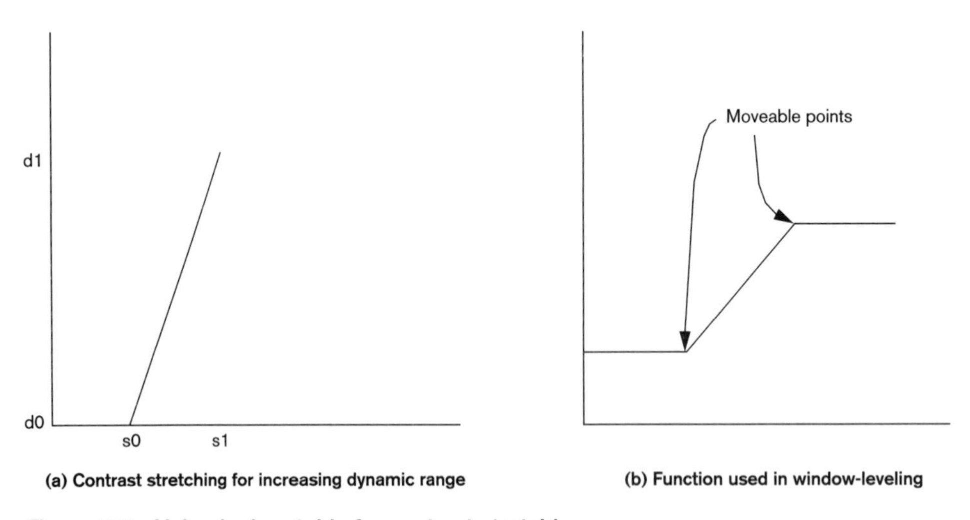

(a) Contrast stretching for increasing dynamic range

(b) Function used in window-leveling

Figure 5.7 Using lookup table for contrast stretching

Besides contrast enhancement, lookup tables are used to perform other kinds of one-to-one mappings on the source pixels. These include functions that can mimic darkroom effects, as well as those that make up for other deficiencies in the image or the viewing system.

The digital negative of an image can be found by applying the following lookup table transformation, where M is the maximum possible pixel value (255 for byte pixels):

$$LUT[x] = M - x \tag{5.3}$$

This yields an image looking exactly like a photographic negative.

A lookup table loaded with a thresholding function creates a binary image that looks like an extremely high-contrast image. The function is:

$$\begin{aligned} LUT[x] &= M, x > T \\ LUT[x] &= 0, x < T \end{aligned} \tag{5.4}$$

The threshold value T is usually found from the histogram.

Several other useful functions can be programmed into a lookup table for image enhancement purposes. Both references 2 and 3 offer a detailed look at these functions. These include a sawtooth function, power law function, rubber band function, the Gaussian error function, and logarithmic transformation. These functions are illustrated in Figure 5.8.

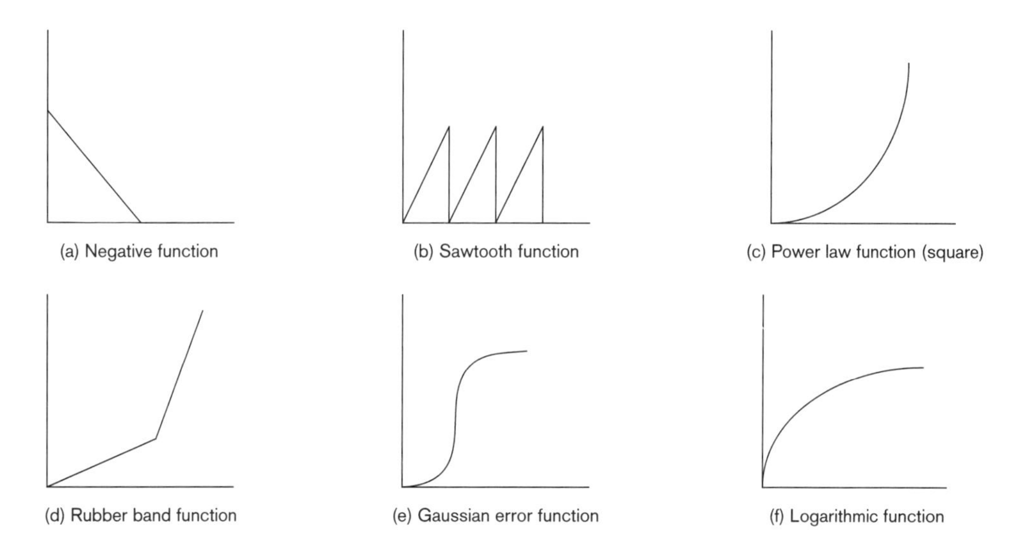

Figure 5.8 Various one-to-one transformations used in lookup table functions

The sawtooth function is useful for displaying an image on a display system that has a smaller dynamic range than the range of pixel values. The power functions are useful for compressing the low-valued pixels and highlighting the nuances in the bright areas of the image. The logarithmic function does just the opposite of this. The rubber band function is a piecewise linear approximation to the power law functions. The Gaussian error function acts like a square function for low pixel values and a square root for high pixel values

Examples of table lookup operations are shown in Figures 5.9 (original), 5.10 (the negative function), and 5.11 (a square function).

Obviously, the quality of the results obtained by these functions depends on the source image itself, and under proper conditions, all of these functions can yield good results.

Figure 5.9 Boat original

Figure 5.10 Boat negative

Figure 5.11 Boat, square lookup table

Other uses of table lookup Besides the previous examples of image enhancement, there are several other uses of table lookup.

One example is data conversion. Often, the data acquisition system that creates the digital image has a dynamic range that is different from that of the display device or the software package. For example, medical images are often acquired at 12 bits per pixel, stored as 16-bit images, and must be displayed on 8-bit devices. Table lookup is useful in transforming between these different data types.

In the following, we summarize the uses of table lookup in other areas that we will encounter in this book:

- In Chapter 6, we will see the uses of table lookup to perform convolution and in morphological image processing.
- In Chapter 7, we will see the uses of table lookup in color image processing—for example, in color quantization using a color cube, as well as in gamma correction.
- In Chapter 8, we will see the uses of table lookup in image geometric modifications. Tables are useful in image scaling using interpolation filters, as well as image modification for special effects.
- In Chapter 9, we will see the uses of table lookups in image data compression. The "splat" algorithm is a table-based approach to computing the Discrete Cosine Transform.

5.4.2 A general table lookup function

In Chapter 3, while discussing performance-tuning, we looked at a table lookup function and tuned it for maximum performance. That example was for a special case that passed a byte image through a 256-entry lookup table to create another byte image. Generalization of this example to other cases is straightforward if lengthy. Therefore, instead of examining more implementations of table lookup, we will concentrate on the specification of a general table lookup function.

This function is offered by the XIL imaging library [4] in the `xil_lookup()` function. In addition to performing staightforward lookup operations, `xil_lookup()` allows the conversion of pixel types (from 8 bit to 16 bit, for example) and the use of multiple lookup tables, which allow the user to convert one-banded images to n-banded images. Finally, individual bands of an n-banded image can be passed through n distinct lookup tables for true multibanded lookups.

Before calling the function `xil_lookup()`, and in accordance with the object-oriented philosophy of XIL, the user must create an object called an XilLookup object, consisting of information about the lookup table as well as the table itself. A lookup is created by calling the function `xil_lookup_create()`, which has the following parameters.

- `state` This is the XIL system state obtained during initialization, and passed to all XIL functions.

- `input_data_type` This is the type of pixel of the source image. It can be unsigned 8-bit, signed 16-bit, or 1-bit deep.
- `output_data_type` This is the type of pixel of the destination image.
- `output_nbands` This is the number of bands in the destination image.
- `num_entries` This is the number of entries in the lookup table. Each entry maps a particular source pixel value or values to a destination pixel value or values.
- `first_entry_offset` The source pixel value for the first entry in the table. This parameter allows "clipping" during lookup operation.
- `data` This is a pointer to the actual entries for the lookup table.

Once these parameters are defined, the XIL lookup table is created by calling `xil_lookup_create()`, which returns a pointer to a lookup table:

```
XilLookup mylookup;

...
/* set up all the parameters and initialize table contents into data */
mylookup = xil_lookup_create(state,input_data_type, output_data_type,
output_nbands, num_entries, first_entry_offset, data);
```

The source image is then passed through the lookup table by calling the function `xil_lookup(src,dest,lookup)`.

Two interesting points stand out here. First, what is the significance of the `first_entry_offset`? Second, how are multibanded lookups (for example, $3 \rightarrow 3$) performed?

The parameter `first_entry_offset` is provided so that the lookup table need not be any larger than it has to be. For example, suppose I want to convert a signed 10-bit image into an unsigned 8-bit image. Since XIL knows only about `BIT`, `BYTE` and `SHORT` images, the 10-bit image is stored as a 16-bit `SHORT` image. However, a true `SHORT`-to-`BYTE` lookup will require a table 65,536 bytes long, whereas a 10-bit-to-`BYTE` lookup requires a table only 1024 bytes long. Therefore, by setting `first_entry_offset` to –512, and `num_entries` to 1024, I can perform this lookup with only a 1024 byte table. This is shown in Figure 5.12.

The second issue is how to perform n-banded-to-n-banded lookup functions. In the call to `xil_lookup()`, one cannot specify the number of bands in the source. The default source image has one band. However, to specify that the source image has more than one band, the parameter `num_entries` is set to zero. Then, the lookup table information, including the number of entries for each table, is embedded into the data part of the lookup table. This enables the user to perform lookups such as those shown in Figure 5.13.

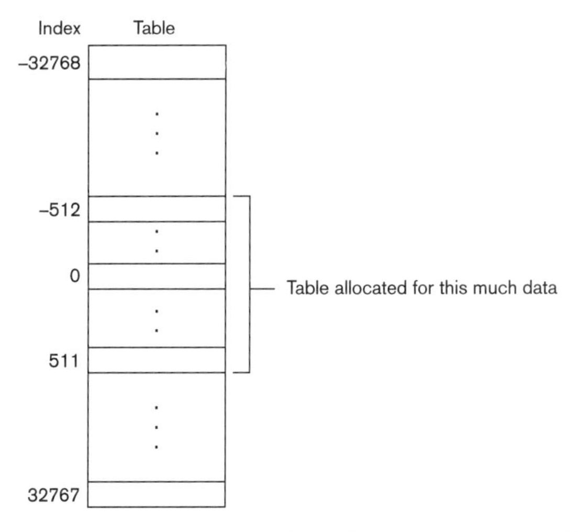

Figure 5.12 The XIL lookup object to convert 10-bit pixels to 8-bit pixels

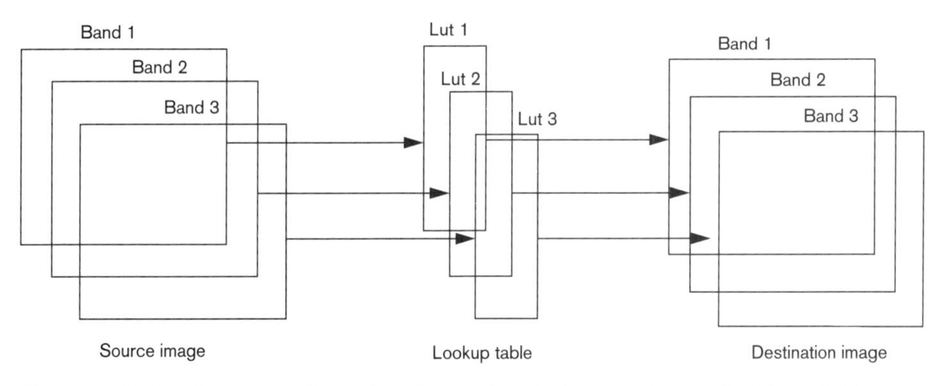

Figure 5.13 Performing a 3 → 3 banded lookup table operation using three tables

5.5 *Histogram-based operations*

So far, we have seen point operations for copying, performing ALU functions, and performing table lookup functions. While these operations are useful in many situations, there is another related class of functions, based on the image statistics generated in a histogram, that is also widely used in imaging, particularly image enhancement. Based on the *histogram*, point operations can be performed for *histogram stretching*, *histogram equalization*, and *adaptive histogram equalization*.

5.5.1 Histogram definition and computation

We now define the histogram of an image. For an image where the pixels can take on N different gray levels (from 0 to $N-1$), the histogram is the array $h(n)$, $n = 0, \ldots, N-1$ where $h(k)$ is the number of pixels having the gray level k.

The histogram is thus the relative frequency distribution of pixel values in the image. The histogram normalized by the total number of pixels (M) in the image, $h(k)/M$, is often used as an approximation to the probability density function of the pixel gray levels of the image.

The histogram of an image is useful for a number of reasons. Given the histogram, a number of useful measurements can be made for the image, including the mean pixel value and other statistical quantities. The histogram tells us the distribution of the pixel values; therefore, by looking at it we can determine if the pixel values of the image need redistribution. Computation of the histogram is shown in the following code for a gray-scale image. The first version of the function is a straightforward approach that yields fair performance.

```
/*
        Function:  pci_histogram
        Operation: Find the histogram of an image
        Input:     src is the source image
                   pointer to the histogram array must be zero'd out
        Output:    histogram is returned in hist[]
*/

void
pci_histogram (pci_image *src, int *hist)
{
        register int srows, scols;
        register unsigned char *spixptr, *spix;
        register unsigned int slb;
        register int i, j;
        register int *lhist = hist;      /* local pointer to hist[] */
        register int bin;

        /* get parameters from the image */
        spix = spixptr = src->data;
        srows = src->height;
        scols = src->width;
        slb = src->linebytes;

        for (i=0; i<srows; i++) {
             for (j=0; j<scols; j++) {
                  *(lhist + *spixptr) += 1;
                  spixptr++;
             }
             spixptr = (spix += slb);
        }
}
```

In a tuned version of the same function, shown in the following code, four pixels are loaded at a time in the integer pixel. Then each byte, that is, each pixel, is stripped from it and the histogram is updated accordingly. This version yields a 33 percent performance gain over the previous example on a processor that can load 32-bit words efficiently. Note that this version works only if the source image is aligned to a 32-bit word.

```
/*
        Function:  pci_histogram2
        Operation: Find the histogram of an image (faster implementation)
        Input:     src is the source image
                   pointer to the histogram (array must be zeroed out)
        Output:    histogram is returned in hist[]
*/

void
pci_histogram2 (pci_image *src, int *hist)
{
        register int srows, scols;
        register unsigned char *spixptr, *spix;
        register unsigned int slb;
        register int i, j;
        register int pix;              /* local pointer to hist[] */
        register int pixel;

        /* get parameters from the image */
        spix = spixptr = src->data;
        srows = src->height;
        scols = src->width;
        slb = src->linebytes;

        for (i=0; i<srows; i++) {
             for (j=0; j<scols>>2; j++) {
                  pixel = *(int *)spixptr;
                  pix = pixel & 0xff;
                  *(hist + pix) += 1;
                  pix = (pixel >> 8) & 0xff;
                  *(hist + pix) += 1;
                  pix = (pixel >> 16) & 0xff;
                  *(hist + pix) += 1;
                  pix = (pixel >> 24) & 0xff;
                  *(hist + pix) += 1;
                  spixptr+=4;
             }
             spixptr = (spix += slb);
        }
}
/*
```

Finally, we note that the C code can probably be speeded up further by moving larger chunks of pixels as allowed by the hardware.

5.5.2 Histogram stretching

In Section 5.4.1, we saw a number of techniques used in image enhancement. These techniques were lookup table based, and they allowed the user to compensate for deficiencies in the way the image was acquired.

The histogram can be used to fine-tune these techniques. One such powerful algorithm is histogram stretching, shown in Figure 5.14.

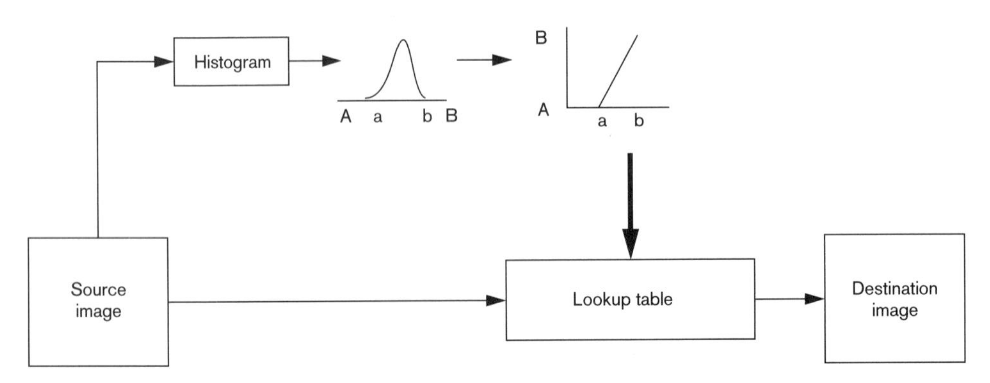

Figure 5.14 Histogram stretching to improve contrast

The histogram of the input image is first computed. For a low-contrast image, the range of the histogram, [*a*,*b*], will lie well within the range of the maximum and minimum possible pixel values [*A*,*B*]. Then a linear ramp function mapping [*a* to *A*] and [*b* to *B*] is loaded in a lookup table. The image is then passed through the lookup table, yielding a higher-contrast result.

5.5.3 Histogram equalization

While a histogram stretch improves contrast for many situations, it does not work well if the histogram of the image is skewed and uneven, with multiple peaks and valleys. For this situation, when it is necessary to enhance details that may be hidden in the picture, histogram equalization is a useful tool. The goal of this operation is to create a destination image whose histogram is *equalized*—that is, it is a flat histogram. Therefore, all the gray levels in the destination image have an approximately equal probability of occurring.

Not surprisingly, histogram equalization is a pointwise mapping operation accomplished using a lookup table. To derive the transformation function, let us consider an analog signal without loss of generality. The extension to digital signals is considered in reference 3. This derivation follows [5].

Let S be the source image and D be the destination image. Let $p_S(s)$ and $p_D(d)$ be the probability density functions of the source gray level s and destination gray level d.

Let the transformation function be $d = T(s)$. Then, from probablity theory, we have

$$p_D(d) = \left[p_S(s)\frac{ds}{dd}\right]_{s = T^{-1}(d)} \tag{5.5}$$

What we need to find is the function $T(s)$, which yields $p_D(d) = 1$. Now consider the transformation function

$$d = T(s) = \int_0^S p_S(w)dw \tag{5.6}$$

which is the cumulative distribution function of the input image. For this function, the derivative $ds/dd = 1/p_S(s)$. If we substitute this into Equation 5.5, we get $p_D(d) = 1$, which is the desired uniform density.

Therefore, the mapping that yields an image with uniform probability density is the cumulative distribution function of the image. When we convert from analog to the digital domain, the output relative frequency is only approximately uniform due to quantization errors.

To implement this digitally, we need to approximate the transformation $T(s)$. This is done by computing the histogram $h(n)$ of the source image, and then accumulating it, as shown in the following equation:

$$T(j) = \sum_{n=0}^{j-1} h(n), j = 0, \ldots, N-1 \tag{5.7}$$

The following code shows the implementation of a histogram equalization routine.

```
/*
        Function : pci_histeq
        Operation: Histogram equalize an image
        Input:     src is the source image
        Output:    dest is the histogram equalized image

*/


void
pci_histeq (pci_image *src, pci_image *dest)
{
        register int i, j;
        int hist[256];
        unsigned char heq_lut[256];
        register int sum = 0;
        register int totpix;

        /* total pixels in the image */
        totpix = src->width * src->height;
```

```
        /* initialize the histogram */
        for (i=0; i<256; i++)
             hist[i] = 0;

        /* compute histogram of the source as before */
        pci_histogram2 (src, hist);

        /* compute table for transforming the image */
        for (i=0;i<256;i++) {
             for (j=0; j< i; j++)
                  sum += hist[j];
             heq_lut[i] = (sum<<8)/totpix;
             sum = 0;
        }

        /* perform histogram equalization using lookup table */
        pci_lookup(src,dest,heq_lut);

}
```

An example of histogram equalization is shown in Figure 5.15 and Figure 5.16.

Figure 5.15 Zabriskie original

Figure 5.16 Zabriskie, histogram equalized

5.5.4 *Adaptive histogram equalization*

Histogram equalization, as described previously, has one disadvantage that renders it sensitive to noise: The mapping function depends on the histogram of the *entire* image. For example, in many medical imaging modalities, the image data have a circular cross-section (that is, the area of interest is circular). However, the format of digital images is rectangular, leaving a large background area with no useful information in it. A straightforward histogram equalization on such an image allows the background to have an unexpected effect on the image.

Adaptive histogram equalization was developed independently by three researchers [6] to remedy this situation. This algorithm consists of moving a window over the image, and performing histogram equalization on the pixel at the center of the window based only on the histogram of the pixels within the window.

This is computationally expensive, since the histogram equalization mapping has to be computed for each pixel in the image. A fast approximation to this algorithm was reported by Pizer, et al. [6]. The simplest form of this approximation is illustrated in Figure 5.17.

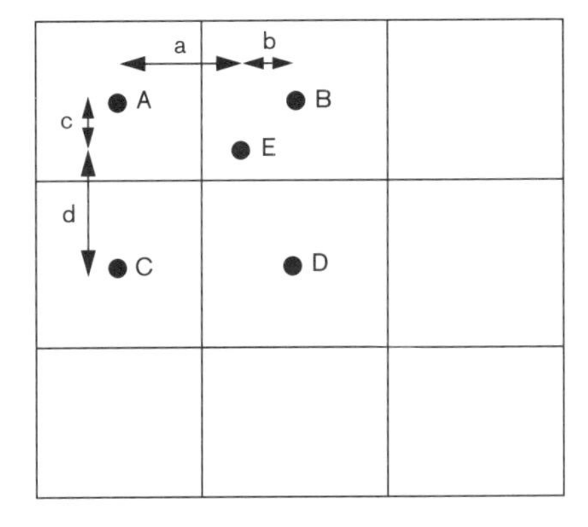

Figure 5.17 Interpolation for adaptive histogram equalization

The image is divided into tiles, and the histogram equalization mapping of the pixels at the center of each tile is computed. Then, the mapping for each pixel in the image is bilinearly interpolated from the four nearest pixels at tile centers. For example, in Figure 5.17, the pixels A, B, C, and D are pixels at tile centers. The mapping $M_A(x)$, which maps an input pixel value x into a histogram-equalized value $M_A(x)$, is computed. This mapping is based only on the pixels within the tile containing A. Similarly, $M_B(x)$, $M_C(x)$, and $M_D(x)$ are also computed. The distances a, b, c, and d between the pixel at E and A, B, C, and D are shown in Figure 5.17. Now the mapping for the pixel E is computed as:

$$M_E(x) = d(bM_A(x) + aM_B(x)) + c(bM_C(x) + aM_D(x)) \tag{5.8}$$

In general, the mapping window and the histogram window need not be the same size. The approximation was found to produce results very close to the true adaptive histogram equalization. The following pseudocode shows how to implement this algorithm. For an implementation using C code, see reference 7.

```
For each tile T in the image
        Compute histogram equalization mapping H(T)
For each tile T in the image
        For each pixel P in tile T
             Find the four closest H(Upper Left), H(Upper Right),
                  H(Lower Left), H(Lower Right)
             Interpolate between these four mappings to get output pixel
```

5.6 Other point operations

In this section, we cover two point operations that do not fall into the categories covered earlier in this chapter, but that are nevertheless useful functions. These are *image compositing using alpha blending* and *image band combination.*

5.6.1 Image compositing using alpha blending

Image compositing is a useful function for both graphics and computer imaging. In graphics, compositing is used for combining several images into one. Typically these images are rendered separately, possibly using different rendering algorithms. For example, the images may be rendered using different types of rendering hardware for different algorithms. In image processing, compositing is needed for any product that needs to merge multiple pictures into one final image. All image editing programs, as well as programs that combine synthetically generated images with scanned images, need this function.

When we have to mix two images, why can't we simply add them? Well, we can, but the resulting picture will have an unnatural look at the places where there are edges. This happens because adding images does not correctly handle the pixels where the two pictures overlap.

Consider the case of an individual pixel on which two pictures, A and B, overlap, as shown in Figure 5.18.

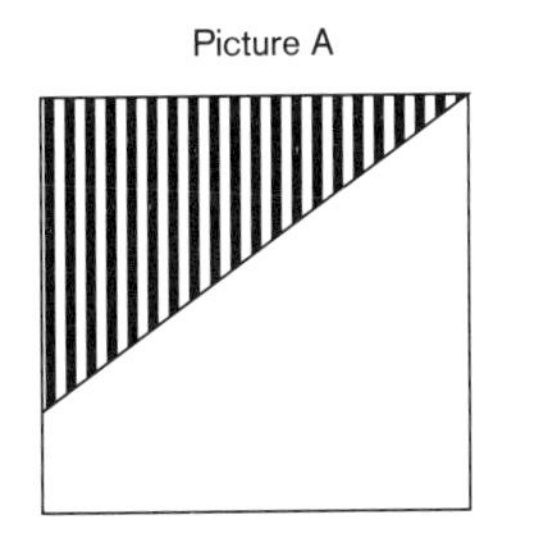

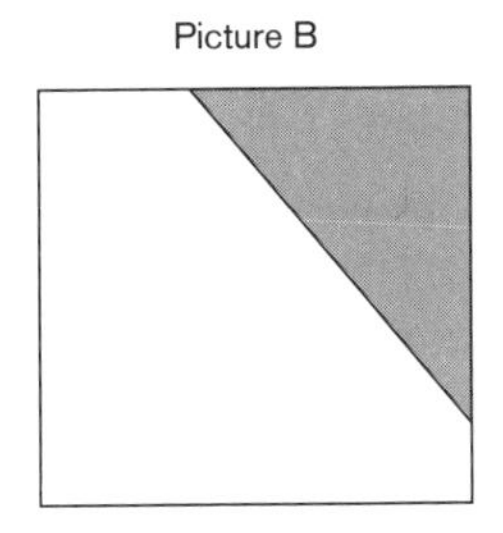

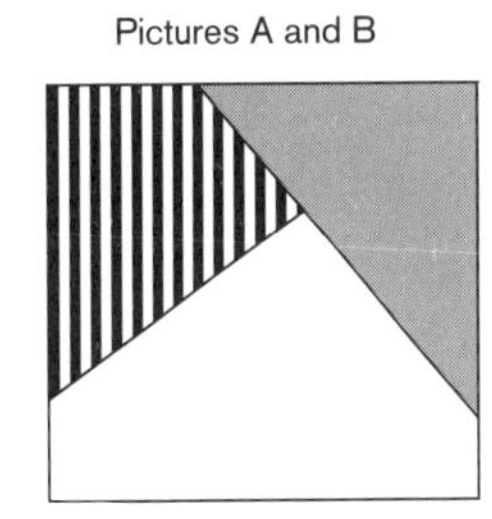

Figure 5.18 Two pictures falling on a pixel

Porter and Duff [8] addressed this scenario in a classic paper. Given this kind of a scenario, the area covered by the pixel can be subdivided into four areas: where none falls, where A only falls, where B only falls, and where both A and B fall. We will refer to these regions as 0, A, B, and AB, respectively. The goal of our blending algorithm is to create an output pixel from the merging of A and B that produces a smooth-looking image.

It turns out that to solve this problem, we need to associate a quantity, $\alpha_N(x,y)$, for the image N falling on the pixel at location (x,y) of the destination image. Alpha represents the coverage of the destination pixel by the image. For example, if we have two images, A and B, then $\alpha_A(x,y)$ and $\alpha_B(x,y)$ represent how much A and B cover the pixel at (x,y).

Given α_A and α_B and the four types of regions within the pixel that we have defined above, we would like to define the operation generating the value of the pixel. This can be specified as a quadruple $\{a,b,c,d\}$, each element representing whether 0, A, or B is used in the corresponding region. For example, the quadruple $\{0,A,B,A\}$ indicates that 0 is left as is, A is placed where the pixel overlaps with A, B is placed where the pixel overlaps with picture B, and B is placed where A and B overlap. There are $1 \times 2 \times 2 \times 3 = 12$ possible quadruples which yield 12 binary functions. (That is, 0 can take on one value, 0; A can take on A or 0; B can take on B or 0; and AB can take on A, B, or 0.) However, for our interests, only three of these are important, as shown in Table 5.3.

Table 5.3 Variations of image compositing

Operation	Quadruple	F_A	F_B
A over B	(0,A,B,A)	1	$1 - \alpha_A$
A in B	(0,0,0,A)	α_B	0
A out B	(0,A,0,0)	$1 - \alpha_B$	0

Table 5.3 is interpreted like this: If the pixel from A has color components (r_A, g_A, b_A, a_A) and the pixel from B has color components (r_B, g_B, b_B, a_B), then the new pixel is (r_C, g_C, b_C, a_C), where r_C is computed as follows:

$$r_C = F_A r_A + F_B r_B \tag{5.9}$$

The other components are computed similarly.

In computer imaging, the term *alpha blend* is an approximation to the operation A over B defined in the above table. It can be defined in terms of two source images, $S1$ and $S2$, and a destination image, D, and an alpha image, α.

$$D(x, y) = (1 - \alpha(x, y))S1(x, y) + \alpha(x, y)S2(x, y) \tag{5.10}$$

The following code example shows how to perform an alpha blend using two source images, a destination image, and a separate alpha image. Note that all the images are assumed to be one-banded byte images. For multibanded images, the extension is straightforward.

```
/*
        Function:  pci_blend
        Operation: Perform alpha blending between two source images
                   using an alpha image
        Input:     Two source images, src1 and src2
                   Alpha image
        Output:    dest is the destination image
*/

void
pci_blend(pci_image *src1, pci_image *src2, pci_image *dest, pci_image
*alpha)
{

        int rows, cols;
        register unsigned char *s1pixptr, *s1pix;
        register unsigned char *s2pixptr, *s2pix;
        register unsigned char *dpixptr, *dpix;
        register unsigned char *apixptr, *apix;
        unsigned int s1lb, s2lb, dlb, alb;
        register int i, j;
        register int result;

        /* get parameters from the image */
        s1pix = s1pixptr = src1->data;
        rows = src1->height;
        cols = src1->width;
        s1lb = src1->linebytes;

        s2pix = s2pixptr = src2->data;
        s2lb = src2->linebytes;

        dpix = dpixptr = dest->data;
        dlb = dest->linebytes;
```

```
        apix = apixptr = alpha->data;
        alb = alpha->linebytes;

        /* do the blend operation over all pixels */
        for (i=0; i<rows; i++) {
                for (j=0; j<cols; j++) {
                  result = *apixptr * (*s2pixptr - *s1pixptr);
                  result += *s1pixptr<< 8;
                  *dpixptr = (unsigned char) (result>>8);
                  apixptr++; dpixptr++; s1pixptr++; s2pixptr++;
                }
                s1pixptr = (s1pix += s1lb);
                s2pixptr = (s2pix += s2lb);
                dpixptr = (dpix += dlb);
                apixptr = (apix += alb);
        }
}
```

An example of blending two images under the control of an alpha image is shown. The original images (Figures 5.19 and 5.20) are blended into Figure 5.22, using the alpha image shown in Figure 5.21.

Figure 5.19 Museum original

Figure 5.20 Bus original

Figure 5.21 Alpha image

Figure 5.22 Images blended with alpha

5.6.2 An image band combination function

When dealing with multibanded images, it is often necessary to produce linear combinations of the individual bands of the image. The need arises, for example, when it is necessary to provide the user with the ability to perform color space conversions, or extract the luminance (gray-level) component of the image. In order to address this need, the XIL imaging library offers a powerful point function called `xil_band_combine()` [4].

The definition of the function is straightforward. For an n-banded source image and an M-banded destination image, a matrix of coefficients A (with $N + 1$ columns and M rows), must be specified. If a_{ij} represents the elements of the matrix, and s_i represents the pixel values at the individual bands of the image, then the destination bands d_i are defined as:

$$\begin{bmatrix} d_0 \\ d_1 \\ \dots \\ d_{M-1} \end{bmatrix} = \begin{bmatrix} a_{00} & a_{01} & a_{02} & \dots & a_{0N} \\ a_{10} & a_{11} & a_{12} & \dots & a_{1N} \\ \dots & \dots & \dots & \dots & \dots \\ a_{M-1,0} & a_{M-1,1} & a_{M-1,2} & \dots & a_{(M-1),N} \end{bmatrix} \begin{bmatrix} s_0 \\ s_1 \\ \dots \\ s_{N-1} \\ 1 \end{bmatrix} \tag{5.11}$$

Thus, for example, the first destination band pixel from Equation 5.11 would be:

$$d_0 = a_{00}S_0 + a_{01}s_1 + \dots + a_{0.N-1}S_{N-1} + a_{0N} \tag{5.12}$$

This function can be used to extract the different bands of a multibanded image. For example, if a 3-banded image is laid out in BGR format, then the following matrix extracts the luminance component of the 3-banded source image into a one-banded destination image:

$$A = \begin{bmatrix} 0.114 & 0.587 & 0.299 & 0 \end{bmatrix} \tag{5.13}$$

As indicated earlier, the band combination function is useful for performing linear color space conversions. For example, the matrix shown in Equation 7.6 may be substituted in place of A to perform an RGB to YIQ conversion. Note that we actually need a 4×3 matrix, where the last column represents bias values added to the resultant pixels.

The implementation of band combination is straightforward. The pseudocode is given below:

```
For each row in the image
        For each column in the image
             For each band in the destination pixel
                  While (matrix_width--)
                       multiply_accumulate source*matrix_element
```

5.7 Conclusion and further reading

Image point operations execute on individual pixels of the source image or images to create pixels of the destination image. Point operations are important operations in imaging. In this chapter we have seen several important classes of point operations, including copying, monadic and dyadic operations, table lookup operations, and histogram-based operations, as well as a set of miscellaneous point operations useful in a variety of applications. We have also examined several coding examples of fast implementations of these operations.

Formal mathematical discussion of image point operations can be found in all the standard textbooks on image processing. I have found references 2 and 3 to be valuable. A discussion of the variations of image blending can be found in reference 8. Clever C implementations can be found in reference 7 as well as other books in the same series.

5.8 References

1 T. Duff, Internet email regarding Duff's device.

2 W. K. Pratt, *Digital Image Processing*, 2nd edition, New York: Wiley, 1991.

3 A. K. Jain, *Fundamentals of Digital Image Processing*, Englewood Cliffs, NJ: Prentice Hall, 1988.

4 *XIL User's Guide*, SunSoft, Inc.

5 R. Gonzalez and P. Wintz, *Digital Image Processing*, 2nd edition, Reading, MA: Addison-Wesley, 1987.

6 S. M. Pizer, et al., "Adaptive Histogram Equalization and Its Variations," *Computer Vision, Graphics, and Image Processing*, Vol. 39, 1987, pp. 355–368.

7 K. Zuiderveld, "Contrast Limited Adaptive Histogram Equalization," *Graphics Gems IV*, P. Heckbert, ed., Cambridge, MA: Academic Press, 1994.

8 T. Porter and T. Duff, "Compositing Digital Images," *Proceedings of ACM SIGGRAPH*, 1984, pp. 253–259.

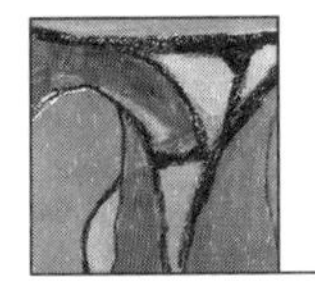

chapter 6

Image neighborhood filtering

6.1 Introduction

This chapter is about neighborhood filters in imaging. Neighborhood filters create a destination pixel based on criteria that depend on the source pixel and the value of pixels in the "neighborhood" surrounding it. Neighborhood filters are widely used in computer imaging. They are used for enhancing and changing the appearance of images by sharpening, blurring, crispening the edges, and noise removal. They are also useful in other application areas such as object recognition, image restoration, and image data compression.

The term *filter* is used in many different senses in imaging. In the desktop publishing and image editing terminology, filter is used in a broad sense to include any operation that automatically changes the pixels of an entire image. These operations include geometric and point operations covered elsewhere in this book. For the purposes of this chapter, the word *filter* refers to *neighborhood* linear and nonlinear filters in the strict sense. That is, we define a filter as a process that changes pixels of the source image based on their values and those of their surrounding pixels. (These filters are called neighborhood filters to distinguish them from *spectral* filters. While neighborhood filters directly use the values of the pixels in the image, spectral filters use Fourier or related methods to transform the image before filtering. In most commercial imaging applications, neighborhood filters are more common.)

When dealing with neighborhood filters, it is useful to distinguish between *linear* and *nonlinear* filters. Linear filters have well-defined properties and their output can be predicted for any input. Nonlinear filters, on the other hand, are more difficult to characterize. In general, nonlinear filters are more computationally expensive to implement but they often yield more striking results. In this chapter we will study both nonlinear and linear filters.

Perhaps the most common imaging filter is the linear filter implemented using a two-dimensional convolution. Although computationally expensive, this filter is useful in a wide range of applications, and one would be hard-pressed to find an imaging product that does not use this ubiquitous operation in some form. Among nonlinear filters, the most commonly used is probably the median filter, which is a powerful filter for removing noise. A related class of nonlinear filters, morphological filters, has also found widespread use. All of these types of filters are considered in this chapter.

We begin in Section 6.2 by comparing linear and nonlinear filters. In Section 6.3, we look at linear filtering of images, paying particular attention to the efficient computation of convolution for linear filtering. Nonlinear filters are the subject of Sections 6.4 and 6.5, where we examine the median filter and related morphological filters and look at efficient implementations of these filters.

6.2 Linear filtering versus nonlinear filtering

Generally speaking, a filter in imaging refers to any process that produces a destination image from a source image. A *linear* filter is illustrated in Figure 6.1. It has the property that a weighted sum of the source images produces a similarly weighted sum of the destination images. Thus, in the figure, the source image *A* produces the destination image *X*, and the source image *B* produces the destination image *Y*. If the image *A* is multiplied by the constant *a*, and *B* is multiplied by the constant *b*, and their sum $aA + bB$ is input into the filter, it produces the destination image $aX + bY$.

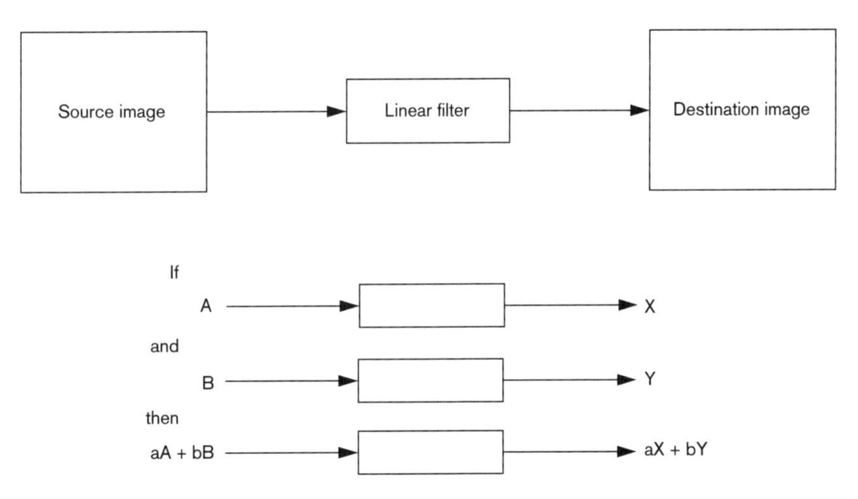

Figure 6.1 A linear fIlter

A special type of linear filter is called a *linear shift-invariant* (LSI) filter. What this means is this: The destination pixel created by the filter depends only on the *values* of the source pixels and not their location. Another way of looking at it is that a given neighborhood of pixels in the source image will yield the same output pixel no matter where in the image it is located. *Convolution* can be used to implement a linear shift-invariant filter on digital signals and images. We will take a detailed look at this operation in Section 6.3.

LSI filters are easy to analyze because they are completely characterized by their *impulse response*. The impulse response of a filter is the output of the filter when an *impulse function* is used. This is shown in Figure 6.2. In one-dimension, an impulse function $\delta(x)$ is zero everywhere except at $x = 0$, where it is 1. In two dimensions, $\delta(x,y)$ is zero everywhere except at $x,y = 0$, where it is 1.

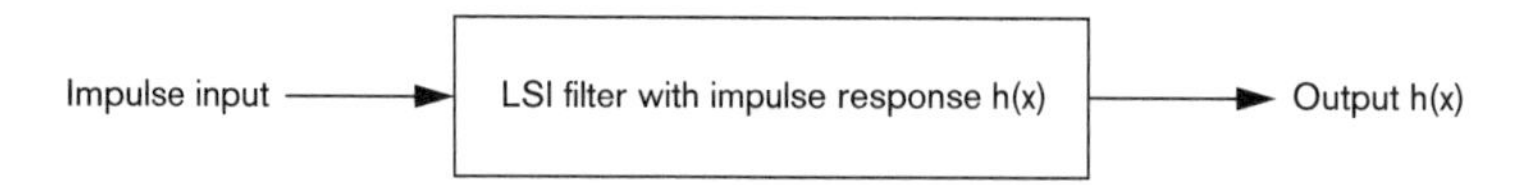

Figure 6.2 Impulse response of a LSI filter

One important property of LSI filters is illustrated by considering the following scenario. Suppose one is given a black box, and the only thing known about the box is that it is an LSI filter. Then, in order to find out the impulse response of the filter, one need only apply an impulse input at the black box and read the output.

What does this mean in terms of images? The black box could be the object module for a C function, `filter (source_image, destination_image)`. Suppose this function is called with `source_image` being an impulse image (that is, an image with zeros everywhere except the center, where it is 1). Then, the `destination_image` returned by the function is the impulse response of the filter, and the filter is completely characterized by this response. This is shown in Figure 6.3.

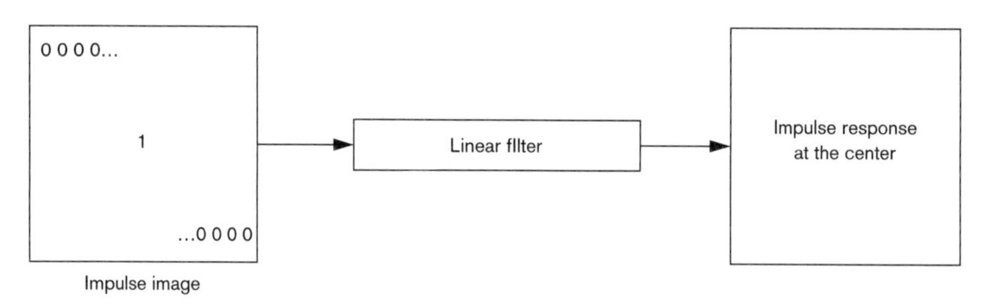

Figure 6.3 Finding the impulse response of a linear fllter

In contrast to linear filters, nonlinear filters are somewhat more difficult to characterize. This is because the output of the filter for a given input cannot be predicted by the impulse response. Nonlinear filters behave differently for different inputs. Therefore, it makes very little sense to talk about their impulse response. However, because of their dynamic nature this also makes nonlinear filters interesting and powerful.

There is another important difference between linear and nonlinear filters. Linear filters allow the user to easily select the frequencies to filter out. For example, we can use a low-pass filter—implemented using convolution—to filter out high-frequency components of the image, thus blurring it, or we can use a high-pass filter—also implemented using convolution—to eliminate the low-frequency components, thus sharpening it. Nonlinear filters, in general, do not allow this kind of selectivity to the user. Examples of blurring and sharpening filters are discussed in Section 6.3.3.

What do all these differences mean in practice? We can contrast the performance of linear and nonlinear filters for a commonly used imaging application: noise removal. The linear filter used in noise removal is a smoothing convolution filter. Such a filter is relatively inexpensive to implement; however, the smoothing filter, while smoothing noise, also smooths the edges in the image, causing substantial loss of high-frequency information. In contrast, the median filter is a nonlinear filter often used for noise removal. The median filter is somewhat more difficult to implement because it requires sorting a series of pixels making up the immediate neighborhood of the pixel. However, median filters, while removing noise, do preserve the edges. In other words, the filter behaves in one manner in the presence of small perturbations in the form of noise, and in a different manner in the presence of large discontinuities represented by edges. For this reason, despite the computational complexity, many users prefer the median filter for noise removal.

Having compared linear and nonlinear filters, we are now ready to take a closer look at their implementation. We begin with linear filters.

6.3 Linear filtering using convolution

6.3.1 One-dimensional discrete convolution

It is instructive to look at discrete convolution in one dimension before going to its two-dimensional generalization. One-dimensional convolution is often useful in imaging, particularly for a class of filters called *separable* filters, which are described later.

One-dimensional convolution is performed between a sequence of numbers called the *source sequence* (which could be a row or a column of an image) and a (typically much smaller) sequence of numbers called the *impulse response* or the *convolution kernel*. The result of the operation is another row of numbers called the *destination sequence*. This is, of course, an implementation of a one-dimensional LSI filter.

One-dimensional convolution is shown in Figure 6.4. In this example, a source sequence is convolved with a 3×1 kernel. The kernel is first flipped about its center. Then, it is slid over the source sequence, and for each position, the source elements overlapping the kernel are multiplied by the corresponding kernel elements. The sum of these products is the result of the convolution of the destination position corresponding to the position of the kernel.

Computation of the fifth element of the destination sequence is shown in Figure 6.4. Note that the elements at the edges of the destination sequence are marked *x*—meaning they are undefined—because the convolution kernel overlaps with only two source elements there.

Mathematically, one-dimensional convolution is expressed as a summation. We assume that the zero element of the source, the (odd-length) kernel, and the destination are all at the left edge of the sequence.

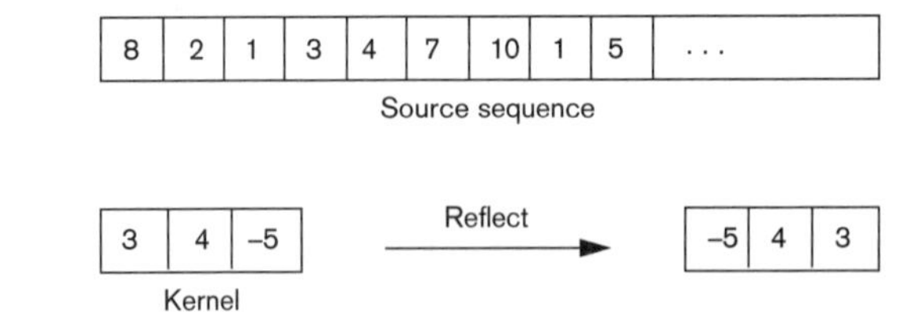

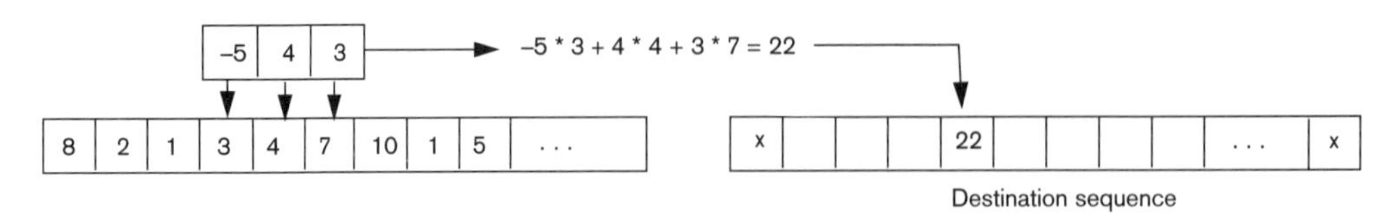

Figure 6.4 Example of one-dimensional convolution (computation of the fifth element)

Then, for a signal $f(m)$ that is M-elements long and an impulse response $h(k)$ that is L-elements long, the destination array $g(m)$ created by the convolution is defined by

$$g(m) = \sum_{x=0}^{L-1} f\left(m + \frac{L-1}{2} - x\right)h(x) \tag{6.1}$$

Note that the pixels in the destination signal created from a complete overlap with the convolution kernel lie in the range:

$$\frac{(L-1)}{2} \le m \le M - \frac{(L-1)}{2}$$

That is, the extent of meaningful data in the destination signal is of size $M - (L - 1)$. For example, in Figure 6.4, since $L = 3$, one pixel at each end of the destination array—those marked x—is invalid because the kernel overlaps with only two source elements.

One-dimensional convolution is useful in imaging, particularly when a two-dimensional convolution kernel can be *separated* into a row and a column kernel to perform *separable convolution*. This is discussed further in Section 6.3.2.

6.3.2 *Two-dimensional discrete convolution*

Two-dimensional discrete convolution is an extension of one-dimensional convolution into two dimensions. In imaging, two-dimensional convolution is the most common way to implement a linear filter. The operation is performed between a source image and a two-dimensional convolution kernel (also called the impulse response) to produce a destination image. As in the one-dimensional case, the convolution kernel is typically much smaller than the source image.

To perform two-dimensional convolution, the kernel is first rotated 180 degrees about its center. (This is the two-dimensional equivalent of flipping the one-dimensional kernel about its center.) Starting at the top, the kernel is moved horizontally over the image, one pixel at a time. Then it is moved down one row and moved horizontally again. This process is continued until the kernel has traversed the entire image. The movement of the kernel is shown in Figure 6.5.

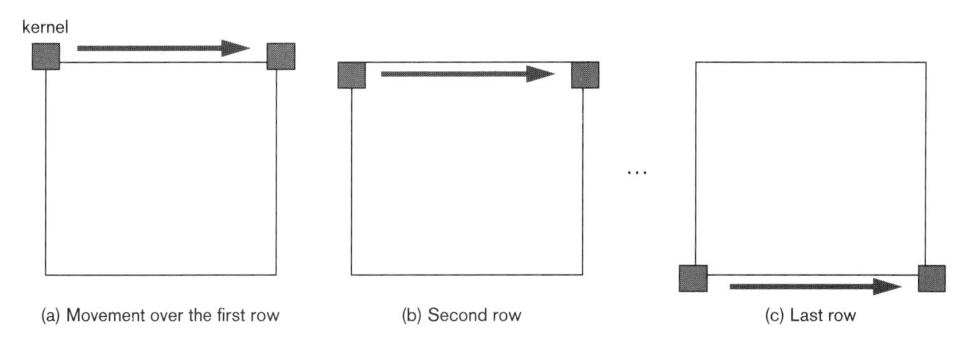

Figure 6.5 Movement of the convolution kernel over the source image

One destination pixel is created every time the kernel moves to a new position. At any position, each element of the kernel is multiplied with the corresponding source pixel that falls underneath it. These products are added to yield the destination pixel corresponding to the current kernel position. This is shown for a 3 × 3 kernel in Figure 6.6, where the destination pixel is computed as $Aa + bb + Cc + Dd + Ee + Ff + Gg + Hh + Ii$.

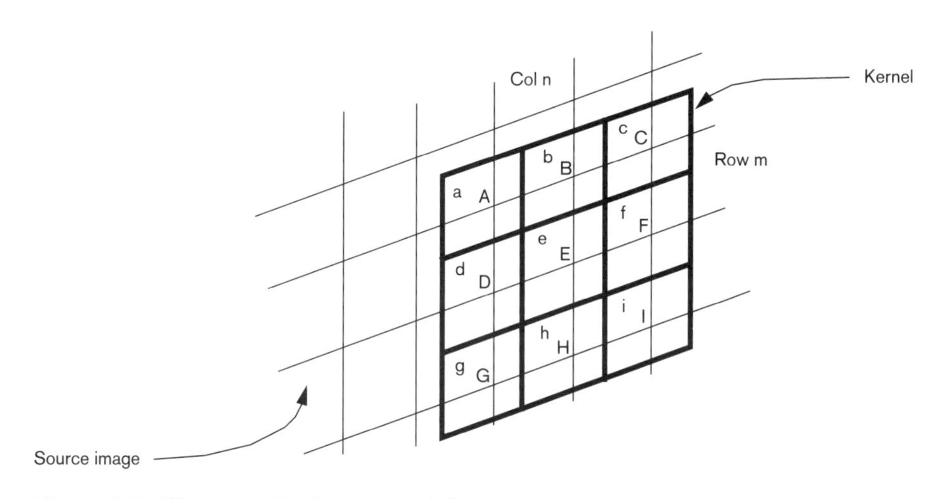

Figure 6.6 Close up of a 3 × 3 convolution

The location of the destination element created for a particular position of the kernel is clarified in Figure 6.7. For the destination pixel at row *m* and column *n*, the kernel is centered at the same location in the source image.

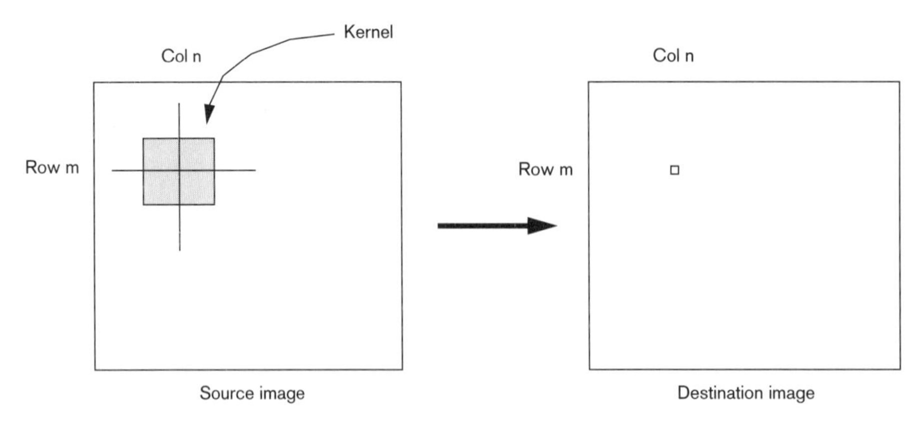

Figure 6.7 Creation of the destination pixel at the location (m,n)

Simply put, two-dimensional discrete convolution is a weighted sum of the pixels in the neighborhood, where the weights are the kernel elements.

Mathematically, two-dimensional discrete convolution is defined as a double summation. Given an $M \times N$ image $f(m,n)$ and a $K \times L$ convolution kernel $h(k,l)$, we define the origin of each to be at the top left corner. We assume that $f(m,n)$ is much larger than $h(k,l)$. Then, the result of convolving $f(m,n)$ by $h(k,l)$ is the image $g(m,n)$ given by

$$g(m, n) = \sum_{x=0}^{K-1} \sum_{y=0}^{L-1} f\left(m + \frac{K-1}{2} - x, n + \frac{L-1}{2} - y\right) h(x, y) \tag{6.2}$$

The pixels of the destination image, $g(m,n)$, are only valid for:

$$\frac{(K-1)}{2} \le m \le M - \frac{(K-1)}{2}$$

and

$$\frac{(L-1)}{2} \le n \le N - \frac{(L-1)}{2}$$

because outside this range, the convolution kernel does not completely overlap with pixels from the source image. This means that the output image has $(M - (K - 1)) \times (N - (L - 1))$ meaningful pixels.

A problem with this computation is that it leaves undefined a band of pixels around the edges, shown in Figure 6.8 for a 3×3 convolution kernel. For this example, only four kernel elements and four source pixels contribute to the destination pixel at the location (0,0), resulting in a destination pixel value that is substantially different from its surroundings.

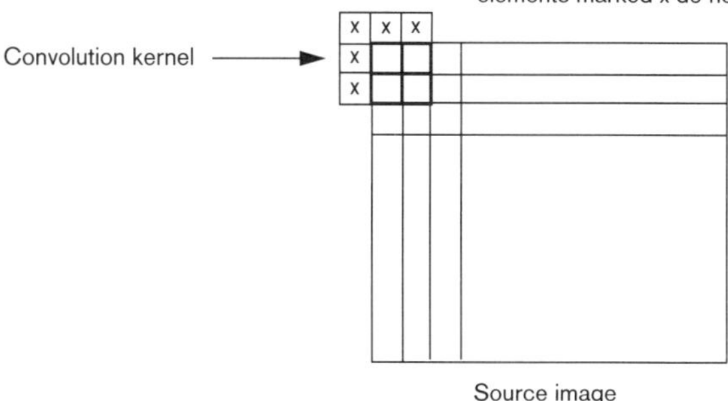

Figure 6.8 Incomplete convolution at the edges

This is not a serious problem if the convolution is applied to the image only once, since a relatively small band around the destination image is affected. However, in many imaging applications, the image is convolved multiple times resulting in a dark band around the image that grows each time it is convolved. The image gives the appearance of slowly shrinking with each iteration of the convolution. Another problem arises when a region of interest within an image is convolved, for example, in an image editing program. Incomplete overlap at the edges can lead to discontinuities around the region of interest.

For these reasons, whenever convolution is being performed, we need to clearly specify the definition of edge pixels. Some common ones are:

- *Zero fill the edges* The destination edge pixels are filled with zeros.
- *Don't write the edges* The destination edge pixels are left untouched.
- *Extend the edges* The edge of the source image is extended, that is, duplicated, by as many pixels as necessary to perform complete convolution at the edges. For example, if a 3×3 convolution is performed on an $M \times N$ image, then an augmented source image, of size $(M + 2) \times (N + 2)$, is created by extending the edge pixels. The augmented image yields a complete $M \times N$ destination image.
- *Reflect the edges* The edge pixels of the source image are reflected along the boundaries to create the augmented image, which is then convolved in the same way as the edge-extend case.

The first edge option is the simplest to implement, and is acceptable in a variety of situations where the image is convolved only once. The second option is used when it is required that no destination pixels be created artificially for the sake of maintaining integrity of the image. The third and fourth options are used whenever it is necessary to seamlessly convolve part of an image, for example, inside a region of interest in an image, or for multiple convolutions.

In all the example programs in this chapter, we have used the zero-fill option on the edges of the destination image.

6.3.3 Implementation of convolution

For a $K \times L$ convolution kernel operating on an $M \times N$ image, a convolution of the image requires approximately $K \times L \times M \times N$ multiply-accumulate operations. Clearly this can be expensive as the image and/or the kernel becomes large. Consider the example of a real-time application where we are required to process 30 images per second. For a 512×512 image and a 3×3 kernel, the processor needs to perform $512^2 \times 9 \times 30 = 70.8$ million multiply-accumulates per second (excluding the loads and stores) in order to meet this requirement.

In this section we will look at several ways to implement convolution. Beginning with a generic two-dimensional convolution function, we will examine ways to optimize two-dimensional convolution in order to reduce the computational complexity. These methods include table-driven convolution, separable convolution, and singular value decomposition/small generating kernel (SVD/SGK) decomposition.

For very large kernels, Fourier methods, which are outside the scope of this book, may be optimal. The reader is referred to reference 1 for a discussion of this.

An $M \times N$ convolution function Our first programming example shows a generic two-dimensional convolution of a $K \times L$ image with an $M \times N$ convolution kernel.

```
/*
        Function:  pci_cnvmxn
        Operation: Convolve an image with an mxn kernel
        Inputs:    src is the source image
                   height, width are kernel height and width
                   kernel is pointer to kernel data
                   The kernel values must be signed 16-bit format
        Output:    dst is the destination image
*/

#define SCA16  (1 << 16)
#define FSCA16  ((float)SCA16)

void
pci_cnvmxn(pci_image *src, pci_image *dst, int height, int width, float
*kernel)
{
        register unsigned char *spixptr, *dpixptr, *spix, *dpix, *sptr;
        register short *lker, *lksav;    /*local kernel */
        register int lw, lh;             /* local copies of ht/wdth*/
        register int result=0;
        int rows, cols;
        int slb, dlb;
        int dw, dh;
        int ccw, ccw_sav, cch;           /* rect of complete coverage */
        int i,j;
        float kerelem;

        /* get the info from the images */
```

```
spixptr = spix = src->data;
slb = src->linebytes;
dpixptr = dpix = dst->data;
dlb = dst->linebytes;
dw = dst->width;
dh = dst->height;

/* copy the kernel, rotate and put into local memory */
lw = width;
lh = height;
lker = (short *)malloc( (lh*lw) * sizeof(short));

lksav = lker;
for (i=0; i<lw*lh; i++)  {
     *lker++ = (short) (*(kernel+lw*lh-i-1) * FSCA16 + 0.5);
}
lker = lksav;


/* width/height of rectangle of complete coverage */
cch = dh - ((lh>>1)<<1);
ccw_sav = ccw = dw - ((lw>>1)<<1);

/* 1st dst pixel to be written is the 1st pixel of above rect. */
dpixptr += (lh>>1)*dlb+(lw>>1);
dpix = dpixptr;

/* perform convolution in rectangle of complete coverage */
while (cch--) {
     while (ccw--) {
          sptr = spixptr;
          while (lh--) {
               while (lw--) {
                    result += *sptr * *lker;
                    lker++;
                    sptr++;
               }
               lw = width;
               sptr = sptr + slb - lw;
          }
          *dpixptr++ = (result)>>16;
          spixptr++;
          lker = lksav;
          lh = height;
          result = 0;
     }
     ccw = ccw_sav;
     dpixptr = dpix += dlb;
     spixptr = spix += slb;
}

/* zero-fill the edges  ...*/
/* ... top edge ... */
dpixptr = dpix = dst->data;
```

```
    for (i=0; i<(lh>>1); i++) {
        for (j=0; j<dw; j++)
            *dpixptr++ = 0;
        dpixptr = dpix += dlb;
    }

    /* ... bottom edge ... */
    dpixptr = dpix = dst->data + (dw-(lh>>1))*dlb;
    for (i=0; i<(lh>>1); i++) {
        for (j=0; j<dw; j++)
            *dpixptr++ = 0;
        dpixptr = dpix += dlb;
    }

    /* ... left edge ... */
    dpixptr = dpix = dst->data;
    for (i=0; i<(lw>>1); i++) {
        for (j=0; j<dw; j++, dpixptr += dlb)
            *dpixptr = 0;
        dpixptr = dpix++;
    }

    /* ... right edge */
    dpixptr = dpix = dst->data + (dw - (lw>>1));
    for (i=0; i<(lw>>1); i++) {
        for (j=0; j<dw; j++, dpixptr += dlb)
            *dpixptr = 0;
        dpixptr = dpix++;
    }

    free(lker);
}
```

A fast 3×3 convolution using table lookup Before hardware multipliers became available, table lookup was used to perform convolution in the early image processors. The idea is to use a table of values, each entry of which corresponds to the product of a pixel value with a kernel value. For a 3×3 kernel whose values are limited to 8 bits of precision, and 8-bit pixels, this table needs to be $3 \times 3 \times 256$ elements long. During the convolution, all products between pixel values and kernel values can then be looked up directly from the table, avoiding the necessity to perform any multiplication.

The following function implements a 3×3 convolution using this technique.

```
/*
    Function:  pci_3x3_lut
    Operation: Convolve an image with 3x3 kernel using table lookup
    Inputs:    src is the source image
               kernel is pointer to kernel data
               The kernel values must be signed 16-bit format
    Output:    dst is the destination image
*/
```

```
#define SCA16  (1 << 16)
#define FSCA16  ((float)SCA16)
#define MAXBYTE   255

void
pci_cnv3x3_lut(pci_image *src, pci_image *dst, float *kernel)
{
        int  i, j;
        int  result;                /* result of convolution sum */
        int  sum;                   /* convolution sum accumulates here */
        int  ker_tab[3][3][256];
        int  src00, src01, src02,   /* source pixel window */
             src10, src11, src12,
             src20, src21, src22;
        int  kx, ky;
        int  slb, dlb, w, h;
        float rot_kernel[9];        /* rotated kernel */

        unsigned char  *sa, *da, *sda, *ssa ;/* save copies */
        unsigned char      *s, *d;  /* image array pointers */

        /* get the data from the image structures */
        da = sda = dst->data;
        sa = ssa = src->data;
        slb = src->linebytes;
        dlb = dst->linebytes;
        w = src->width;
        h = src->height;

        /* rotate the kernel */
        for (i=0; i<9; i++)
             rot_kernel[i] = *(kernel+8-i);

        /* initialize the kernel table */
        for (i = 0; i < 9; i++) {
        kx = i % 3;
        ky = i / 3;
        for (j = 0; j < 256; j++)
             ker_tab[ky][kx][j] = (int) (j * rot_kernel[i] *
                  FSCA16 + 0.5);
        }

        /* 1st and last row are excluded - handled at the end */

        /* advance source and destination pointers to row 1 */
        da += dlb;
        sa += slb;

        for (j = 1; j < h-1; j++) {
             d = (unsigned char*)da;
             s = (unsigned char*)sa;
```

```
/* advance source pointer to col 1 */
++s;

src10 = s[-slb-1];
src11 = s[-1];
src12 = s[slb-1];

src20 = s[-slb];
src21 = s[0];
src22 = s[slb];

/* perform zero-fill at edge */
 d[0] = 0;

/* advance destination pointer to col 1 */
++d;

for (i = 1; i < w-1; i++) {

     /* copy over source pixels towards the left column */
     src00 = src10;
     src10 = src20;
     src01 = src11;
     src11 = src21;
     src02 = src12;
     src12 = src22;

     /* read in the new values */
     ++s;
     src20 = s[-slb];
     src21 = s[0];
     src22 = s[slb];

     /* perform full convolution */
     sum = ker_tab[0][0][src00]+ ker_tab[0][1][src01]+
           ker_tab[0][2][src02]+ ker_tab[1][0][src10]+
           ker_tab[1][1][src11]+ ker_tab[1][2][src12]+
           ker_tab[2][0][src20]+ ker_tab[2][1][src21]+
           ker_tab[2][2][src22];

     result = sum >> 16;

     /* clamp */
     if (result & ~MAXBYTE) {
          if (result < 0) result = 0;
          else result = MAXBYTE;
     }
d[0] = result;
++d;
}

/* copy over source pixels towards the left column */
src00 = src10;
src10 = src20;
```

```
            src01 = src11;
            src11 = src21;
            src02 = src12;
            src12 = src22;

            /* perform zero-fill at edge */
             d[0] = 0;

            da += dlb;
            sa += slb;
        }

        /* perform zero-fill at top and bottom edges */

        /* handle the topmost row */
        j = 0;
        d = (unsigned char*)sda;
        s = (unsigned char*)ssa;

        /* zero-fill on left */
        src10 = src11 = src12 = 0;

        /* zero-fill on top */
        src21 = s[0];
        src22 = s[slb];

        for (i = 0; i < w; i++) {
                *d++ = 0;
        }

        /* handle the bottommost row */
        j = h-1;
        d = (unsigned char*)sda + j*dlb;
        s = (unsigned char*)ssa + j*slb;

        /* zero-fill on left */
        src10 = src11 = 0;

        /* zero-fill on bottom */
        src20 = s[-slb];
        src21 = s[0];
        for (i = 0; i < w; i++) {
            d[0] = 0;
            d++;
        }
}
```

Separable convolution using packed lookup tables For square convolution kernels of size $L \times L$, if the two-dimensional convolution can be expressed as a separable convolution, then the complexity drops from $L^2 \times M \times N$ to $2L \times M \times N$. Mathematically, this requires that the convolution kernel $h(k,l)$ can be expressed as the outer product of two one-dimensional vectors $r(k)$ and $c(l)$. This operation is shown in Figure 6.9.

(a) Separate 2-d kernel into the outer product of two 1-d vectors

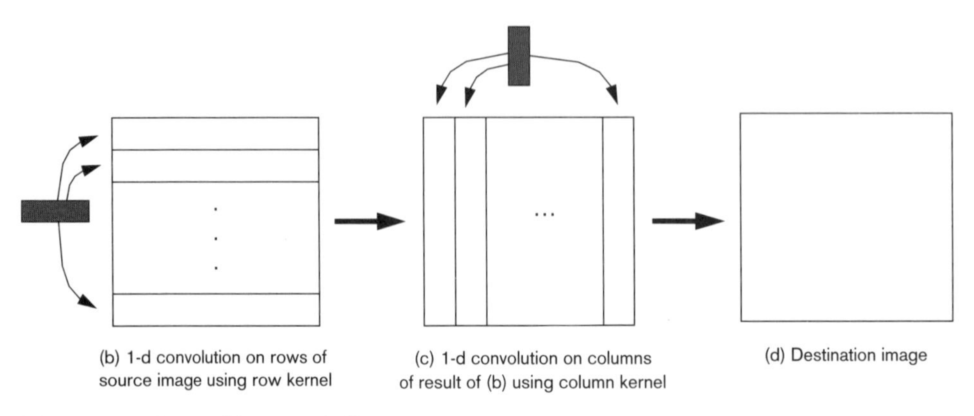

Figure 6.9 Separable convolution

The rows of the source image are then convolved with row kernel (*b* in Figure 6.9), and the columns of the result are convolved with the column vector (*c* in Figure 6.9). Assuming that enough intermediate precision is maintained, the resulting destination image is identical to the one obtained by performing two-dimensional convolution on the source image by the two-dimensional kernel. In this case, substantial computation savings can be made.

The idea of separable convolution, when combined with using table lookup during convolution, can be taken one step further in the case where the one-dimensional vectors are symmetric (a common case). Wolberg and Massalin [2] have reported an algorithm for performing convolution that results in faster computation than the straightforward approach.

To illustrate the algorithm, we consider the case shown in Figure 6.10, where a seven-point one-dimensional convolution is computed over a set of input pixels. The kernal values are $k_1, \ldots, k_7$.

The outputs at the locations *D*, *E*, and *F* are as follows:

- $\text{Out}(D) = k3A + k2B + k1C + k1E + k2F + k3G + k0D$
- $\text{Out}(E) = k3B + k2C + k1D + k1E + k2F + k3G + k0E$
- $\text{Out}(F) = k3C + k2D + k1E + k1F + k2G + k3H + k0F$

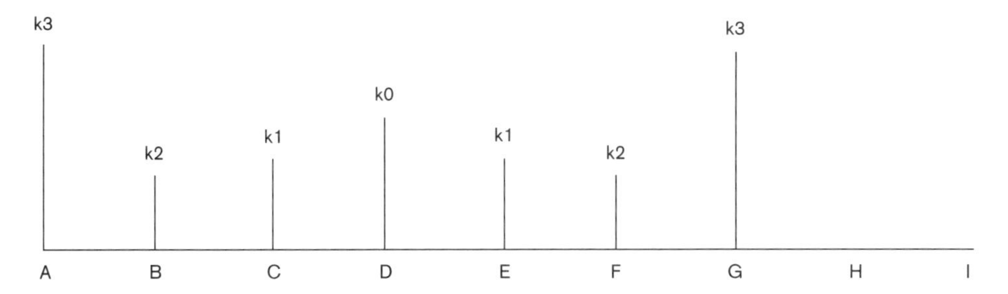

Figure 6.10 One-dimensional convolution

Using a straightforward table lookup approach would require the storing of all the values ki*pixel in the table. Then, we would look up the values and add them up, as shown in the previous programming example.

Taking this idea a step further, we can pack the lookup table values for each pixel multiplied by *k*1, *k*2, and *k*3 and store them in a 30-bit long word. This results in the table shown in Figure 6.11. The flow of the algorithm can take advantage of the partial sums being reused.

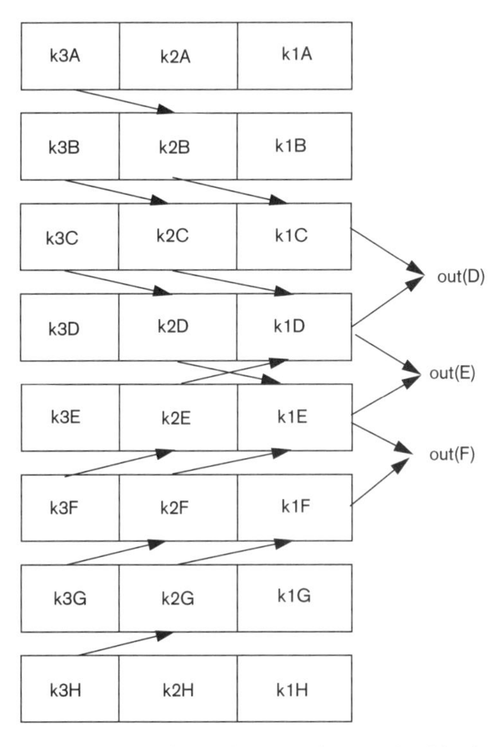

Figure 6.11 Convolution using packed lookup tables

Now, let us define two (30-bit) variables, `fwd` and `rev`, as follows. Each 10-bit field of `fwd` is created by right-shifting and adding the table value of the previous table entry. For example, here are the first few values of `fwd`:

Bits 20–29	**10–19**	**0–9**
$k3B$	$k3A + k2B$	$k2A + k1B$
$k3C$	$k3B + k2C$	$k3A + k2B + k1C$
$k3D$	$k3C + k2D$	$k3B + k2C + k1D$
$k3E$	$k3D + k2E$	$k3C + k2D + k1E$

and so forth. Similarly, the values of `rev` are computed from the bottom up:

Bits 20–29	**10–19**	**0–9**
$k3E + k2D$	$k2E + k1D$	$k1E$
$k3F + k2E + k1D$	$k2F + k1E$	$k1F$
$k3G + k2F + k1E$	$k2G + k1F$	$k1G$
$k3H + k2G + k1F$	$k2H + k1G$	$k1K$

Now we notice that the output at D, out(D), is simply the sum of the bits 0–9 of `fwd` and 20–29 of `rev` in the second row of values. Similarly the output at E is the same summation using the third row of values.

Therefore, to compute the one-dimensional convolution, all that one needs to do is to initialize `fwd` and `rev` and update them every time through the loop using the following pseudocode:

```
Initialize fwd, rev, table
Initialize srcptr, dstptr
While (input_len--) {
        fwd <- (fwd>>10) + table[srcptr]
        rev <- (rev<<10) + table[srcptr+6]
        *outptr <- (fwd & 0x3ff) + ((rev>>20) & 0x3ff) + k0*srcptr
}
```

This algorithm can now be used for performing separable convolution on the image. The code for this algorithm is given in reference 2, therefore we will not repeat it here.

Convolution with large kernels In the previous examples, we have seen several software implementations of convolution that work reasonably well for relatively small kernel sizes. As the kernel becomes larger, convolution becomes more unwieldy because the number of multiply-accumulate operations is directly proportional to the square of the kernel size.

There are several ways of efficiently implementing convolution with large kernels. These include Singular Value Decomposition/Small Generating Kernel (SVD/SGK) convolution, Fourier transform-based filtering, and recursive filtering. In this section we will look at the first method. Discussion of the other two methods, which are beyond the scope of this chapter, can be found in reference 1.

The motivation for the SVD/SGK [1] came from the desire to implement convolution with large kernels using hardware convolvers that accelerated 3×3 convolutions. One such convolver was the Vicom image processor, which is discussed in some detail in Chapter 3. However, the principles of SVD/SGK are applicable to either hardware or software convolvers that have been optimized for smaller convolution kernels. For the sake of clarity, we assume that fast 3×3 convolutions are available to us.

Suppose we want to convolve an $M \times M$ image $F(j,k)$ with an $L \times L$ kernel $H(j,k)$, where L is larger than 3. Then, the SVD/SGK algorithm effectively decomposes $H(j,k)$ into a sequence of 3×3 kernels with which the image is sequentially convolved.

This is done in two steps. The first step is to write the matrix $H(j,k)$ as a sum of r matrices, $H_0(j,k) \ldots H_{r-1}(j,k)$, each separable and $L \times L$ in size. This step can be done with the singular value decomposition for the matrix H. Here the index r is the number of unique singular values of H (also called the rank of H).

The decomposition allows us to write

$$H(j, k) = \sum_{i=0}^{r-1} H_i(j, k) \qquad (6.3)$$

Using this equation, the convolution of $F(j,k)$ by $H(j,k)$ can be realized, as shown in Figure 6.12.

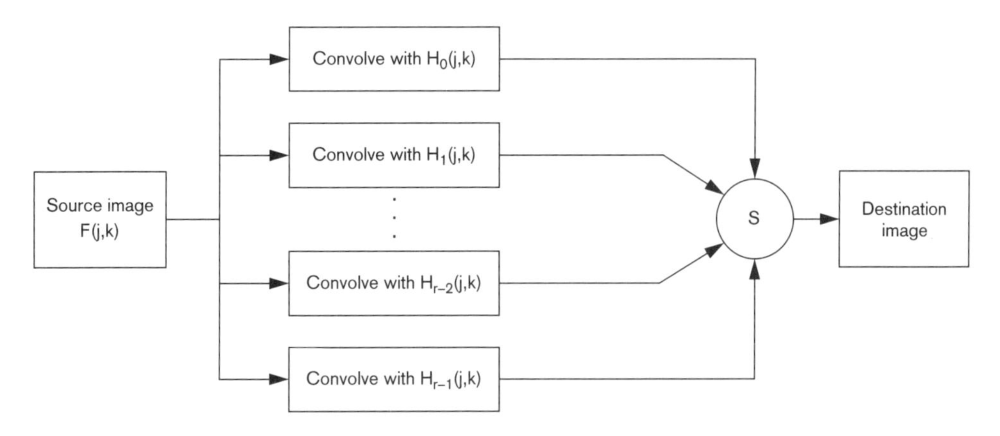

Figure 6.12 SVD/SGK convolution

In the second step, each H_i matrix is decomposed into a sequence of 3×3 matrices. First, the matrix is separated into the outer product of a column vector and a row vector. Then, each of these vectors is decomposed into a sequence of 3×1 and 1×3 vectors. Finally, the 3×1s and 1×3s are paired to give 3×3 vectors. This process is shown graphically in Figure 6.13.

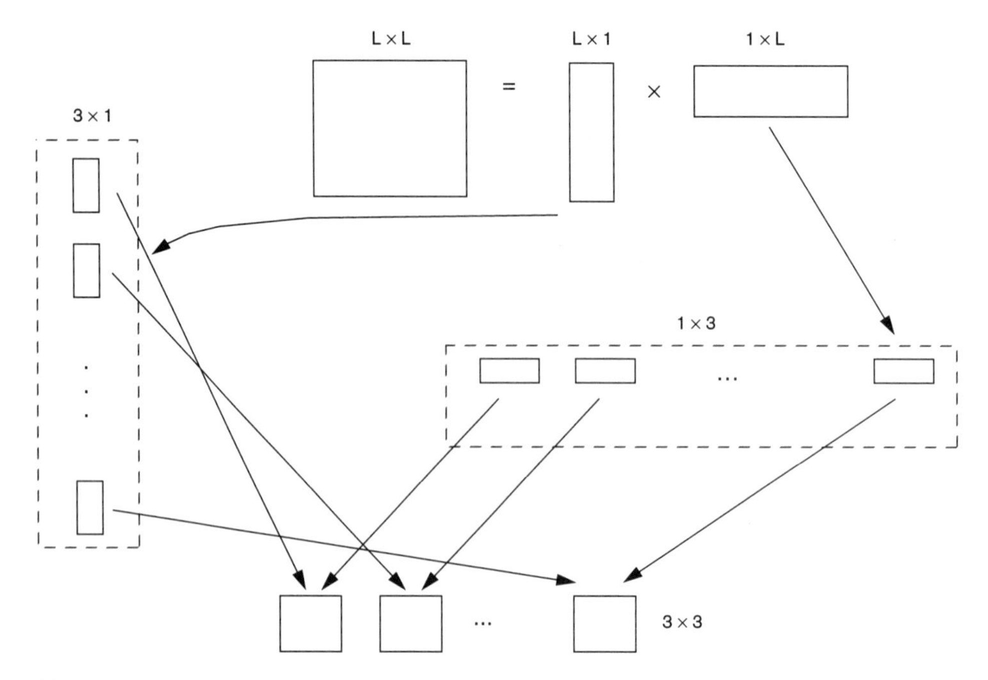

Figure 6.13 The SGK algorithm applied to an L × L kernel matrix

Given that the 3×3 convolution is of high speed, the performance of the SVD/SGK algorithm depends on two factors: the number r, which is the rank of the matrix, and the number of 3×3 matrices generated from each $L \times L$ matrix. Note that given the size $L \times L$ of $H(j,k)$, the number of 3×3 matrices generated is a constant at $(L - 1) / 2$. The number r, on the other hand, is the rank of $H(j,k)$, and the computational requirement of this algorithm is directly proportional to r.

6.3.4 Versatility of convolution

By now, the reader has a clear understanding of the convolution operation. It is essentially a weighted summation done within a moving window. The values being summed are the pixels within the window, and the weights are the values in the convolution kernel.

Much of the information content of pictures comes from the relative pixel values within small localized neighborhoods. Therefore, linear filtering using convolution is a powerful image processing technique. We briefly look at some of the applications of convolution here.

Smoothing filters Perhaps the most straightforward application of the convolution operation is in smoothing filters. As the window moves, it replaces the center pixel with the average of the pixels within the window. This is a smoothing filter because if there is noise, it gets smoothed out. In areas of uniform values, the pixels do not change.

The following convolution kernels perform smoothing in a 3×3 window:

$$L1 = \frac{1}{9}\begin{bmatrix} 1 & 1 & 1 \\ 1 & 1 & 1 \\ 1 & 1 & 1 \end{bmatrix}; L2 = \frac{1}{10}\begin{bmatrix} 1 & 1 & 1 \\ 1 & 2 & 1 \\ 1 & 1 & 1 \end{bmatrix}; L3 = \frac{1}{16}\begin{bmatrix} 1 & 2 & 1 \\ 2 & 4 & 2 \\ 1 & 2 & 1 \end{bmatrix}$$

The kernel $L1$ takes an unweighted average of all the pixels within the window, while the kernels $L2$ and $L3$ assign more weight to more important "central" pixels.

While smoothing filters are simple to implement, they have two major drawbacks: They blur edges, and they are unable to wipe out noise pixels that are large in value compared to the neighborhood pixels. The effect is to smear the noise pixels. As we will see in the next section, median filter makes a more effective filter.

Another use of smoothing filters is to low-pass filter the image before subsampling it. For example, an image pyramid (also called a MIPmap) is built by repeatedly smoothing the image and subsampling it. For this purpose a Gaussian convolution kernel, obtained by sampling a two-dimensional Gaussian function, is often used.

Sharpening filters As the name suggests, sharpening filters are used to sharpen images by crispening the edges. This is done by subtracting the surrounding pixels from a weighted version of the center pixel. The following kernels are commonly used for sharpening filters:

$$S1 = \begin{bmatrix} 0 & -1 & 0 \\ -1 & 5 & -1 \\ 0 & -1 & 0 \end{bmatrix}; S2 = \begin{bmatrix} -1 & -1 & -1 \\ -1 & 9 & -1 \\ -1 & -1 & -1 \end{bmatrix}; S3 = \begin{bmatrix} 1 & -2 & 1 \\ -2 & 5 & -2 \\ 1 & -2 & 1 \end{bmatrix}$$

Unsharp masking is also frequently used for sharpening images. It subtracts a smoothed version of the image from the original image to yield a sharper version of the image. Scale factors are used to adjust the amount of sharpening:

unsharp masked image = $K1$ * original image – $K2$ * blurred image

where $(K1 + K2) = 1$. Typically, the ratio of $K1$:$K2$ ranges from 1.5:1 to 5:1 [1].

Edge detection Edges are very important features of images because the human visual system uses edges for understanding and recognizing objects. Therefore, the detection of edges is an important part of many image processing systems, particularly those that analyze the image.

Here we briefly look at edge detection based on computing the rate of change of pixel values, that is, the *image gradient*. There are two such classes of edge detectors: directional and direction-invariant. Directional edge detectors are based on finding the image gradient along the

x and y directions; direction-invariant detectors, as the name suggests, are independent of the direction of the edges.

Among the directional edge detectors, a simple yet powerful one is the Sobel edge detector, which is shown in Figure 6.14.

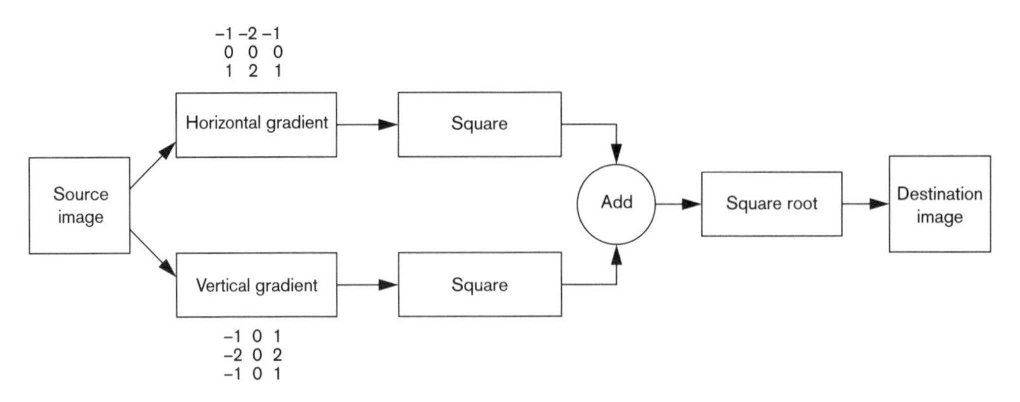

Figure 6.14 The Sobel edge detector

The result of the Sobel operator, $D(x,y)$, is defined as

$$D(x, y) = \sqrt{(SH(x, y))^2 + (SV(x, y))^2} \quad (6.4)$$

where $SH(x,y)$ and $SV(x,y)$ are the horizontal and vertical gradient images generated from the source image by convolving it with the following kernels:

$$H = \begin{bmatrix} -1 & -2 & -1 \\ 0 & 0 & 0 \\ 1 & 2 & 1 \end{bmatrix}; V = \begin{bmatrix} -1 & 0 & 1 \\ -2 & 0 & 2 \\ -1 & 0 & 1 \end{bmatrix}$$

The square and square-root operations are applied to each pixel and are typically implemented using a table lookup. The Sobel edge detector may be implemented in multiple passes or in a single pass through the image. However, if multiple passes are made, care must be taken to ensure the precision of the intermediate image values.

While the Sobel edge detector works fine on "normal" images, it is susceptible to noise, which gets amplified by the gradient process. (When you take the derivative of noise, you get more noise.)

A more advanced directional edge detector is the Canny edge detector [3]. This detector convolves the image with a kernel based on the first derivative of the Gaussian function, and then performs some edge-linking operations to create the final edge image. The first derivative of the Gaussian essentially smooths the image, reducing noise, before taking the gradient of the image.

A simple direction-invariant edge detector is the 3×3 approximation to the Laplacian operator, which approximates the magnitude of the second derivative at any point. The kernel used is:

$$S = \begin{bmatrix} -1 & -1 & -1 \\ -1 & 9 & -1 \\ -1 & -1 & -1 \end{bmatrix}$$

As before, this kernel is susceptible to noise. A more sophisticated edge detector is the Marr-Hildreth edge detector, which uses the Laplacian of the Gaussian function as the convolution kernel. This operator has the benefit that by choosing the standard deviation of the Gaussian function appropriately, it can be tuned to smooth out image features at any given scale.

There is a large body of literature on edge detection, much of which is beyond the scope of this book. The reader is referred to Chapter 16 of reference 1 for more details on the subtleties of edge detection, as well as another class of edge detection, based on model-fitting, that we have not discussed here.

Miscellaneous uses of convolution Besides filtering and edge detection, convolution has some often unexpected uses in imaging.

The following convolution kernel translates the image to the left by one pixel:

$$T = \begin{bmatrix} 0 & 0 & 0 \\ 1 & 0 & 0 \\ 0 & 0 & 0 \end{bmatrix}$$

The following convolution kernel can be used for bilinear interpolation, provided the rows and columns are interlaced with zeros:

$$B = \begin{bmatrix} 0.25 & 0.5 & .25 \\ 0.5 & 1 & 0.5 \\ 0.25 & 0.5 & 025 \end{bmatrix}$$

An example is shown in Figure 6.15.

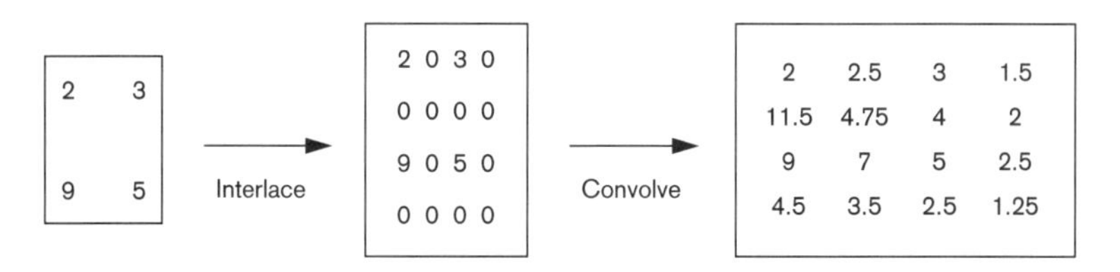

Figure 6.15 Bilinear interpolation using convolution

Convolution is also used in morphological filters to bitstack the pixels; this is explored in more detail in Section 6.5.

Examples of image convolution are shown in Figures 6.16 through 6.20 (Sobel edge detector).

Figure 6.16 **Bus original**

Figure 6.17 **Bus, 3×3 blur**

Figure 6.18 **Bus, 5×5 blur**

Figure 6.19 Bus, 3 × 3 high pass

Figure 6.20 Bus, Sobel edge detector

6.4 Nonlinear filtering I: the median filter and its variations

As we saw in Section 6.2, nonlinear filters differ substantially from linear filters. While in many instances they are more powerful than linear filters, they are also more difficult to model as well as to implement. In this section and the next, we study two classes of nonlinear filters that have found use in many imaging applications. Among these filters is the median filter, which is a powerful noise-removal filter. The properties and implementations of the median filter have been studied in great detail. We also look at several variations of the median filter.

6.4.1 *Definition and properties*

The median filter was initially proposed for filtering one-dimensional signals by Tukey [4]. It was extended to images by Pratt [1].

The filter is implemented by moving a window over the image. For each position of the window, the center pixel is replaced by the median of the pixels covered by the window. To find the median of N numbers, we sort them and pick the one with rank $(N + 1) / 2$.

There are several possible shapes for the window. These are shown in Figure 6.21. The choice of these windows is discussed shortly.

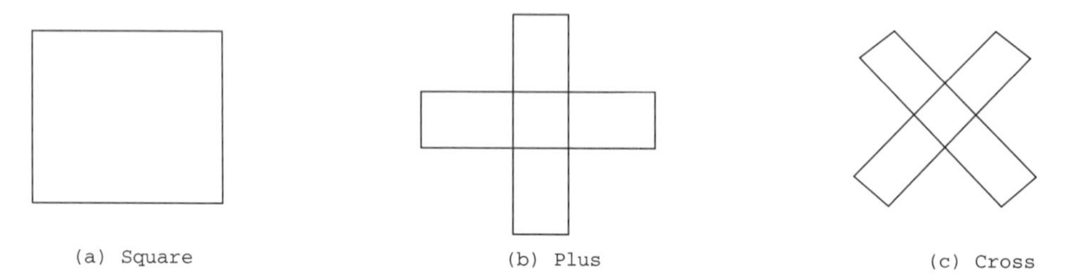

Figure 6.21 Windows used for median filter

Several properties of the median filter are noteworthy. First, it is a nonlinear filter. This is because if we are given two sequences $f(m)$ and $g(m)$ then median($(f(m) + g(m))$ does not, in general, equal median($f(m)$)+ median($g(m)$). Thus, one cannot use the principle of superposition to predict the output of median filter based on the transfer function, because this filter has no "transfer function." Studies have shown that the transfer function-like characteristics of the median filter actually change with the input.

Second, it is quite effective in removing impulse noise, unlike linear filters, which smooth edges and are unable to eliminate impulse noise completely. This is shown in Figure 6.22, where a white dot in a dark area of the image is removed by the filter. Note that a linear noise-smoothing filter would have smeared this dot over the surrounding pixels.

Third, the median filter does not blur the edges of objects in an image. This is critical because edges contain the most information. The median filter is different from linear filters in this respect because the latter tend to smooth edges. This is shown in Figure 6.23, where an edge represented by the transition of the pixel values from a to b is left untouched.

Fourth, the median filter affects certain other features of the image. It tends to wipe out thin lines and rounds out corners of images, and larger window sizes introduce a "blockiness" in the image. The plus- and cross-shaped windows shown Figure 6.21 perform better with corners and thin lines than do the square-shaped windows.

A *separable* median filter has also been defined [5]. This filter requires two passes through the image. The first pass consists of filtering all the rows of the image over a horizontal window. The columns of the resultant image are then filtered by a vertical median filter. This filter, while

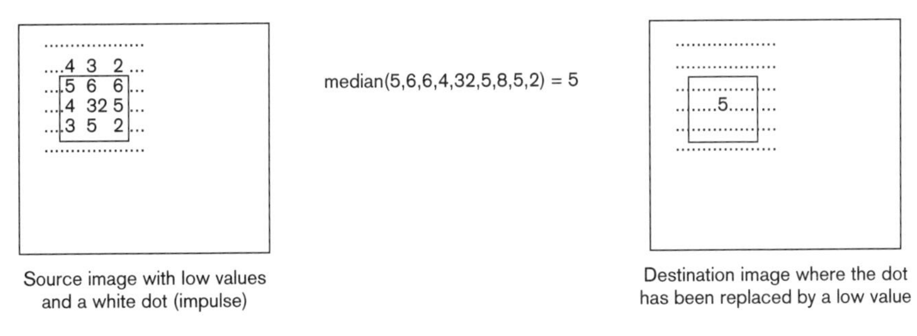

Figure 6.22 Median filter removes impulse noise from images

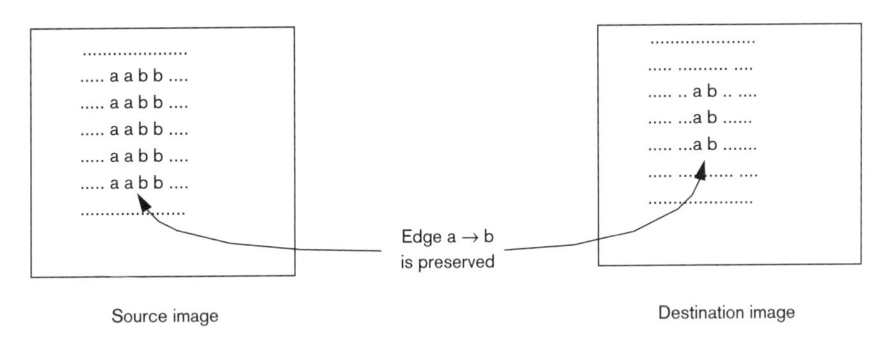

Figure 6.23 Edge preservation by a 3 × 3 square-shaped median filter

being less expensive to implement, has properties that are very similar to the nonseparable median filters.

6.4.2 Implementation and coding examples

Implementation of the median filter requires sorting. For this reason, it can be quite time-consuming. Several algorithms have been developed for the efficient implementation of the median filter. In this section we describe two such algorithms. The first one, proposed by Huang et al. [6], is based on the histogram of the image. The second one, reported by Paeth [7], is similar to classical compare-and-exchange algorithms (see, for example, reference 8).

Brute-force implementation In a brute-force implementation, the entire window is sorted using a quicksort or similar algorithm. The time required depends on the sorting algorithm used. For example, for an $m \times m$ window, quicksort requires an average of m^2 comparisons. The pseudocode is shown in the following code segment.

```
for all columns j in image
        for all rows i in image {
             compute median of window centered at source[i,j]
             replace destination [i,j] with this median
        }
```

Histogram algorithm Consider a median filter with an $m \times m$ rectangular window. In the brute-force approach above, when the window is moved from the position $[i,j]$ to $[i,j + 1]$, there are $(m - 1)m$ pixels that are already sorted. The histogram algorithm takes advantage of this fact and processes the new pixels to enter the window. It works best for square-shaped windows, although it can also be applied for plus-shaped and cross-shaped windows.

The algorithm relies on computing the variable `mdn` (the median) and the array `hist[]` (the histogram within the window) at the start of each row. As the window is moved to the right, both `hist[]` and `mdn` are updated. The pseudocode of the algorithm is as follows.

```
1. Compute the following quantities for pixels in the first window:
        a. hist[], the gray-level histogram
        b. mdn, the median
        c. ltmdn, the no. of pixels having gray level < mdn.

2. Move the window right one column. Each pixel, g, of the leftmost column is
deleted. hist and ltmdn are updated.
        hist[g] <- hist[g]-1
        ltmdn <- ltmdn -1 if g <mdn
Each pixel, g, of the rightmost column is added. hist and ltmdn are updated.
        hist[g] <- hist[g]+1
        ltmdn <- ltmdn +1 if g <mdn
Now hist has the histogram of the current window. ltmdn is the number of pix-
els in the current window having gray level less than the median of the pre-
vious window.

3. Define th = (m^2 - 1)/2. Compare ltmdn with th. There are two cases.
Case 1: ltmdn > th This indicates that mdn is greater than the current
median. Update ltmdn and mdn as follows
        while (ltmdn > th) {
             ltmdn <- ltmdn - hist[mdn]
             mdn <- mdn -1
        }

Case 2: ltmdn<= th This indicates that mdn is less than or equal to the
current median. Update ltmdn and mdn as follows
TOP:    if (ltmdn + hist[mdn] <= th) {
             ltmdn <- ltmdn + hist[mdn]
             mdn <- mdn +1
             go to TOP
        }
        else current median = mdn
```

The algorithm is implemented for a 3×3 window in C and shown in the following code.

```
/*
        Function:  pci_mdn3x3_histo
        Operation: Perform 3x3 median filter using Huang's histogram
                   algorithm
        Input:     src is the source image
        Output:    dst is the filtered image
*/

#define TH 4
void
pci_mdn3x3_histo(pci_image *src, pci_image *dst)
{
        register unsigned char *spixptr, *dpixptr, *spix, *dpix, *sptr;
        register unsigned char *s0ptr, *s1ptr, *s2ptr;
        register unsigned char median=0;
        int rows, cols;
        int slb, dlb;
        int dw, dh;
        int lw, lh;
        int i;
        int hist[256];                 /* 8-bit pixels */
        int ltmdn=0;
        unsigned char first[9];        /* contents of the 1st window */

        /* get the info from the images */
        spix = src->data;
        slb = src->linebytes;
        dpixptr = dpix = dst->data;
        dlb = dst->linebytes;
        dw = dst->width;
        dh = dst->height;

        s0ptr = spix;
        s1ptr = s0ptr+slb;
        s2ptr = s1ptr+slb;

        /* init the histgram */
        for (i=0; i<256; i++)
             hist[i] = 0;

        /* zero out the top row */
        for (i=0; i<dw; i++)
             *dpix++ = 0;

        /* start at the beginning of the second row */
        dpixptr += dlb;
        dpix = dpixptr;

        /* perform median using histogram algo. */
        lh = dh - 2;
        while (lh--) {

             /* handle left edge */
             *dpixptr++ = 0;
```

```
/* get first nine pixels */
first[0]=*s0ptr; first[1]=*(s0ptr+1); first[2]=*(s0ptr+2);
first[3]=*s1ptr; first[4]=*(s1ptr+1); first[5]=*(s1ptr+2);
first[6]=*s2ptr; first[7]=*(s2ptr+1); first[8]=*(s2ptr+2);

/* compute histgram of the 1st window */
for (i=0; i<9; i++)
     hist[first[i]]++;

/* compute the median */
qsort(first,9,1,intcompare);
median = first[4];

/* compute ltmdn */
for (i=0; i<4; i++)
     if (first[i] < median) ltmdn++;

*dpixptr++ = median;

/* now move right, applying algorithm along one row */
lw = dw - 3;
while (lw--) {
     /* first subtract the first 3 pixels... */
     /* ...by updating hist[]... */
     hist[*s0ptr]--;
     hist[*s1ptr]--;
     hist[*s2ptr]--;

     /* ... and updating ltmdn */
     if (*s0ptr < median)
          ltmdn--;
     if (*s1ptr < median)
          ltmdn--;
     if (*s2ptr < median)
          ltmdn--;

     /* bring in new pixels by updating source pointers */
     s0ptr++;
     s1ptr++;
     s2ptr++;

     /* add three new pixels ... */
     /* ... by updating hist[]... */
     hist[*(s0ptr+2)]++;
     hist[*(s1ptr+2)]++;
     hist[*(s2ptr+2)]++;

     /* ... and updating ltmdn */
     if (*(s0ptr+2) < median)
          ltmdn++;
     if (*(s1ptr+2) < median)
          ltmdn++;
     if (*(s2ptr+2) < median)
          ltmdn++;
```

```
                /* find the median */
                if (ltmdn > TH) {
                     do {
                          median--;
                          ltmdn -= hist[median];
                     } while (ltmdn > TH);
                }
                else {
                     while ((ltmdn + hist[median]) <= TH) {
                          ltmdn += hist[median];
                          median++;
                     }
                     /* else current median is the new median */
                }
                *dpixptr++ = median;
           }

           /* handle the right edge */
           *dpixptr++ = 0;

           /* clean out the row-sensitive variables */
           for (i=0; i<256; i++)
                hist[i]=0;
           median = 0;
           ltmdn = 0;

           /* update the three source pointers */
           spix += slb;
           s0ptr = spix;
           s1ptr = spix + slb;
           s2ptr = s1ptr + slb;

           dpixptr = dpix += dlb;
      }

      /* fix the bottom row */
      for (i=0; i<dw; i++)
           *dpixptr++ = 0;
}
```

Paeth's median-finding algorithm Paeth [7] has proposed a fast algorithm for finding the median of nine numbers, representing the pixels in a 3×3 square window. The algorithm is based on the premise that finding the median of a set of nine numbers effectively divides the set into three smaller sets: a set of four numbers, whose values are less than (or equal to) the median; a set consisting solely of the median; and a set of four numbers, whose values are greater than (or equal to) the median. The median is found by successively taking pairs of numbers out of the nine numbers and placing them into the nonmedian sets until the median remains.

The algorithm will be shown here in pseudocode; C code can be found in reference 7. The basis of the implementation is the macro `s2(a,b)`, which places the letters *a* and *b* in ascending order:

```
s2(a,b): swap a and b if a > b
```

We can now build up the operators `minmaxk(a0, ..., ak-1)` from this. The operator takes k elements $a_0, \dots, a_{k-1}$ and places the minimum of these in a_0 and the maximum of these in a_{k-1}. For example:

```
minmax3(a,b,c): s2(b,c), s2(a,c), s2(a,b)
minmax4(a,b,c,d): s2(a,b); s2(c,d), s2(a,c), s2(b,d)
```

and so on. Then, the median of the nine elements of the array `a[9]` can be found by

```
minmax6( a[0], a[1], a[2], a[3], a[4], a[5]);
minmax5( a[1], a[2], a[3], a[4], a[6]);
minmax4( a[2], a[3], a[4], a[7]);
minmax3( a[3], a[4], a[8]);
```

This algorithm is similar to the classical compare-and-exchange (CEX) algorithm [8] used to compute the median filter. The algorithm, shown in Figure 6.24 for an example array of five elements, consists of successively comparing two elements and exchanging them to place them in ascending order. Each box in the figure represents a CEX, which compares the two input values and places the smaller at one output and the larger at the other. Note that the comparison at the bottom right is not strictly necessary for computation of the median.

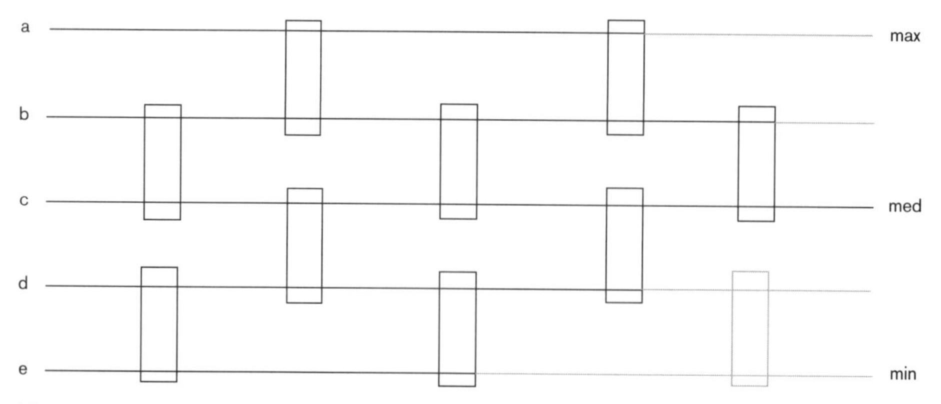

Figure 6.24 Compare and exchange to find median of five elements (a,b,c,d,e)

Examples of median filtering are shown in Figures 6.25 through 6.28.

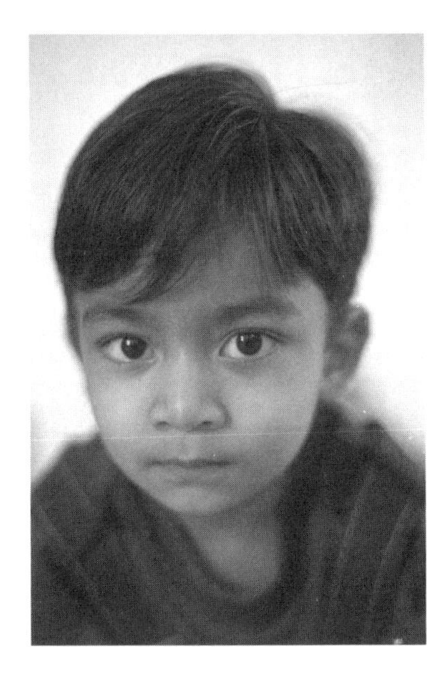

Figure 6.25 Ihsan original

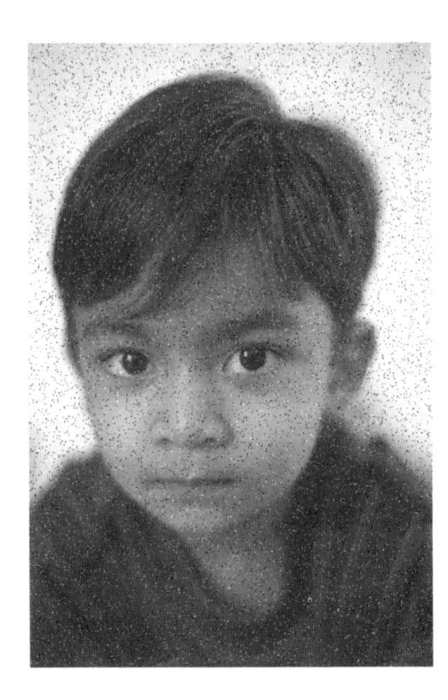

Figure 6.26 Ihsan, noise added

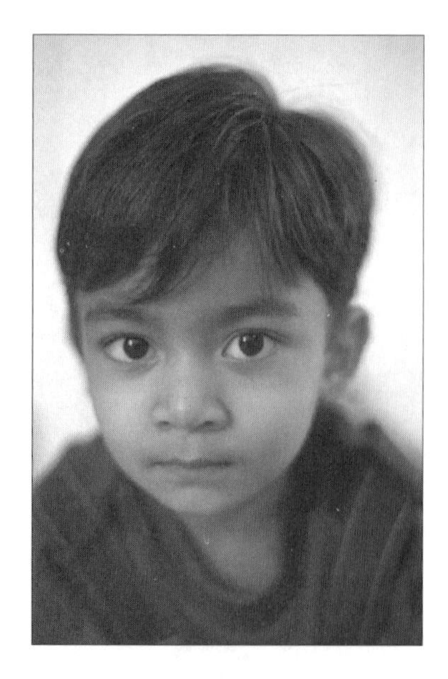

Figure 6.27 Ihsan, median filtered

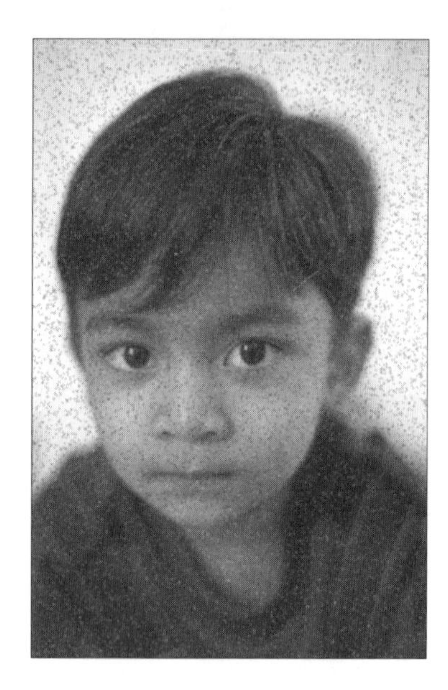

Figure 6.28 Ihsan, 3×3 low-pass filtered using moving average convolution

6.4.3 *Variations of the median filter: pseudomedian and weighted median*

Numerous variations of the median filter have been proposed in the literature. These include the pseudomedian filter [8], the weighted median filter [9], and hybrid filters combining the rank-ordering with linear filtering.

The pseudomedian filter is a computationally efficient variation of the median filter that has properties similar to the median filter. Here we briefly present the idea behind this filter by considering one-dimensional arrays. Extension to two dimensions is straightforward and left to the reader.

Given three elements $\{a,b,c\}$, their median can be written as

$$med(a, b, c) = max(min(a, b), min(b, c), min(c, a)) \tag{6.5}$$

Alternatively,

$$med(a, b, c) = min(max(a, b), max(b, c), max(c, a)) \tag{6.6}$$

Similarly, for five elements $\{a,b,c,d,e\}$, the median can be written as

$$med(a, b, c, d, e) = max(A, B, C, D, E, F, G, H, I, J) \tag{6.7}$$

where $A = \min(a,b,c)$, $B = \min(a,b,d)$, $C = \min(a,b,e)$, $D = \min(a,c,d)$, $E = \min(a,c,e)$, $F = \min(a,d,e)$, $G = \min(b,c,d)$, $H = \min(b,c,e)$, $I = \min(b,d,e)$, and $J = \min(c,d,e)$.

The pseudomedian, *pmed*, of these two sequences requires fewer comparisons. For the three-element case, it is:

$$pmed(a, b, c) = \frac{1}{2}max(min(a, b), min(b, c)) + \frac{1}{2}min(max(a, b), max(b, c)) \tag{6.8}$$

For the five-element case, it is:

$$pmed(a, b, c, d, e) = \frac{1}{2}max(A, B, C) + \frac{1}{2}min(A, B, C) \tag{6.9}$$

where $A = \min(a,b,c)$, $B = \min(b,c,d)$, and $C = \min(c,d,e)$.

The one-dimensional filter is extended to two dimensions by stringing together the elements in a two-dimensional window. This filter can be implemented substantially faster than a median filter, particularly on pipeline image processors, such as the Vicom. Images filtered using the pseudomedian filter have their corners preserved.

Another variation of the median filter is the *weighted median filter* (WMF) reported by Brownrigg [9]. It has found use in astronomical image processing. This filter attempts to overcome some of the constraints of the median filter. In particular, the square-shaped median filter smooths out corners (shown in Figure 6.29) and wipes out lines of one-pixel width, whereas a plus-shaped median filter preserves corners but also preserve lines of one-pixel width. However, in astronomical image processing, one often requires that corners be preserved and single-pixel lines be wiped out. Neither a square- nor a plus-shaped median filter is able to do this.

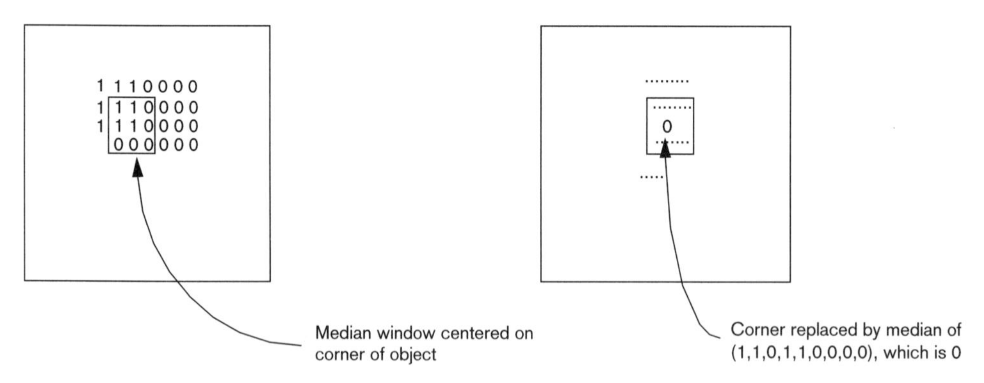

Figure 6.29 A square-shaped median filter smooths out corners

The WMF, which is an attempt to fine-tune the response of the median filter in order to meet this kind of need, is defined by a two-dimensional array, *W*, of nonnegative integers, such that the sum of the array elements is odd. The filter is implemented in a way similar to the median—a window is moved over the image and pixels falling in the window are sorted—except that the number of copies of a pixel that are in the list is the corresponding number from *W*. For example, if the WMF is the following array, then it is really a 3×3 square-shaped median filter.

$$W1 = \begin{bmatrix} 1 & 1 & 1 \\ 1 & 1 & 1 \\ 1 & 1 & 1 \end{bmatrix}$$

For the WMF below,

$$W2 = \begin{bmatrix} 1 & 1 & 1 \\ 1 & 3 & 1 \\ 1 & 1 & 1 \end{bmatrix}$$

the pixels in a 3×3 window are sorted, but three copies of the center pixel are used in the list. Then, the median of 11 values is used to replace the center pixel. Brownrigg has shown that the filter *W2* above is able to preserve corners while eliminating single-pixel lines.

While several other variations of the median filter are available, these are outside the scope of this book. We now proceed to look at another type of powerful nonlinear filters.

6.5 *Nonlinear filtering II: morphological filters*

Morphological image processing is an area of study concentrating on the *shape* of objects in an image. Morphological filters, which are based on these concepts, modify shapes so that features can be extracted and the images can be analyzed. The structure of the objects can be greatly simplified by using these filters.

The theory of morphological image processing was developed by Serra [10] using set-theoretic principles. Some of the earlier ideas used in morphology came from Minkowski [11] and from Matheron [12]. While these theoretical principles form a rich area of study, morphological filters based on empirical observations have also been implemented by various individuals. These include Rosenfeld [13], Pavlidis [14], and Pratt and Kabir [15]. Pratt [1] has shown that these filters based on the empirical observations can be derived from the theory developed by Serra and others.

Morphological filters apply to binary and to gray-scale images. The use of morphological filters for binary images is widespread. In image analysis applications, where the goal is to automatically identify and recognize objects in an image, these filters are used to detect features, to smooth objects, and to simplify the shape of an object into its essential structure.

6.5.1 *Binary morphology*

In binary images, where all pixel values are either 0 or 1, morphological filters, such as those described in reference 1, are based on *hit-or-miss transformations*. A hit-or-miss transformation is defined by three parameters: the *window size*, the *hit table*, and the *rule*.

Given a particular (x,y) location in the source image, a *hit* occurs if the pattern of 0s and 1s in the neighborhood window centered at (x,y) matches one of the patterns in the hit table. When this happens, the destination pixel at the same location is set according to the rule we have defined. The rule typically changes the state of the pixel (from 0 to 1 or vice versa).

An example is shown in Figure 6.30. In this example, the window size is 3×3, and the patterns in the hit table are shown on the left. In the source image, the location of hits is outlined. The rule we have chosen is that the destination pixels where hits occur are set to 0.

If we restrict ourselves to a 3×3 window operating on binary images, then the window can be represented as a set of numbers x, $x0$, …, $x7$, as shown in Figure 6.31.

Since the pixels are all binary, this pixel nomenclature can be used to represent a set of patterns using a logical equation. For example, the patterns used in the example in Figure 6.30 can be represented as (x & ~$x0$ & ~$x1$ & ~$x2$ & $x3$ & ~$x4$ & ~$x5$ & ~$x6$ & ~$x7$), and (x & ~$x0$ & $x1$ & ~$x2$ & ~$x3$ & ~$x4$ & ~$x5$ & ~$x6$ & ~$x7$), where & represents *logical and* and ~ represents *not*.

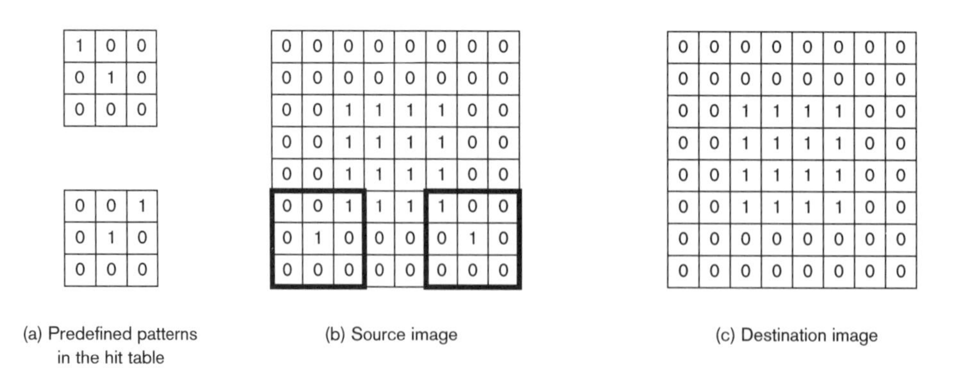

(a) Predefined patterns in the hit table

1	0	0
0	1	0
0	0	0

0	0	1
0	1	0
0	0	0

(b) Source image

0	0	0	0	0	0	0	0
0	0	0	0	0	0	0	0
0	0	1	1	1	1	0	0
0	0	1	1	1	1	0	0
0	0	1	1	1	1	0	0
0	0	1	1	1	1	0	0
0	1	0	0	0	0	1	0
0	0	0	0	0	0	0	0

(c) Destination image

0	0	0	0	0	0	0	0
0	0	0	0	0	0	0	0
0	0	1	1	1	1	0	0
0	0	1	1	1	1	0	0
0	0	1	1	1	1	0	0
0	0	1	1	1	1	0	0
0	0	0	0	0	0	0	0
0	0	0	0	0	0	0	0

Figure 6.30 An example of a hit-or-miss transformation

x3	x2	x1
x4	x	x0
x5	x6	x7

Figure 6.31 A 3 × 3 window used in morphological filters

Basic binary morphological operations Having defined hit-or-miss transformations, we can now look at specific morphological filters. There are more than a dozen basic operators. Here we concentrate on five of these: *erosion, dilation, opening, closing,* and *interior pixel removal.* The other operators are similar in principle and are derived in reference 1.

The *erosion* operation peels away a layer of pixels from the outer boundary of the objects in the image. An example of erosion is shown in Figure 6.32.

The rule for generating the hit table for erosion is simple: If at least one of the eight neighbors of the pixel is 1, then the pixel is set to 0. Therefore, the hit table consists of the set of patterns that satisfy the following logical expression x & (x0 or *x*1 or *x*2 or *x*3 or *x*4 or *x*5 or *x*6 or *x*7).

Dilation is the inverse operation of erosion. It adds a layer of pixels one pixel wide to the outline of the object. An example is shown in Figure 6.33.

The rule for generating the hit tables for dilation is also simple: If the pixel is 0 and at least one of the eight neighbors is 1, then set the pixel to 1. The patterns in the hit table satisfy the following logical expression ~*x* & (x0 or *x*1 or *x*2 or *x*3 or *x*4 or *x*5 or *x*6 or *x*7).

Based on erosion and dilation, we can define two new operators called *opening* and *closing.* Opening is an erosion followed by a dilation. It has an overall smoothing effect on the object shapes, and is useful for removing pixel noise from the image. It is useful for finding flaws in an

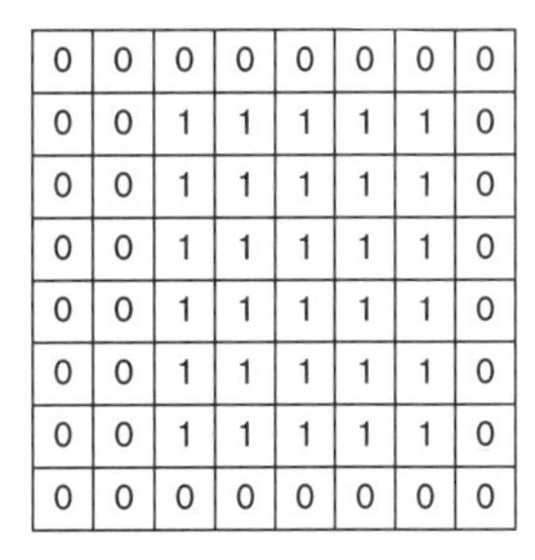

0	0	0	0	0	0	0	0
0	0	1	1	1	1	1	0
0	0	1	1	1	1	1	0
0	0	1	1	1	1	1	0
0	0	1	1	1	1	1	0
0	0	1	1	1	1	1	0
0	0	1	1	1	1	1	0
0	0	0	0	0	0	0	0

(a) Source image

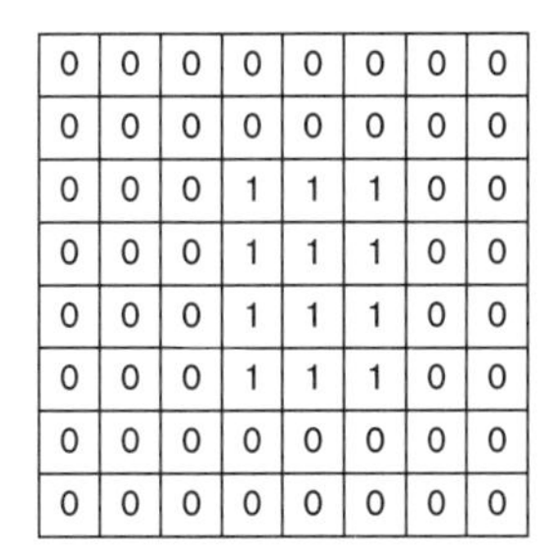

0	0	0	0	0	0	0	0
0	0	0	0	0	0	0	0
0	0	0	1	1	1	0	0
0	0	0	1	1	1	0	0
0	0	0	1	1	1	0	0
0	0	0	1	1	1	0	0
0	0	0	0	0	0	0	0
0	0	0	0	0	0	0	0

(b) Destination image

Figure 6.32 Morphological erosion

0	0	0	0	0	0	0	0
0	0	1	1	1	1	1	0
0	0	1	1	1	1	1	0
0	0	1	1	1	1	1	0
0	0	1	1	1	1	1	0
0	0	1	1	1	1	1	0
0	0	1	1	1	1	1	0
0	0	0	0	0	0	0	0

(a) Source image

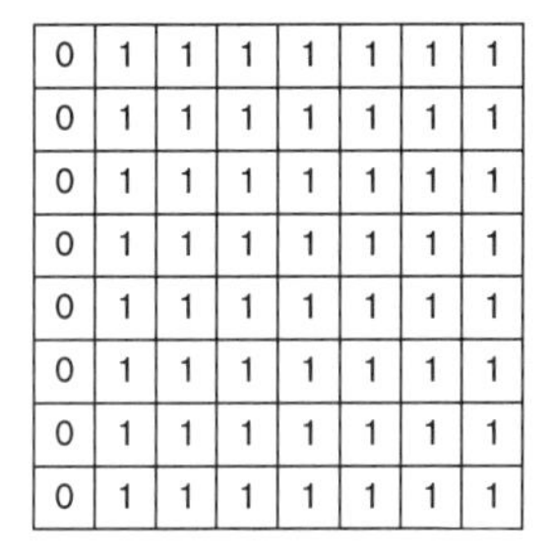

0	1	1	1	1	1	1	1
0	1	1	1	1	1	1	1
0	1	1	1	1	1	1	1
0	1	1	1	1	1	1	1
0	1	1	1	1	1	1	1
0	1	1	1	1	1	1	1
0	1	1	1	1	1	1	1
0	1	1	1	1	1	1	1

(b) Destination image

Figure 6.33 Morphological dilation

image (for example, if a straight line has a nick on it, opening will break the line into two lines of similar width). Closing is a dilation followed by an erosion. It can be used to close gaps in the image.

The last type of basic morphological filter is called *interior pixel removal*. The operation of this filter is shown in Figure 6.34. It yields the outline, or the edge of the object. This is very useful for object recognition, where often one needs to make feature measurements on the object for recognition purposes. The corresponding logical equation for interior pixel removal is x & (~$x0$ or ~$x2$ or ~$x4$ or ~$x6$).

Implementation of binary morphological operations A brute-force implementation of any of these functions requires scanning the image and checking each 3×3 neighborhood against all the patterns in the pattern table. Clearly, this can be time consuming.

If one has access to fast implementations of convolution and table lookup, then the implementation of erosion, dilation, and interior pixel removal can be done using a *pixel stacker*.

0	0	0	0	0	0	0	0
0	0	1	1	1	1	1	0
0	0	1	1	1	1	1	0
0	0	1	1	1	1	1	0
0	0	1	1	1	1	1	0
0	0	1	1	1	1	1	0
0	0	1	1	1	1	1	0
0	0	0	0	0	0	0	0

0	0	0	0	0	0	0	0
0	0	1	1	1	1	1	0
0	0	1	0	0	0	1	0
0	0	1	0	0	0	1	0
0	0	1	0	0	0	1	0
0	0	1	0	0	0	1	0
0	0	1	1	1	1	1	0
0	0	0	0	0	0	0	0

Figure 6.34 Morphological interior pixel removal

Consider an image being convolved with the convolution kernel shown in Figure 6.35. Note that the source pixels are all 0 or 1. The convolution result is as shown in the figure, provided that the destination pixel is 9 bits or wider.

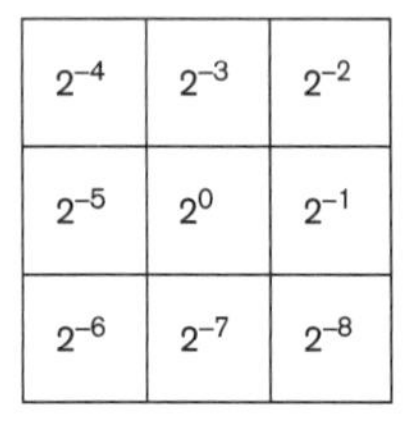

(a) Convolution kernel

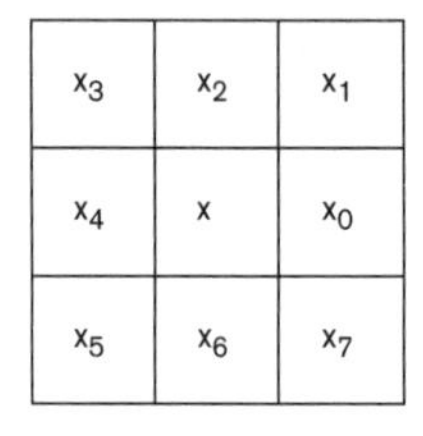

(b) Source neighborhood

(c) Bits of the destination pixel

Figure 6.35 Pixel stacking operation

In essence, we have stacked the 3×3 neighborhood of the source pixel x into the bits of one destination pixel. Now consider the sequence of operations shown in Figure 6.36 [1].

In this implementation, the convolution operation is used to stack the neighborhood of the (binary) source image into an intermediate image. In other words, the 3×3 neighborhood of each source pixel is encoded into the corresponding pixel of the intermediate image. The (9-bits-in, 9-bits-out) lookup table is programmed so that all entries corresponding to the hit patterns are changed according to the corresponding rule. All entries corresponding to missed patterns are passed through without change. This is an efficient way to perform basic binary morphological operations. For example, to perform erosion, the lookup table is composed of two parts, as shown in Figure 6.37.

Thus, all entries of value 255 or less correspond to source neighborhoods where the central pixel, x, is 0. Therefore, the destination pixel is 0. The entry at 256 corresponds to an isolated pixel neighborhood, which is not erased. The entries between 257 and 510 correspond to neighborhoods where, in addition to the central nonzero pixel, there is at least one other pixel, which

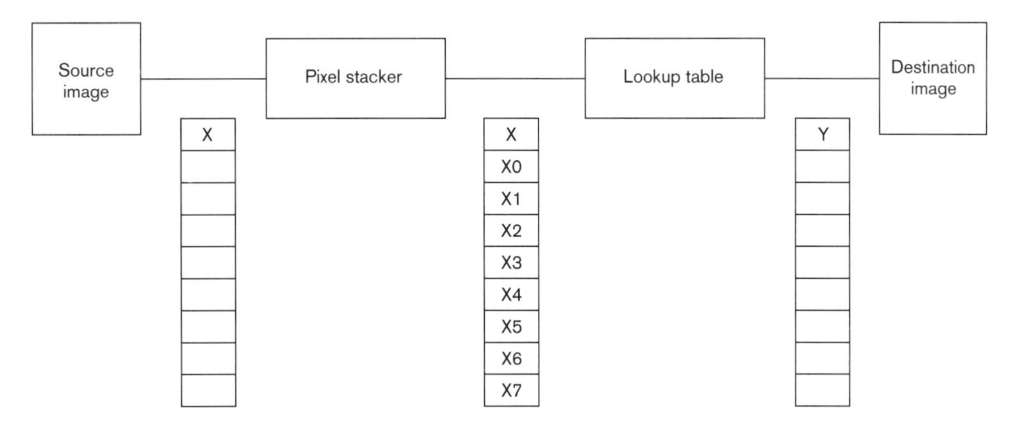

Figure 6.36 Block diagram of hit-or-miss transformation using convolution and table lookup

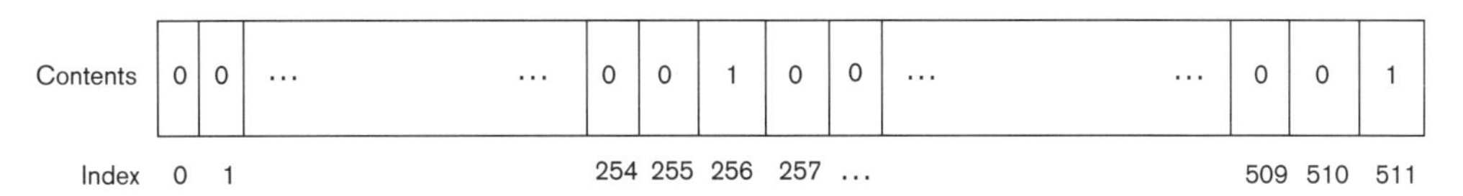

Figure 6.37 The lookup table loaded for an erosion operation

is also nonzero. Therefore, the destination pixel is set to 0 for these neighborhoods. Finally the last entry, 511, corresponds to a neighborhood where all the pixels are 1. The source pixel is left untouched here.

Next we provide C implementation of table-driven morphological filters. In order to optimize the subroutine, the pixel stack is defined in a slightly different manner from our earlier definition, as shown in Figure 6.38. When the 3×3 window is moved right horizontally, this configuration allows updating the pixel stack by simply shifting to the left by three places and or-ing in the new pixels, as shown in Figure 6.39.

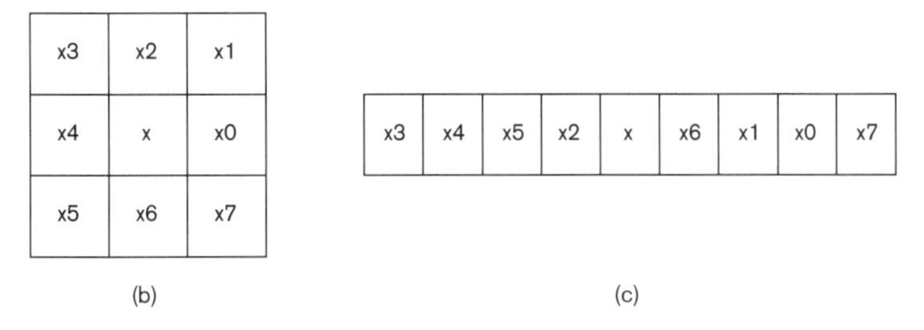

Figure 6.38 Pixel stacking used in the C program

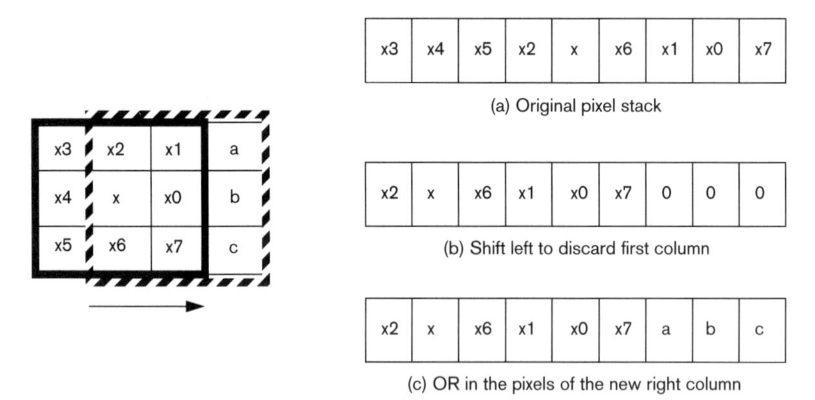

(a) Original pixel stack

(b) Shift left to discard first column

(c) OR in the pixels of the new right column

Figure 6.39 Updating the pixel stack as the window moves

The two subroutines shown in the following code generate the hit tables for erosion and interior pixel removal, respectively.

```
/*
          Function:   pci_gen_erode
          Operation:generates a morphology table for "erosion" operation.
          Input:      none
          Output:     pointer to table
*/

void
pci_gen_erode(unsigned short *table)
{
          register int i;

          for (i=0; i<512; i++)
                    table[i] = 0;
          table[511] = 1;
          table[16] = 1;
          return;
}

/*
          Function:   pci_gen_ipr
          Operation:generates a morphology table for "interior pixel
                      remove" operation.
          Input:      none
          Output:     pointer to table
*/

void
pci_gen_ipr(unsigned short *table)
{
          register int i;
          for (i=0; i<512; i++)
                    table[i] = 0;
```

```
	for (i=0; i<512; i++) {
		if ((i&0x10) == 0x10) /* i.e., the center pixel is 1 */
			table[i] = 1;

		if (((i&0x2)==2) && ((i&0x8)==8) && ((i&0x10)==0x10)
		  && ((i&0x20) == 0x20) && ((i&0x80)==0x80))
			table[i] = 0;
		}

	return;
}
```

The following subroutine goes over the image and performs the morphological filter specified by the table `m_table`.

```
/*
	Function: pci_morph
	Operation:perform unconditional table driven morphology
	Input:    src is the source image which is binary (ie, 0 or 1)
	          only lsb of the 8-bit pixel is considered
	          m_table is the morhpology table (see text for details)
	Output:   dst is the destination image
*/

void
pci_morph(pci_image *src, pci_image *dst, unsigned short *m_table)
{
	register unsigned char *s0ptr, *s1ptr, *s2ptr;
	register unsigned char *spixptr, *dpixptr, *spix, *dpix, *sptr;
	register unsigned int bitstack;
	register unsigned short *l_table; /* local copy of table */
	unsigned char first[9];
	int rows, cols;
	int slb, dlb;
	int dw, dh;
	int lw, lh;
	int i;

	/* get the info from the images */
	spix = src->data;
	slb = src->linebytes;
	dpixptr = dpix = dst->data;
	dlb = dst->linebytes;
	dw = dst->width;
	dh = dst->height;

	l_table = m_table;

	s0ptr = spix;
	s1ptr = s0ptr+slb;
	s2ptr = s1ptr+slb;
```

```
/* zero out the top row */
for (i=0; i<dw; i++)
     *dpix++ = 0;

/* start at the beginning of the second row */
dpixptr += dlb;
dpix = dpixptr;

/* perform morphological filter using table-driven algorithm*/
lh = dh - 2;
while (lh--) {

     /* handle left edge */
     *dpixptr++ = 0;

     /* get first nine pixels */
     first[0]=*s0ptr; first[1]=*(s0ptr+1); first[2]=*(s0ptr+2);
     first[3]=*s1ptr; first[4]=*(s1ptr+1); first[5]=*(s1ptr+2);
     first[6]=*s2ptr; first[7]=*(s2ptr+1); first[8]=*(s2ptr+2);

     /* stack the bits of the first window */
     bitstack = 0;
     bitstack |= first[0]<<8;
     bitstack |= first[1]<<5;
     bitstack |= first[2]<<2;
     bitstack |= first[3]<<7;
     bitstack |= first[4]<<4;
     bitstack |= first[5]<<1;
     bitstack |= first[6]<<6;
     bitstack |= first[7]<<3;
     bitstack |= first[8];

     *dpixptr++ = l_table[bitstack];

     s0ptr += 2;
     s1ptr += 2;
     s2ptr += 2;

     /* now move right, applying algorithm along one row */
     lw = dw - 3;
     while (lw--) {
          /* first subtract the first 3 pixels... */
          /* ...turning off the bits moved out of window... */
          bitstack <<= 3;
          bitstack &= 0x01ff;

          /* bring in new pixels by updating source pointers */
          s0ptr++;
          s1ptr++;
          s2ptr++;

          /* turn on corresponding bits for new pixels */
          bitstack |= (*s0ptr)<<2;
          bitstack |= (*s1ptr)<<1;
          bitstack |= (*s2ptr)<<0;
```

```
                /* copy entry from table */
                *dpixptr++ = l_table[bitstack];
            }

            /* handle the right edge */
            *dpixptr++ = 0;

            /* clean out the row-sensitive variables */
            bitstack = 0;

            /* update the three source pointers */
            spix += slb;
            s0ptr = spix;
            s1ptr = spix + slb;
            s2ptr = s1ptr + slb;

            dpixptr = dpix += dlb;
        }

        /* fix the bottom row */
        for (i=0; i<dw; i++)
            *dpixptr++ = 0;
}
```

Before proceeding to conditional morphological operations, we present some example images. Figure 6.40 (original) has been thresholded into a binary image, Figure 6.41. The erosion of this image using two iterations is shown in Figure 6.42. Figure 6.43 shows two iterations of dilation, and Figure 6.44 shows the result of interior pixel removal.

Figure 6.40 Snog Rock original

Figure 6.41 Snog Rock, thresholded

Figure 6.42 Snog Rock, eroded 2×

Figure 6.43 Snog Rock, dilated 2×

Figure 6.44 Snog Rock, interior pixels removed

Conditional operations and implementation The preceeding algorithm provides an efficient way to perform basic morphological operations. These operations are useful for a number of reasons, as outlined above. In certain cases, however, we wish to perform morphological operations that preserve the structure of the objects. For example, we might be interested in a skeleton representation of the object, or we might want to thin down a rectangle into a line one-pixel wide.

In these cases, it is important to preserve the connectivity of the objects in the image. The basic algorithms described above fail to do this, as shown in the following example of erosion (Figure 6.45).

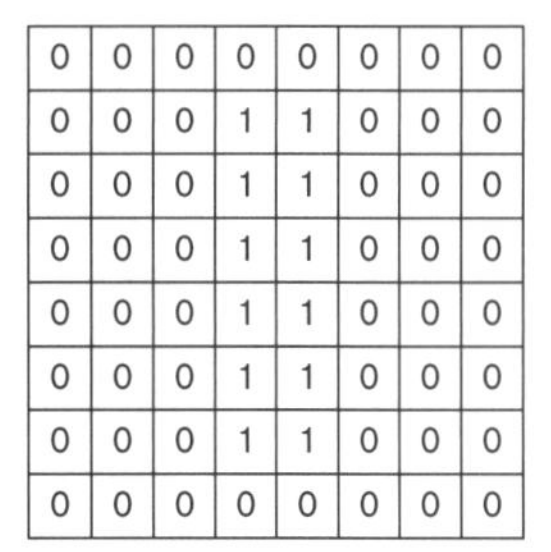

0	0	0	0	0	0	0	0
0	0	0	1	1	0	0	0
0	0	0	1	1	0	0	0
0	0	0	1	1	0	0	0
0	0	0	1	1	0	0	0
0	0	0	1	1	0	0	0
0	0	0	1	1	0	0	0
0	0	0	0	0	0	0	0

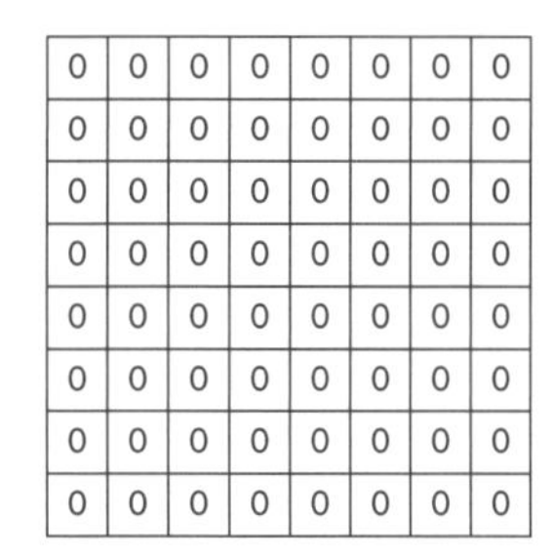

0	0	0	0	0	0	0	0
0	0	0	0	0	0	0	0
0	0	0	0	0	0	0	0
0	0	0	0	0	0	0	0
0	0	0	0	0	0	0	0
0	0	0	0	0	0	0	0
0	0	0	0	0	0	0	0
0	0	0	0	0	0	0	0

Figure 6.45 Erosion of a line that is two pixels wide

In this case, we need to be more clever if we want to preserve the two-pixel-wide line. For this reason, we introduce the concept of *conditional* morphological operations [1].

In conditional morphological operations, two passes are made through the image. In the first pass, pixels whose neighborhoods match patterns in the hit table are marked for change. The marked pixel neighborhoods are compared to a second hit table for changing. This is shown in Figure 6.46.

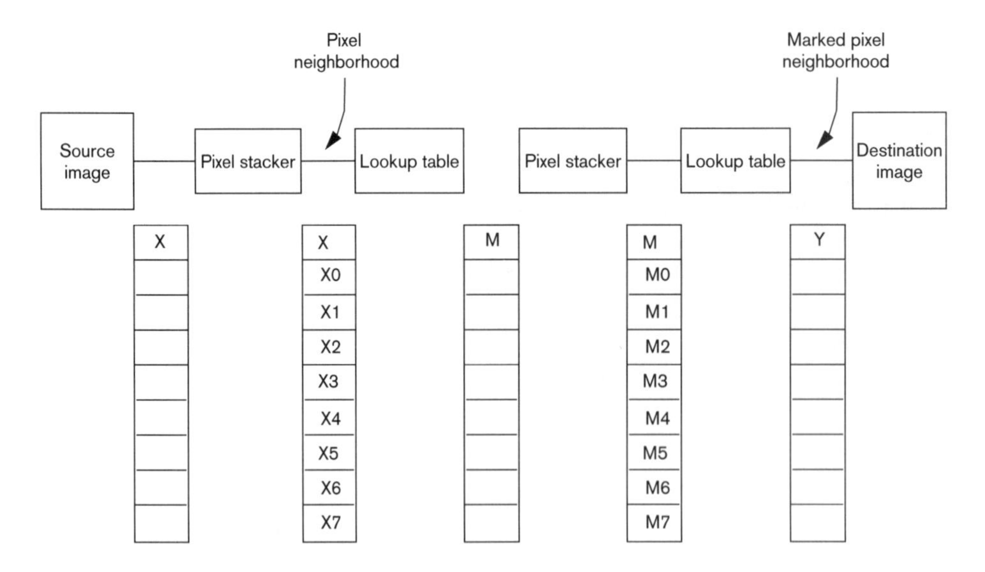

Figure 6.46 Conditional binary morphological operations

The three operations of interest that are performed using this algorithm are *shrink*, *thin*, and *skeletonize*. These operations are defined as follows for objects without holes:

- *Shrinking* An object erodes it to a single pixel at or near its center of mass.
- *Thinning* An object erodes it to a minimally connected stroke, one or two pixels wide, located equidistant from its nearest outer boundaries.
- *Skeletonizing* An object results in a stick figure representation of the object consisting of the set of points that are equally distant from two close points of an object boundary.

The hit tables for these operations are shown in Tables 6.1, 6.2, and 6.3. Conditional morphological operations are implemented using a similar technique as the basic morphological operations.

Table 6.1 Shrink, thin, and skeletonize conditional mark patterns—M = 1 if hit

Table	Bond	Pattern							
		0 0 1	1 0 0	0 0 0	0 0 0				
S	1	0 1 0	0 1 0	0 1 0	0 1 0				
		0 0 0	0 0 0	1 0 0	0 0 1				
		0 0 0	0 1 0	0 0 0	0 0 0				
S	2	0 1 1	0 1 0	1 1 0	0 1 0				
		0 0 0	0 0 0	0 0 0	0 1 0				
		0 0 1	0 1 1	1 1 0	1 0 0	0 0 0	0 0 0	0 0 0	0 0 0
S	3	0 1 1	0 1 0	0 1 0	1 1 0	1 1 0	0 1 0	0 1 0	0 1 1
		0 0 0	0 0 0	0 0 0	0 0 0	1 0 0	1 1 0	0 1 1	0 0 1
		0 1 0	0 1 0	0 0 0	0 0 0				
T K	4	0 1 1	1 1 0	1 1 0	0 1 1				
		0 0 0	0 0 0	0 1 0	0 1 0				
		0 0 1	1 1 1	1 0 0	0 0 0				
S T K	4	0 1 1	0 1 0	1 1 0	0 1 0				
		0 0 1	0 0 0	1 0 0	1 1 1				
		1 1 0	0 1 0	0 1 1	0 0 1				
S T	5	0 1 1	0 1 1	1 1 0	0 1 1				
		0 0 0	0 0 1	0 0 0	0 1 0				
		0 1 1	1 1 0	0 0 0	0 0 0				
S T	5	0 1 1	1 1 0	1 1 0	0 1 1				
		0 0 0	0 0 0	1 1 0	0 1 1				
		1 1 0	0 1 1						
S T	6	0 1 1	1 1 0						
		0 0 1	1 0 0						
		1 1 1	0 1 1	1 1 1	1 1 0	1 0 0	0 0 0	0 0 0	0 0 1
S T K	6	0 1 1	0 1 1	1 1 0	1 1 0	1 1 0	1 1 0	0 1 1	0 1 1
		0 0 0	0 0 1	0 0 0	1 0 0	1 1 0	1 1 1	1 1 1	0 1 1

Table 6.1 Shrink, thin, and skeletonize conditional mark patterns—M = 1 if hit (continued)

Table	Bond	Pattern							
		1 1 1	1 1 1	1 0 0	0 0 1				
S T K	7	0 1 1	1 1 0	1 1 0	0 1 1				
		0 0 1	1 0 0	1 1 1	1 1 1				
		0 1 1	1 1 1	1 1 0	0 0 0				
S T K	8	0 1 1	1 1 1	1 1 0	1 1 1				
		0 1 1	0 0 0	1 1 0	1 1 1				
		1 1 1	0 1 1	1 1 1	1 1 1	1 1 1	1 1 0	1 0 0	0 0 1
S T K	9	0 1 1	0 1 1	1 1 1	1 1 1	1 1 0	1 1 0	1 1 1	1 1 1
		0 1 1	1 1 1	1 0 0	0 0 1	1 1 0	1 1 1	1 1 1	1 1 1
		1 1 1	1 1 1	1 1 1	1 0 1				
S T K	10	0 1 1	1 1 1	1 1 0	1 1 1				
		1 1 1	1 0 1	1 1 1	1 1 1				
		1 1 1	1 1 1	1 1 0	0 1 1				
K	11	1 1 1	1 1 1	1 1 1	1 1 1				
		0 1 1	1 1 0	1 1 1	1 1 1				

Table 6.2 Shrink and thin unconditional mark patterns—$P(M,M_0,M_1,M_2,M_3,M_4,M_5,M_6,M_7) = 1$ if hit

Pattern			
Spur			
0 0 M	M 0 0		
0 M 0	0 M 0		
0 0 0	0 0 0		
Single 4-connection			
0 0 0	0 0 0		
0 M 0	0 M M		
0 M 0	0 0 0		
4-connected offset			
0 M M	M M 0	0 M 0	0 0 M
M M 0	0 M M	0 M M	0 M M
0 0 0	0 0 0	0 0 M	0 M 0
Spur corner cluster			
0 M M	M M 0	0 0 M	M 0 0
0 M M	M M 0	M M 0	0 M M
M 0 0	0 0 M	M M 0	0 M M

Table 6.2 Shrink and thin unconditional mark patterns— $P(M,M_0,M_1,M_2,M_3,M_4,M_5,M_6,M_7) = 1$ if hit (continued)

Pattern							
Corner cluster							
M M D							
M M D							
D D D							
Tee branch							
D M 0	0 M D	0 0 D	D 0 0	D M D	0 M 0	0 M 0	D M D
M M M	M M M	M M M	M M 0	M M 0	M M 0	0 M M	0 M M
D 0 0	0 0 D	0 M D	D M 0	0 M 0	D M D	D M D	0 M 0
Vee branch							
M D M	M D C	C B A	A D M				
D M D	D M B	D M D	B M D				
A B C	M D A	M D M	C D M				
Diagonal branch							
D M 0	0 M D	D 0 M	M 0 D				
0 M M	M M 0	M M 0	0 M M				
M 0 D	D 0 M	0 M D	D M 0				

Note: $A \cup B \cup C = 1$, $D = 0 \cup 1$.

Table 6.3 Skeletonize unconditional mark patterns— $P(M,M_0,M_1,M_2,M_3,M_4,M_5,M_6,M_7) = 1$ if hit

Pattern			
Spur			
0 0 0	0 0 0	0 0 M	M 0 0
0 M 0	0 M 0	0 M 0	0 M 0
0 0 M	M 0 0	0 0 0	0 0 0
Single 4-connection			
0 0 0	0 0 0	0 0 0	0 M 0
0 M 0	0 M M	M M 0	0 M 0
0 M 0	0 0 0	0 0 0	0 0 0
L corner			
0 M 0	0 M 0	0 0 0	0 0 0
0 M M	M M 0	0 M M	M M 0
0 0 0	0 0 0	0 M 0	0 M 0
Corner cluster			
D M M	D D D	M M D	D D D
D M M	M M D	M M D	D M M
D D D	M M D	D D D	D M M

Table 6.3 Skeletonize unconditional mark patterns— $P(M,M_0,M_1,M_2,M_3,M_4,M_5,M_6,M_7) = 1$ if hit (continued)

Pattern			
Tee branch			
D M D	D M D	D D D	D M D
M M M	M M D	M M M	D M M
D 0 0	D M D	D M D	D M D
Vee branch			
M D M	M D C	C B A	A D M
D M D	D M B	D M D	B M D
A B C	M D A	M D M	C D M
Diagonal branch			
D M 0	0 M D	D 0 M	M 0 D
0 M M	M M 0	M M 0	0 M M
M 0 D	D 0 M	0 M D	D M 0

Note: $A \cup B \cup C = 1$, $D = 0 \cup 1$.

Generalized binary dilation and erosion The preceeding algorithms are implementations of binary morphological operations based on hit-or-miss operations using 3×3 neighborhoods. A more general definition of dilation and erosion also exists. First, some definitions are needed.

- A source image $F(i,j)$ is said to be dilated or eroded by a *structuring element* $H(i,j)$—$H(i,j)$ is rather like a convolution kernel.
- Dilation is performed by translating $F(i,j)$ by all nonzero elements in $H(i,j)$ and then taking their logical union. That is, the image $F(i,j)$ is translated by the coordinate (relative to the origin) of each nonzero element of $H(i,j)$, The logical union of all of these translated versions of $F(i,j)$ yields the dilation of $F(i,j)$ by $H(i,j)$.
- Erosion is performed by translating $F(i,j)$ by all the elements in $H(i,j)$ and then taking their logical intersection. That is, the image $F(i,j)$ is translated by the coordinate (relative to the origin) of each nonzero element of $H(i,j)$, The logical intersection of all of these translated versions of $F(i,j)$ yields the dilation of $F(i,j)$ by $H(i,j)$.

An example of the dilation of an object by a vertical structuring element two pixels high is shown in Figure 6.47. The origin of the structuring element is marked with an arrow. An example of the erosion of an object by a slightly different structuring element, with origin similarly marked, is shown in Figure 6.48.

Pratt [1] has shown that the unconditional erosion and dilation performed using hit-or-miss transformations is equivalent to generalized erosion and dilation using a 3×3 structuring element. For other structuring elements, the implementation follows from the definition of the operation. That is, the image is translated and or-ed or and-ed with itself as many times as necessary according to the structuring element.

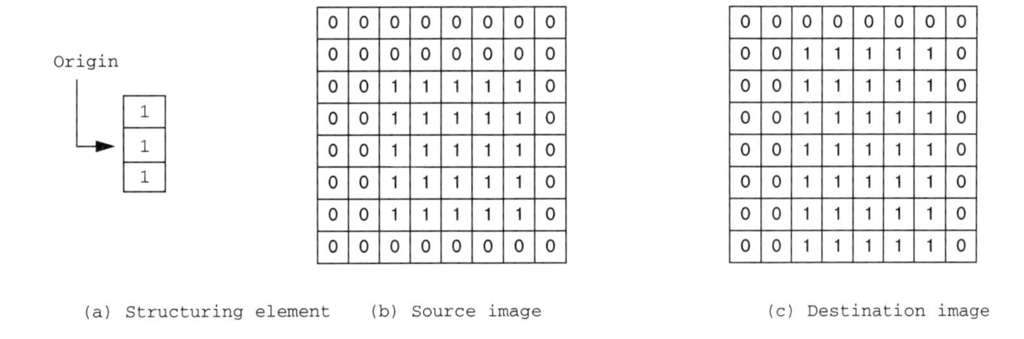

Figure 6.47 Dilation of an image

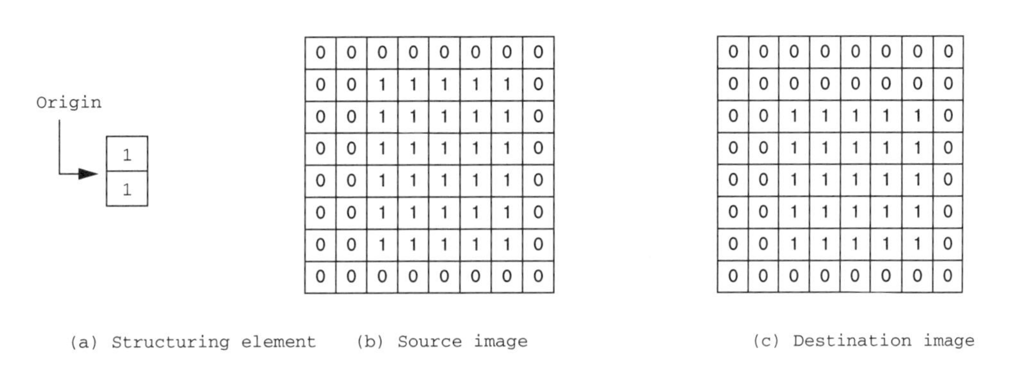

Figure 6.48 Erosion of an object by a structuring element

6.5.2 Gray-scale morphology

The concepts developed in binary image morphology can be extended to gray-scale images. These operations are useful in the analysis of gray-scale images, where, for example, measurements of object sizes can be made using erosion.

Erosion and dilation in gray-scale images are implemented as neighborhood minimum and neighborhood maximum operations, respectively. Taking the minimum effectively shrinks a bright object that is standing out from its surroundings, and taking the maximum expands it.

Opening and closing are also defined for gray-scale images. As before, opening is an erosion (minimum) followed by a dilation (maximum), and closing is the reverse operation.

6.6 Conclusion and further reading

Neighborhood filters are used in many ways in computer imaging. They come in linear and nonlinear flavors. In this chapter we have studied various image neighborhood filters. We have seen efficient implementations of both linear filters, such as convolution, as well as nonlinear filters, such as median filters and morphological filters.

More mathematical details on neighborhood filters can be found in reference 1, where edge detection and mathematical morphology are also discussed. Details on the use of neighborhood filters can be found in reference 16. A systematic and mathematically rigorous approach to edge detection can be found in reference 3. Discussions of efficient implementations can be found in references 2, 7, and 15.

6.7 References

1. W. K. Pratt, *Digital Image Processing*, 2nd edition, New York: Wiley, 1993.
2. G. Wolberg and H. Massalin, "Fast Convolution with Packed Lookup Tables," *Graphics Gems IV*, Cambridge, MA: Academic Press, 1994.
3. J. Canny, "A Computational Approach to Edge Detection," *IEEE Transactions on Pattern Analysis and Machine Intelligence*, Vol. 8, No. 6, November 1986, pp. 679–698.
4. J. W. Tukey, *Exploratory Data Analysis*, Reading, MA: Addison-Wesley, 1974.
5. P. M. Narendra, "A Separable Median Filter for Image Noise," *Proceedings of the IEEE Conference on Pattern Recognition and Image Processing*, IEEE Computer Society Press, 1978.
6. T. S. Huang, G. T. Yang, and G. Y. Tang, "A Fast Two-Dimensional Median Filtering Algorithm," *IEEE Trans. Accoust., Speech and Signal Processing*, Vol. ASSP-27, February 1979, pp. 13–18.
7. A. Paeth, "Median Finding on a 3 × 3 Grid," *Graphics Gems I*, Cambridge, MA: Academic Press, 1992.
8. W. K. Pratt, T. J. Cooper, and I. Kabir, "Pseudomedian Filter," *Proceedings SPIE*, Vol. 534, 1985, pp. 34–43.
9. D. R. K. Brownrigg, "The Weighted Median Filter," *Communications of the ACM*, Vol. 27, No. 8, August 1984.
10. J. Serra, *Image Analysis and Mathematical Morphology*, London: Academic Press, 1982.
11. H. Minkowski, "Volumen und Oberflache," *Math. Ann.*, Vol. 57, 1903, pp. 447–459.
12. G. Matheron, *Random Sets and Integral Geometry*, New York: Wiley, 1975.
13. A. Rosenfeld and A.C. Kak, *Digital Picture Processing*, 2nd edition, Vol. 2, New York: Academic Press, 1982.
14. T. Pavlidis, "A Thinning Algorithm for Discrete Binary Images," *Computer Graphics and Image Processing*, Vol. 13, No. 2, 1980, pp. 142–157.
15. W. K. Pratt and I. Kabir, "Morphological Binary Image Processing with a Local Neighborhood Pipeline Processor," *Computer Graphics*, Tokyo, 1984.
16. J. C. Russ, *The Image Processing Handbook*, Boca Raton, FL: CRC Press, 1993.

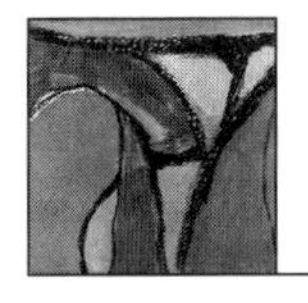

chapter 7

Color in computer imaging

7.1 Introduction

Color plays an important role in our lives. In particular, color enhances our perception of the world in many ways. The importance of color in perception carries over from the physical world to the world of pictures. In computer imagery, color serves both aesthetic and utilitarian purposes by making pictures more pleasing and by packing more information into them.

This chapter is about the use of color in computer imaging operations. In the last few years, due to the advances in image acquisition and display technologies, color image processing has become increasingly important in computer imaging. In order to keep pace with these advances, imaging (and graphics) systems must frequently perform some common operations related to color. In this chapter, we examine these operations from the user's and the implementor's point of view.

The operations covered in this chapter include the representation of color images in numerical terms via color spaces, conversion between different color spaces, quantization of color images for display, and gamma correction. This set of operations provides a robust and useful set of tools for many color imaging applications.

We should, however, note that color science is a far-reaching field of study. The cause, perception, and measurement of color have been studied for centuries by scientists in many disciplines, including physics, philosophy, psychology, physiology, and engineering. Color science has been covered in excellent detail in a number of texts including references 1 and 2. In this chapter we are primarily interested in a relatively small set of algorithmic issues in color as they relate to computer imaging. Therefore, we will keep the discussion of the theoretical aspects of color to a minimum, except where such discussions clarify otherwise confusing issues.

Finally, we should point out that the algorithms and implementations presented in this chapter are well known from the public domain. Many color operations—especially those in more sophisticated applications such as precise color reproduction or color management systems—require proprietary knowledge, particularly about device and ink characteristics. Also, skilled artisans are often employed to perform color matching and related tasks. Because of the subjective nature of color perception, large amounts of black magic pervade the world of color. Coverage of these topics is beyond the scope of our discussion.

What, then, *does* our discussion cover? We begin, in Section 7.2, with fundamentals and motivational examples. In Section 7.3, we define and discuss color spaces, and show how to convert between them. We also attempt to sort out, in a meaningful way, the large number of color spaces that can be both confusing and intimidating. Important color spaces (both device-independent and device-dependent) are covered, and programming examples in C are provided. In Section 7.4, we discuss algorithms related to the display of color images. These include gamma correction and color quantization algorithms. Gamma correction is a way to compensate for the nonlinearities present in video display monitors. Color quantization is used to display high-precision pixels (e.g., 24-bit pixels) in a lower-precision display device (e.g., an 8-bit display).

7.2 Fundamentals and motivating examples

The theoretical foundation of color reproduction, as used in computer imaging, lies in *tristimulus color theory*, which says that the sensation of a particular color can be reproduced by combining the appropriate amounts of three primary colors. According to this theory, which was first shown by Thomas Young in his classic work on color [3], any color can be decomposed into a combination of three primary colors.

Results of physiological experiments support tristimulus color theory. The human visual system has been shown to have three types of receptors (called cones) in the retina of the eye. Cones are receptive to three bands of colors: red, green, and blue. The spectral response—that is, the brightness sensations by lights of different colors—for each of these cones is shown in Figure 7.1.

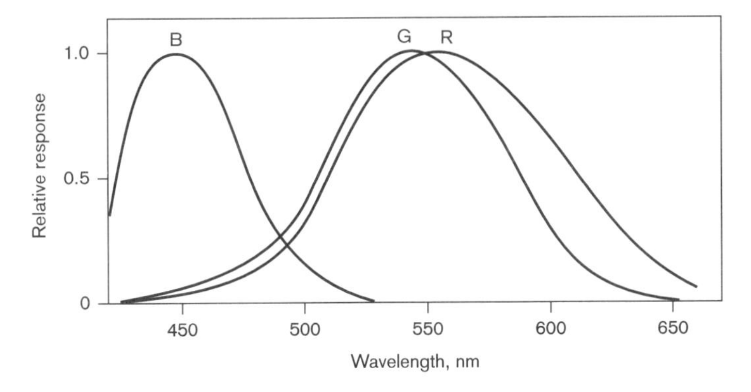

Figure 7.1 Spectral sensitivity of cones

These curves intuitively tell us that visible colors can be decomposed into three primary colors. A *color space* is a multidimensional space formed by the coordinate axes, which represent primary colors. Any point in this space represents a color. As we will see, all but one of the color spaces considered in this chapter are three dimensional.

Note, however, that a given set of primary colors cannot reproduce all visible colors. All we are saying is that, given a color, it is possible to decompose it into a set of primaries. Therefore, all points in the color space may not be physically realizable. We will see the implications of this later in this chapter.

Tristimulus color theory is useful in a number of ways. For instance, color matching is an important problem that tries to answer the following question: Given a color and a particular set of primaries, how much of each primary does one need to mix in order to match that color? To answer this question we need to conduct the following (possibly hypothetical) experiment.

Given a white surface, consider the situation where light of the given color shines on one half of the surface and a mixture of lights of the three primary colors is projected onto the other half of the surface. The amounts of the primary colors being mixed in this setup can be varied. These amounts are adjusted until the light from the primary mixture, shining on one half of the surface, looks exactly like the colored light shining on the other half of the surface. The amounts of the primaries required to match the given color are called the *tristimulus values* of the color.

Extensive measurements of tristimulus values have been conducted by the Comission Internationale d'Eclairage (CIE), an international body devoted to setting up color references. For instance, in 1931, the CIE performed some experiments on a well-defined set of primary colors. This set was {red at 700 nm, green at 546.1 nm, blue at 435.8 nm}. The experiments characterized the amounts of each primary that were needed by a *standard observer* to match any given light. These characterizations, called the *CIE spectral matching curves*, are shown in Figure 7.2.

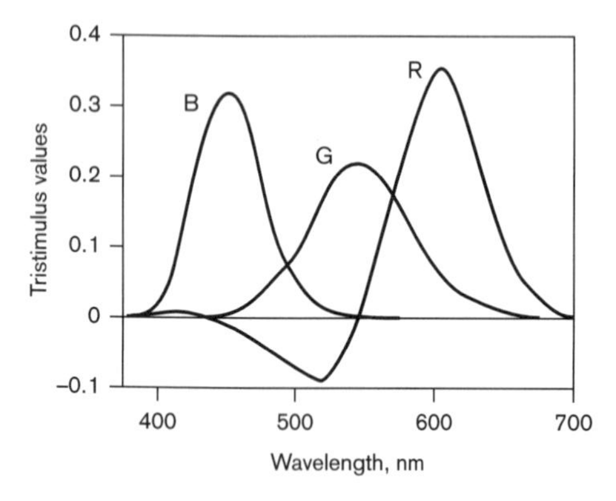

Figure 7.2 Tristimulus color matching functions for CIE reference RGB color space

The negative values in the curve correspond to colors that are not physically realizable. A negative tristimulus value may be interpreted in terms of our experiment above: It means that the primary must be added to the color being matched. To overcome the inconvenience of these negative numbers, the CIE defined a theoretical color space called XYZ, whose spectral matching curves are all positive. XYZ and its related spaces L*a*b* and L*u*v* are not realizable by any physical device. However, they are useful for colorimetric purposes.

Besides the tristimulus values, another concept from color science that is important for computer imaging is that of the *chromaticity diagram*. Any point in a (three-dimensional) color coordinate space represents a color; however, as we have seen, all of these colors are not visible or physically realizable. The set of points representing the visible colors form a solid in the space. If one coordinate is held constant (at unity, for example) then the visible colors form an area in two-dimensional plane. This area is the *chromaticity diagram*. An example of the chromaticity diagram (which, of course, depends on the primaries) is shown in Figure 7.3.

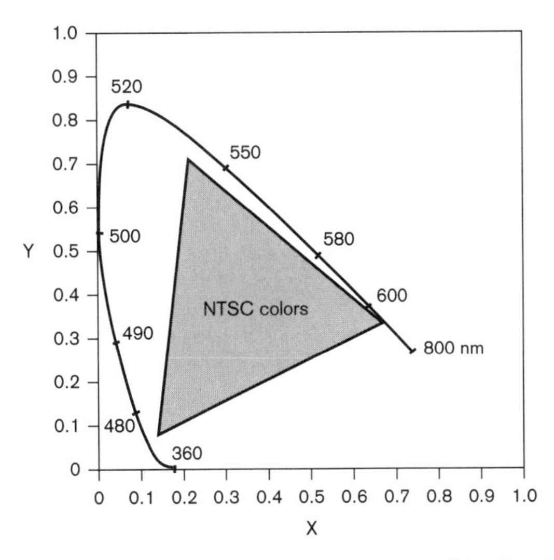

Figure 7.3 Chromaticity diagram for CIE primaries

7.2.1 Devices that produce color

In physical color reproduction devices, primaries are mixed in an *additive* or a *subtractive* manner to produce color. In an additive color system, the primary colors (red, green, and blue) are added to reproduce color. The amount of each primary color is controlled to produce the desired color.

In a subtractive system, the primaries are subtracted from white light to reproduce the desired color. This subtraction is performed by filters of complementary colors that attenuate the primary colors. A cyan filter attenuates red light and passes the other colors; similarly, magenta and yellow attenuate green and blue colors, respectively. The amount of attenuation can be controlled to produce the desired color.

These two methods of mixing colors are shown in Figure 7.4.

A television or a color monitor are examples of an additive color system (*a* in Figure 7.5). The colors seen on a monitor are generated when electrons, emitted by three electron guns, excite the phosphor layer in the picture tube. The voltages applied to the electron guns are proportional to the amounts of primaries that need to be added to produce the desired color; for a digital display system, these voltages are generated from the pixel values in the display memory of a video display card using D/A converters.

The phosphor is laid out as dots or stripes, and a mask is used to ensure that each gun excites a phosphor of only one color. The light emitted from these dots is blended by the eye to produce the sensation of color.

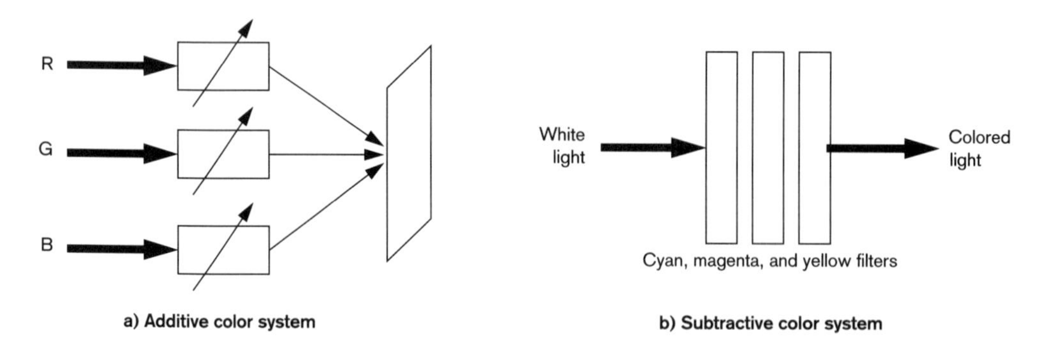

Figure 7.4 Methods of color reproduction

A photographic enlarger is a subtractive color system (*b* in Figure 7.5). Light from a condenser source is passed through adjustable cyan, magenta, and yellow filters—which subtract colors from the white light—before it passes through the negative and is projected into the photographic paper layered with light-sensitive dyes.

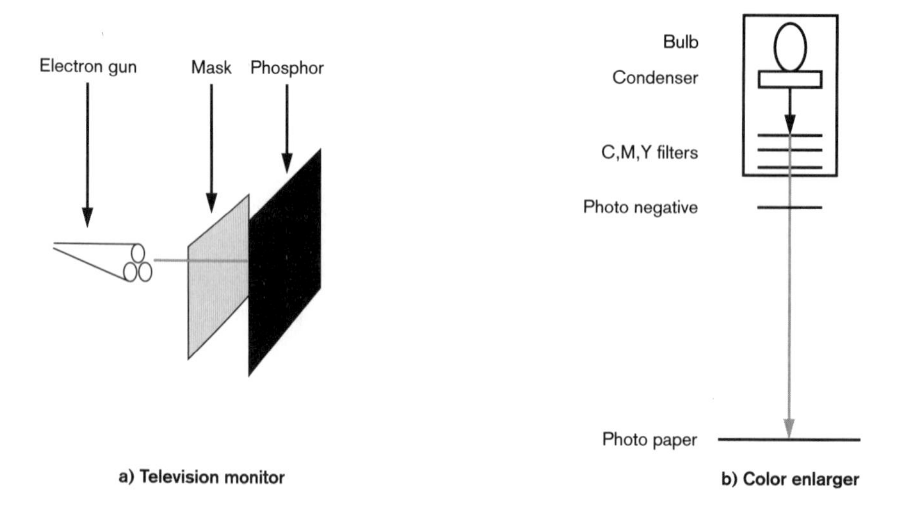

Figure 7.5 Additive and subtractive color systems

Color printing is also a subtractive color system. Color is reproduced by depositing appropriate amounts of cyan, magenta, and yellow inks on paper. These inks subtract color from the light that is reflected from the paper and seen by the eye.

The above examples demonstrate the fundamental physical principles used in creating color. However, the real-life situations encountered in color applications in computer imaging applications require more elaborate methods, which we now briefly summarize.

In many color imaging applications, more than one color device may be involved. For instance, in desktop publishing, color scanners, monitors, and printers are used to acquire, display, and print color images. In order to closely reproduce the appearance of colors in these devices, the color properties of the processes used by these devices must be quantified. This process is called *device characterization* [4], and it consists of measurement of the color properties and finding a transformation to map from the device color space and a standard color space such as CIElab. Characterization of devices is closely coupled to their *calibration*, that is, proper adjustment of the equipment to comply with specifications. Thus, the characterization of a color monitor, for example, depends, among other things, on the calibration of its white-point, the tonal response curves for each color primary, and the chromaticities of the RGB colors.

More generally, device characterization must be done using lookup tables to model every possible color that the device can reproduce. Since this is prohibitively expensive, a subsampled set of colors is used in conjunction with interpolation along multiple axes.

The range of colors that can be reproduced by a device is called the *gamut* of that device. The gamuts of several color devices are shown in Figure 7.6 [5]. The range of colors that our eye can see is larger than the gamut of any of these devices.

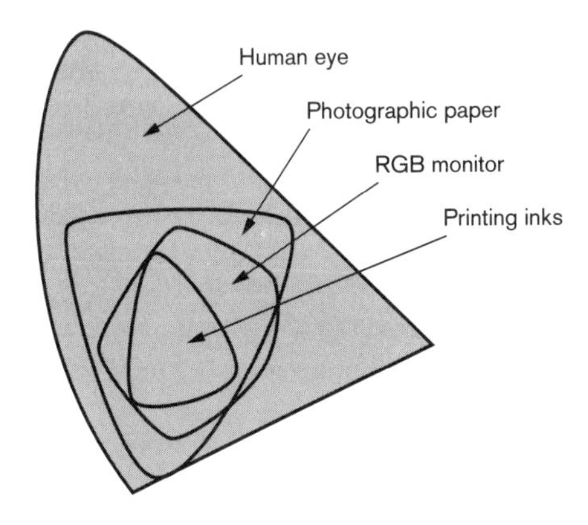

Figure 7.6 Gamuts of different devices

7.2.2 *Examples and Applications*

The techniques and algorithms studied in this chapter are important for several practical reasons: the *measurement and matching* of colors; the *reproduction* of color in the different types of devices used in computer imaging systems; the *transmission* and *compression* of color information; and the *display* of color images.

Color spaces are fundamental to general color science. They are useful in *colorimetry*, where the goal is to measure the characteristics of color. Color spaces are used to study color, and to characterize perception of color. For instance, to characterize the sensitivity of color changes to the human eye, CIE color spaces are used in conjunction with two-dimensional figures known as *MacAdam ellipses* [6].

Color spaces and their interchange are also useful in the *reproduction* of color in different devices. Cameras and scanners for acquisition, monitors for display, and color printers all use different physical methods for creating color. For a particular device, colors must be described in terms related to the physical device parameters that create and control the color.

In Figure 7.7 we show some example devices and the color spaces commonly associated with them. The JPEG compressor/decompressor (discussed in detail in Chapter 9) commonly uses a video color space, such as the CCIR Rec. 601 YCbCr color space. PhotoCD images are stored in the PhotoYCC color space. A color printer "understands" color in terms of the primary colors of its inks, for instance, cyan, magenta, and yellow (CMY). High Definition Television Systems (HDTVs) have been specified to work in a newer video color space called CCIR Rec. 709 YCbCr. Monitors and scanners work in RGB color spaces.

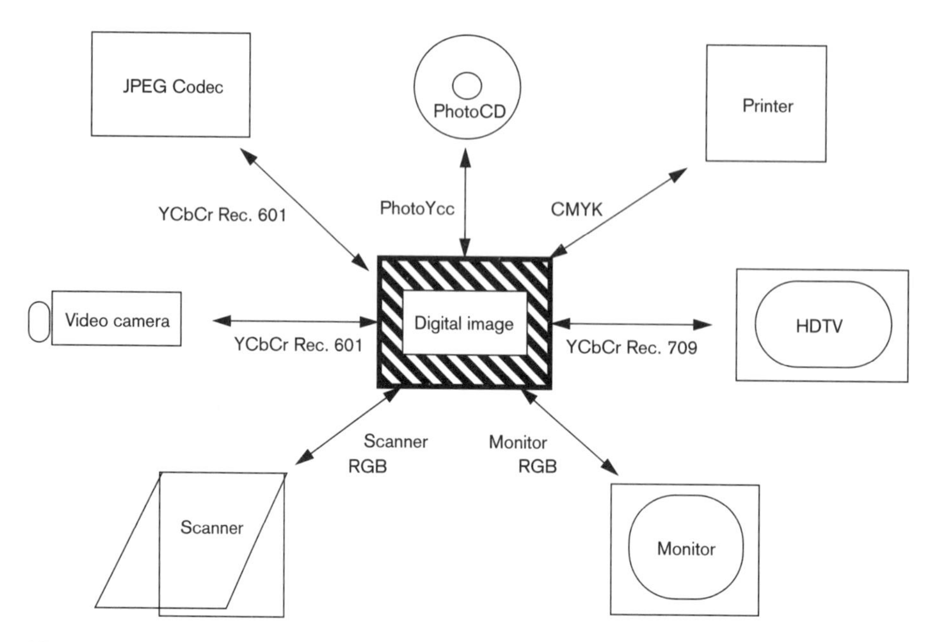

Figure 7.7 Use of color spaces on different devices in computer imaging

In order to move color images from one device to another, we need to be able to interchange between all these different color spaces. Thus, for example, a JPEG decompressed image usually needs to be converted from YCbCr (601) to RGB after it has been uncompressed and before it can be displayed on a monitor. Similarly, a PhotoCD image must be converted into CMY or the related CMYK color space before it can be printed.

The presence of so many color spaces can confuse the reader. Note, however, that many of them are closely related. In Section 7.3, we have attempted to sort out the profusion of color spaces.

A third use of color in computer imaging is in *transmitting* color information during broadcasting. By judiciously selecting the color space, we can transmit color information efficiently over finite bandwidth channels. This is done by converting the primary RGB color space into a video color space such as YUV that separates brightness information (Y) from color information (U and V). In color terminology, Y is called the *luminance* and U and V are called the *chrominance* of the image.

This decomposition has two advantages. First, since the human eye is less sensitive to chrominance than it is to luminance, the U and V can be subsampled without appreciable loss of image quality, resulting in compression of the image. In analog applications, such as television broadcasting, compression is achieved by assigning a lower bandwidth to U and V. In digital images, U and V are assigned less storage (typically half or quarter) than Y. The second advantage is that extraction of Y, which is the black-and-white equivalent of the image, enables backward compatibility with existing monochrome television sets.

Once we have an image in the desired color space, there are some other *display*-related considerations. Often, color images are represented using 8 bits for each of red, green, and blue channels. However, many display systems use indexed 8-bit color. If the image is in 24-bit color, how does one display it on an 8-bit color display? This problem is resolved using *color quantization* techniques. Another issue is the correction of nonlinearities in the display device. *Gamma correction* is often required before a perceptually correct color image can be seen on the monitor.

7.3 Working with color spaces

The preponderance of color spaces almost always confuses and intimidates the novice—and sometimes even the experienced—engineer working on computer imaging. The large number of color spaces, inconsistent definitions, conflicting nomenclature, and everyone's claim to be a "standard" all contribute to this confusion.

In the following figures we attempt to sort out the color spaces based on their use. All color spaces can be categorized into device-independent or device-dependent color spaces (Figure 7.8).

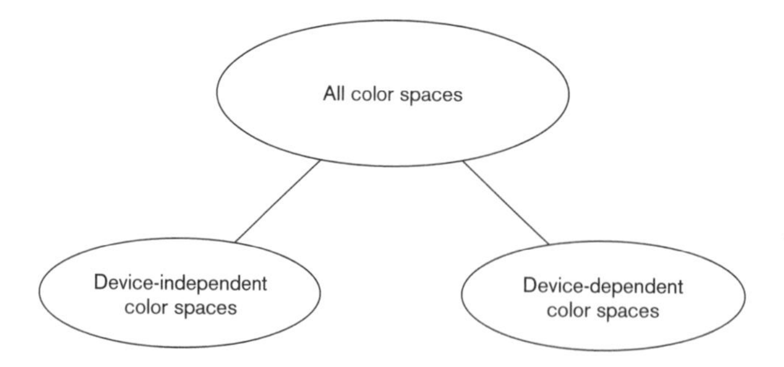

Figure 7.8 Classification of color spaces

Device-independent color spaces can be divided into various CIE-defined color spaces, such as XYZ, L*a*b*, and L*u*v* (Figure 7.9). As we have mentioned, these are primarily useful for colorimetric purposes. They are also used in color management systems.

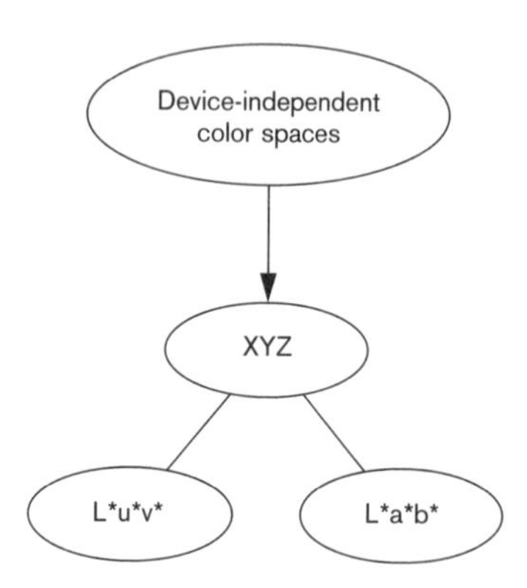

Figure 7.9 Device-independent color spaces

Device-dependent color spaces can be divided into three broad classes: printing color spaces, video color spaces, and monitor color spaces (Figure 7.10). The printing color spaces, CMY and CMYK, are based on the colors of inks used in printing and photography. Monitor color space is basically variants of RGB. Video color spaces—all of them closely related to the YUV luminance-chrominance color space—divide into several individual color spaces depending on the application.

Some important properties of the color spaces we have covered in this chapter are presented in Table 7.1.

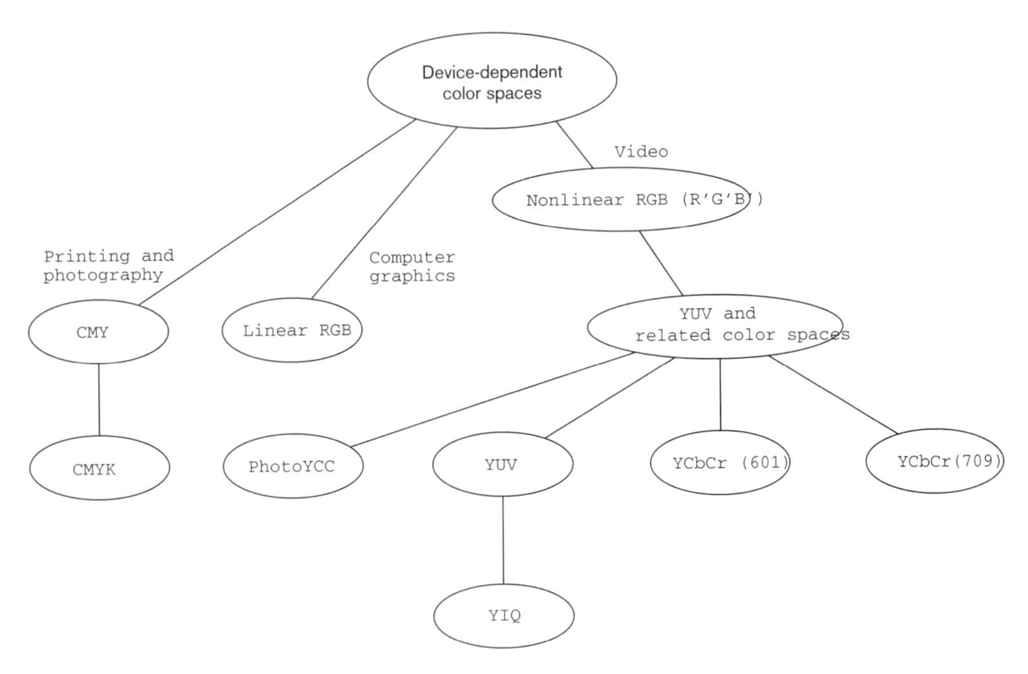

Figure 7.10 Device-dependent color spaces

Table 7.1 Applications of various color spaces

Color space	Device or medium	Applications
L*a*b*	Device-independent	Color management systems
XYZ	Device-independent	Colorimetry, color management systems
L*u*v*	Device-independent	Colorimetry
Linear RGB and variants	Monitor	Computer graphics
R'G'B'(RGB 709)	Monitor	Monitors
YUV	Television	SECAM, PAL
YIQ	Television	NTSC
YCbCr 601	Component video	NTSC, video cameras
YCbCr 709	Component video	HDTV
PhotoYCC	PhotoCD	Wide variety of applications
CMY	Printing ink, photographic dye	Printing, color photography
CMYK	Printing ink	Printing

Having sorted out the profusion of color spaces, we are now ready to take a closer look at the individual color spaces.

7.3.1 *Device-independent color spaces*

As we have mentioned earlier, the CIE sets the standards for color references for colorimetry purposes. In this section, we will look at device-independent color spaces defined by the CIE. These spaces are primarily useful for colorimetric purposes and as reference color spaces.

The XYZ color space The CIE reference RGB tristimulus curves, as shown in Figure 7.2, cause some awkwardness because of their negative values. The XYZ color space is an abstract color space defined to have tristimulus matching curves that are all positive, thus avoiding this awkwardness. The computation of XYZ from RGB depends on the monitor white-point. An equation of the following form is used if the white-points of the RGB and the XYZ space are the same.

$$\begin{aligned} X &= R_x R + G_x G + B_x B \\ Y &= R_y R + G_y G + B_Y B \\ Z &= R_Z R + G_Z G + B_Z B \end{aligned} \tag{7.1}$$

The chromaticity diagram and tristimulus matching curves of the XYZ color space are shown in Figure 7.11.

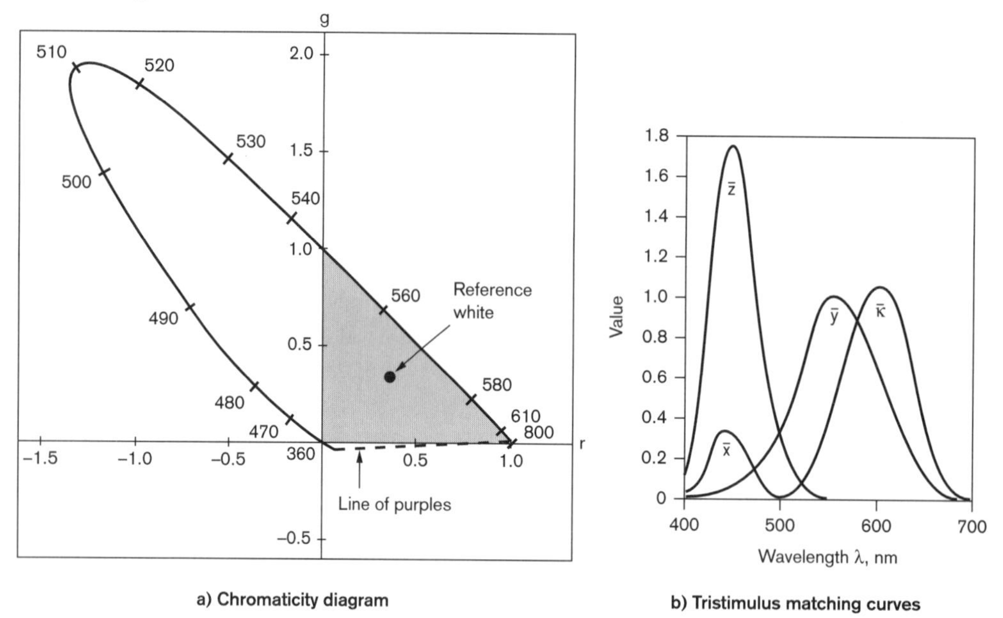

a) Chromaticity diagram b) Tristimulus matching curves

Figure 7.11 Chromaticity diagram and tristimulus matching curves for CIE XYZ primaries

While the XYZ color space is useful for colorimetric calculations, a disadvantage of this color space is that it is not perceptually uniform. In other words, traversing the same distance in the chromaticity diagram does not necessarily yield the same differences in color perception. The L*a*b* color space addresses this problem.

*The L*a*b* and L*u*v* color spaces* The CIE introduced the perceptually uniform L*a*b* (sometimes called lab or CIElab) and L*u*v* color spaces, derived from the XYZ color space, in 1976. In these color spaces, just noticeable difference (JND) between two colors has been formally defined to be a Euclidean distance of 1.

L*a*b* are computed from the XYZ values using the following equations:

$$L^* = 25\left(100\frac{Y}{Y_0}\right)^{1/3} - 16, \quad \frac{Y}{Y_0} \geq 0.01$$

$$L^* = 903.3\frac{Y}{Y_0}, \quad \frac{Y}{Y_0} < 0.01$$

$$a^* = 500\left(\left(\frac{X}{X}\right)^{1/3} - \left(\frac{Y}{Y_0}\right)^{1/3}\right) \qquad (7.2)$$

$$b^* = 200\left(\left(\frac{X}{X_0}\right)^{1/3} - \left(\frac{Z}{Z_0}\right)^{1/3}\right)$$

Here, X_0, Y_0, and Z_0 are the tristimulus values of the reference white being used in the system. L^* is very close to the perceptual brightness of the color; a^* and b^* measure the colorfulness of the color.

In the L*u*v* system, L^* is the same as before; u^* and v^* are computed in a slightly different manner. The reader is referred to reference 5 for more details.

Both the L*a*b* and L*u*v* color spaces are useful for colorimetric calculations and for color management systems where device-specific color spaces are first converted into L*a*b* before being converted into color spaces for other devices.

7.3.2 Device-dependent color spaces

We will look at three classes of device-dependent color spaces and conversions between them. These are the RGB primaries and their variations, the YUV primaries and their variations, and the CMY (and CMYK) primaries and their variations.

RGB color primaries and their variations The linear RGB model is the most common device-dependent color space used in computer imaging and graphics computations. Each of the linear RGB components is defined to be in the range of 0 to 1, where 0 represents black and 1 represents white.

While the linear RGB color space is used widely in the display of color images on monitors, it does have two disadvantages. First, it is not perceptually uniform; and second, it does not account for the nonlinearity of the voltage-brightness relationship of the cathode ray tube. These deficiencies were addressed by the definition of the CCIR Rec. 709 standard for gamma corrected RGB, henceforth referred to as R'G'B'. The R' component of the R'G'B' color space is derived from the linear R using the following equations:

$$R' = \begin{cases} 0.45R, & R < 0.018 \\ 1.099R^{0.45} - 0.099, & R \geq 0.018 \end{cases} \tag{7.3}$$

The green and blue components, G' and B', are computed using identical equations with R replaced by G and B, respectively.

This mapping is shown in Figure 7.12. The discontinuity in the equation at 0.018 is needed to limit the slope of the curve at small values (since the derivative of a power function goes to infinity at 0 for powers less than 1).

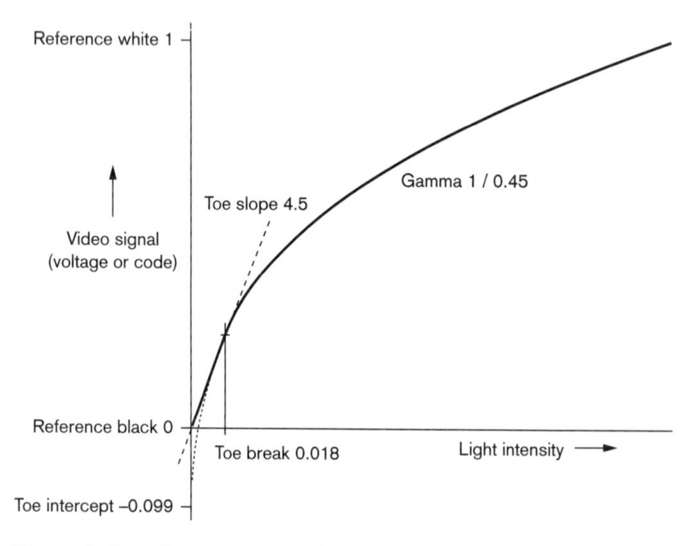

Figure 7.12 Gamma-correction curve for CCIR Rec. 709

YUV and related color spaces and conversions The YUV color space and its variations YIQ, YCbCr-601, YCbCr-709, and PhotoYCC, are used in broadcast television, video, and other applications where two needs must be satisfied: bandwidth or storage must be conserved, and the black-and-white image corresponding to the color image may be needed. In all of these color spaces, the RGB values are expressed in terms of a set of luminance/chrominance color coordinates for reasons having to do with efficiency, backward compatibility with monochrome television systems, and increased dynamic range.

As we briefly saw in Section 7.2.2, luminance (Y) is a measure of image brightness, and chrominance (U and V) is a measure of the color present in the image. Luminance represents the black-and-white equivalent of the color image, and the eye is more sensitive to it because, among other things, it contains edge information. On the other hand, the eye is not as sensitive to chrominance, which contains the color information. These facts are taken advantage of in YUV and its related colorspaces.

The YUV color space The YUV color space is the basic color space used by the Phase Alternating Line (PAL), NTSC (some), and SECAM composite video standards. It is a color space for composite video.

YUV was designed so that color television signals could be shown on existing black-and-white television as monochrome signals. Taking advantage of the eye's insensitivity towards chrominance, YUV separates the luminance component of the picture from the chrominance component. The chrominance components can then be assigned lower bandwidths, since the eye is able to make up for degradations in them. The Y component is used as the black-and-white signal.

The computation of YUV from R'G'B' (that is, the gamma-corrected RGB as defined by CCIR Rec. 709) is shown in Equation 7.4. Since green is represented most strongly in the luminance component, the chrominance components are computed by removing (i.e., subtracting) the luminance Y from R' and B'. The conversion equations are:

$$\begin{aligned} Y &= 0.299R' + 0.587G' + 0.114B' \\ U &= 0.4949(B' - Y) \\ V &= 0.877(R' - Y) \end{aligned} \tag{7.4}$$

and

$$\begin{aligned} R' &= Y + 1.14V \\ G' &= Y - 0.395U - 0.58V \\ B' &= Y + 2.032U \end{aligned} \tag{7.5}$$

Note that the values of all the color components in these equations are normalized between 0 and 1.

The YIQ color space The YIQ color space is a variation on the YUV color space, which can reduce color bandwidth even more. Y remains the same as before; I and Q are found by rotating (by 33 degrees) and axis-flipping the U and V coordinates. The corresponding equations are:

$$\begin{aligned} Y &= 0.299R' + 0.587G' + 0.114B' \\ I &= 0.596R' - 0.274G' - 0.322B' \\ Q &= 0.211R' - 0.523G' + 0.312B' \end{aligned} \tag{7.6}$$

Conversion from YIQ to R'G'B' is accomplished by the following equations:

$$\begin{aligned} R' &= Y + 0.956I + 0.621Q \\ G' &= Y - 0.272I - 0.647Q \\ B' &= Y - 1.106I + 1.703Q \end{aligned} \tag{7.7}$$

Note, once again, that the values of the color components in these equations are normalized between 0 and 1.

The CCIR Rec. 601 YCbCr color space The YCbCr-601 color space specifies digital coding for component video. The coding is specified to be in 8 bits, that is, the values are between 0 and 255. This color space is used by component digital video equipment (including D1 digital video tape recorders) and by JPEG and MPEG compression/decompression standards (see Chapter 9).

In the YCbCr-601 YCbCr color space, conversion equations from R'G'B', are as follows. If R', G', and B' are 8-bit pixels, then they must be normalized.

$$\begin{aligned} Y_{601} &= 16 + 219y \\ Cb_{601} &= 128 + 126(B' - y) \\ Cr_{601} &= 128 + 160(R' - y) \end{aligned} \tag{7.8}$$

where

$$y = (0.299R' + 0.587G' + 0.114B') \tag{7.9}$$

These equations place black at $Y = 16$ and white at $Y = 235$. The *footroom* and *headroom*, as shown in Figure 7.13, refer to the values left over. Headroom and footroom are used in video to accommodate extreme values of color, or intermediate values that are generated during video processing.

Figure 7.13 Headroom and footroom in CCIR Rec. 601

The conversions from YCbCr (601) to R'G'B', also assuming unsigned 8-bit values, are:

$$\begin{aligned} R' &= 1.164Y_{601} + 1.596Cr_{601} - 222.9 \\ G' &= 1.164Y_{601} - 0.392Cb_{601} - 0.813Cr_{601} + 135.57 \\ B' &= 1.164Y + 2.03Cb_{601} - 278.77 \end{aligned} \tag{7.10}$$

For C code implementation of these equations, see the programming examples in Section 7.3.3.

The CCIR Rec. 709 YCbCr color space The YCbCr-709 is a calibrated nonlinear color space for worldwide HDTV production and program exchange. It was agreed on in April 1990.

The YCbCr-709 space reflects that CRT phosphors have undergone substantial change since the introduction of the YUV color space in the early 1950s. Hence the Y component is computed somewhat differently.

The conversions from R', G', and B' are as follows. As before, if R', G', and B' are 8-bit pixels, then they must be normalized:

$$\begin{aligned} Y_{709} &= 16 + 219y \\ Cb_{709} &= 128 + 21(B' - y) \\ Cr_{709} &= 128 + 142(R' - y) \end{aligned} \tag{7.11}$$

where

$$y = 0.2125R' + 0.7154G' + 0.0721B' \tag{7.12}$$

The inverse conversions are:

$$\begin{aligned} R' &= 1.164Y_{709} + 1.795Cr_{709} - 248.48 \\ G' &= 1.164Y_{709} - 0.212Cb_{709} - 0.533Cr_{709} + 76.88 \\ B' &= 1.164Y + 2.1Cb_{709} - 288.38 \end{aligned} \tag{7.13}$$

The Kodak PhotoYCC color space The Kodak PhotoYCC color space was intended for use in a device-independent manner. Most of the color spaces discussed so far have been designed for one kind of device. However, the goal of PhotoYCC is to work in both photographic, broadcast, printing, and computer industries, and to minimize the effort required to interchange color images between devices in these disparate media.

The color space is very similar to the YCbCr-601 color space, except that Y is coded with lots of headroom and no footroom. Cb and Cr are scaled in a different manner to allow for a wider color gamut.

The most interesting aspect of PhotoYCC is its extended dynamic range. For example, by allowing plenty of headroom in the definition of Y, it enables the coding of "whiter-than-white" colors. These situations arise in photography where specular highlights can arise, for example, from reflections from chrome surfaces.

PhotoYCC also allows for certain negative values of RGB. While these values cannot be reproduced by a CRT monitor, they can be reproduced by certain inks. Therefore, they need to be represented digitally for desktop publishing applications.

Due to the possible negative values of R, G, and B, the computation of R'G'B' needs to be extended from the original CCIR Rec. 709 definition, which we have followed so far:

$$\begin{aligned} R' &= 1.099R^{0.45} - 0.099, \ R \geq 0.018 \\ R' &= 4.5R, \ 0.018 \geq R \geq -0.018 \\ R' &= -1.099|R|^{0.45} - 0.099, \ R \leq -0018 \end{aligned} \tag{7.14}$$

Similar equations apply for G' and B'.

To use the above equations for 8-bit color components, the color values must be normalized (i.e., divided by 255) before the conversion and denormalized (i.e., multiplied by 255) after.

The PhotoYCC coordinates are computed from R', G', and B' as follows [7]. As before, if R', G', and B' are 8-bit pixels, then they must be normalized.

$$\begin{aligned} Y_{\text{photo}} &= \frac{255}{1.402}y \\ Cb_{\text{photo}} &= 156 + 111.4(B' - y) \\ Cr_{\text{photo}} &= 137 + 135.64(R' - y) \end{aligned} \tag{7.15}$$

where

$$y = (0.299R' + 0.587G' + 0.114B') \tag{7.16}$$

For digital video RGB values, based on the SMPTE 240M digital specifications for broadcast TV, the photoYCC values are first normalized:

$$\begin{aligned} Y_{\text{photo}} &= 1.3584Y_{\text{photo}} \\ Cb_{\text{photo}} &= 2.2179(Cb_{\text{photo}} - 156) \\ Cr_{\text{photo}} &= 1.8215(Cr_{\text{photo}} - 137) \end{aligned} \tag{7.17}$$

R, G, and B are now computed using:

$$\begin{aligned} R &= Y_{\text{photo}} + Cr_{\text{photo}} \\ G &= Y_{\text{photo}} - 0.194Cb_{\text{photo}} - 0.509Cr_{\text{photo}} \\ B &= Y_{\text{photo}} + Cb_{\text{photo}} \end{aligned} \tag{7.18}$$

CMY and CMYK color spaces and conversions The CMY and CMYK (cyan, magenta, yellow, and black) color spaces are useful in color printing and photography. Cyan, magenta, and yellow are subtractive primaries, and are the complements of additive primaries red, green, and blue. Subtractive primaries create a given color by subtracting (or removing) colors from white. For example, cyan, being the complement of red, removes red from white.

The inks used in color printing and color filters used in color photography have subtractive properties. For example, cyan ink applied to white paper prevents the reflection of red from the paper. Similarly, increasing the amount of cyan in a color photographic filter reduces the amount of red in the final photographic print.

Just as equal amounts of red, green, and blue produce shades of gray, similar proportions of cyan, magenta, and yellow in theory produce degrees of blackness. In practice, this is true in color photography, but not in color printing, where equal amounts of cyan, magenta, and yellow inks actually produce dark brown shades instead of shades of true black. Therefore, in printing, it is desirable to substitute black ink for the mixed black portions of a color. This is done by computing the black component, represented by K in the CMYK nomenclature.

The equations for generating CMY from RGB are:

$$\begin{aligned} C &= 1 - R \\ M &= 1 - G \\ Y &= 1 - B \end{aligned} \tag{7.19}$$

The inverse equations are:

$$\begin{aligned} R &= 1 - C \\ G &= 1 - M \\ B &= 1 - Y \end{aligned} \tag{7.20}$$

For the CMYK system, C, M, and Y are generated as above. K is generated as the minimum of C, M, and Y. K must then be subtracted from C, M, and Y:

$$\begin{aligned} K &= min(C, M, Y) \\ C &= C - K \\ M &= M - K \\ Y &= Y - K \end{aligned} \tag{7.21}$$

In a digital image, where pixel values are represented by unsigned bytes, the 1s above are replaced by 255.

7.3.3 Programming examples in C

In this section, we will look at example programs for color space conversion. These examples are based on 24-bit color images, where each color component is represented as an unsigned byte. The images are assumed to be in the pci format introduced in Chapter 1.

Linear RGB to nonlinear RGB conversions The conversion of linear-light RGB into nonlinear RGB using the CCIR Rec. 709 is shown in Figure 7.10. The first step is to generate the table as follows:

```
/*
        Function: pci_gentab_l2nl
        Operation:Generate a table for converting linear RGB
                  into (Rec 709) gamma-corrected RGB
        Conversion Equations:
                  rgamma = .45rlinear (rlinear < 0.018)
                    = 1.099rgamma^0.45 - 0.099 (rlinear>= 0.018)
                  Similarly for b and g.
*/

#define NORM 256
unsigned char pci_l2nltab[256];
void pci_gentab_l2nl()
{
        int i;
        float in, out;
        for (i=0; i<256; i++) {
             in = (float) i/NORM;
             if (in < 0.018)
                  out = 4.5 * in;
             else
                  out = 1.099 * pow(in,0.45) - 0.099;
             pci_l2nltab[i] = (unsigned char ) (out * MAXBYTE);
        }
        return;
}
```

Now the conversion is reduced to passing the three bands of the image through the table `pci_l2709[]`:

```
/*
Function:     pci_rgbl2rgb709
Operation:    convert a linear rgb image into a gamma-corrected rgb image
              using nonlinear mapping contained in the table pci_l2709tab
Conversion Equations:
              rgamma = .499rlinear (rlinear < 0.018)
                     = 1.099rgamma^0.45 - 0.099 (rlinear>= 0.018)
              Similar equations for blue and green
Input:        src image, where 24-bit pixels are laid out like this:
              BGRBGRBGR... and each B, G, and R is an unsigned char
Output:       dest image, laid out in a similar manner to src image
*/

void pci_rgbl2rgb709(pci_image *src, pci_image *dest)
{
        register unsigned char *l2709tab = &pci_l2709tab[0];
        register unsigned char *spix, *dpix, *spixptr, *dpixptr;
        register unsigned int pixel_stride, rows, ccols, slb, dlb;
        int cols;
```

```
        /* get parameters from the images */
        spix = spixptr = src->data;
        dpix = dpixptr = dest->data;
        pixel_stride = src->pixel_stride;
        ccols = cols = src->width;
        rows = src->height;
        slb = src->linebytes;
        dlb = dest->linebytes;

        while (rows--) {
             while (ccols--) {
                  *dpixptr = *(l2709tab+*spixptr);
                  *(dpixptr+1) = *(l2709tab+*(spixptr+1));
                  *(dpixptr+2) = *(l2709tab+*(spixptr+2));
                  spixptr += pixel_stride;
                  dpixptr += pixel_stride;
             }
             ccols = cols;
             spixptr = (spix += slb);
             dpixptr = (dpix += dlb);
        }
}
```

For the inverse conversion, the table `pci_7092l[]` must be generated using the inverse equation. Then, the same table lookup method can be applied:

```
/*
        Function: pci_gentab_nl2l
        Operation:Generate a table for converting nonlinear RGB
                  into linear RGB
        Conversion Equations:
                  rlinear = .222*rlinear (rlinear < 0.081)
                          = (.9099*rgamma+.09)^gamma) (rlinear>= 0.018)
                  Similarly for b and g
*/

unsigned char pci_nl2ltab[256];
void pci_gentab_nl2l()
{
        int i;
        float in, out;
        for (i=0; i<256; i++) {
             in = (float) i/NORM;
             if (in < .081)
                  out = .222*in;
             else
                  out = pow((.9099*in + .09),2.22);
             pci_nl2ltab[i] = (unsigned char ) (out * MAXBYTE);
        }
}
```

YCbCr-601 conversions In this section, we will consider several variations of conversions between the CCIR Rec. 601 YCbCr color space and linear and CCIR Rec. 709 nonlinear RGB color space.

The first example converts from YCbCr-601 to RGB-709. The macro `CLIP(x,y)` forces the computed floating-point value `x` into the range 0 to 255 required for unsigned byte representation of pixels.

```
#define CLIP(x,y) \
        if (x<0) y=0; \
        else if (x>255) y = 255; \
        else y = (unsigned char) x;

/*
        Function: pci_ycc6012rgb709
        Operation:Convert a YCbCr 601 image into an RGB 709 image
        Input:    src image, where 24-bit pixels are laid out like this:
                  YCbCrYCbCrYCbCr... and each Y, Cb, Cr is an unsigned char
        Conversion Equations:
                  r709 = 1.169Y          +  1.6Cr   - 223.5
                  b709 = 1.169Y - .394Cb - .8144Cr + 136
                  g709 = 1.169Y + 2.03Cb            - 278.77
        Output:   dest image, laid out in BGR format
*/

#define K00 1.169
#define K02 1.6
#define K03 -223.5
#define K11 -.394
#define K12 -.8144
#define K13 136.
#define K21 2.03
#define K23 -278.77
void pci_ycc6012rgb709(pci_image *src, pci_image *dest)
{
        register unsigned char *spixptr, *dpixptr;
        register unsigned char y,cb,cr;
        register float partial, result;
        register unsigned int pixel_stride, rows, ccols;
        int cols;

        /* get parameters from the images */
        spixptr = src->data;
        dpixptr = dest->data;
        pixel_stride = src->pixel_stride;
        ccols = cols = src->width;
        rows = src->height;
        while (rows--) {
             while (ccols--) {
                  y = *spixptr;
                  cb = *(spixptr+1);
                  cr = *(spixptr+2);

                  partial = K00*(float)y;
```

```
                        result = partial + K02*(float)cr + K03;
                        CLIP(result,*(dpixptr+2));
                        result = partial +K11*(float)cb + K12*(float)cr + K13;
                        CLIP(result,*(dpixptr+1));
                        result = partial + K21*(float)cb + K23;
                        CLIP(result,*(dpixptr));

                        spixptr += pixel_stride;
                        dpixptr += pixel_stride;
                }
                ccols = cols;
        }
}
```

The second example converts from YCbCr-601 to linear RGB. The macro `CLIP_TAB(x,y)` forces the computed floating-point value `x` into the range 0 to 255 required for unsigned byte representation of pixels; however, the value of `x` is used to look up the `z` `(pci_70921_tab[])` first. The same constants `K00..K23` as the previous program are used.

```
#define CLIP_TAB(x,y,z) \
        if (x<0) y=0; \
        else if (x>255) y = 255; \
        else y = *(z+(unsigned char) x);
```

The code is identical to the YCbCr-601-to-RGB-709 conversion, except for the inner loop, where `CLIP_TAB` is invoked:

```
/*
        Function: pci_ycc6012rgbl
        Operation: Convert a YCbCr 601 image into an RGB linear image
        Input:    src image, where 24-bit pixels are laid out like this:
                  YCbCrYCbCrYCbCr... and each Y, Cb, Cr is an unsigned char
        Conversion Equations:
                  r709 = 1.169Y          +  1.6Cr  - 223.5
                  b709 = 1.169Y - .394Cb - .8144Cr + 136
                  g709 = 1.169Y + 2.03Cb           - 278.7
        Followed by 709->linear conversion (nonlinear->linear)

        Output:   dest image, laid out in BGR format
*/

void pci_ycc6012rgbl(pci_image *src, pci_image *dest)
{
        register unsigned char *tabptr = &pci_nl2ltab[0];
        register unsigned char *spixptr, *dpixptr;
        register unsigned char y,cb,cr;
        register float partial, result;
        register unsigned int pixel_stride, rows, ccols;
        int cols;

        /* get parameters from the images */
        spixptr = src->data;
```

```
        dpixptr = dest->data;
        pixel_stride = src->pixel_stride;
        ccols = cols = src->width;
        rows = src->height;

        while (rows--) {
             while (ccols--) {
                  y = *spixptr;
                  cb = *(spixptr+1);
                  cr = *(spixptr+2);

                  partial = K00*(float)y;
                  result = partial + K02*(float)cr + K03;
                  CLIP_TAB(result,*(dpixptr+2),tabptr); /* --R */
                  result = partial +K11*(float)cb + K12*(float)cr + K13;
                  CLIP_TAB(result,*(dpixptr+1),tabptr); /* -GR */
                  result = partial + K21*(float)cb + K23;
                  CLIP_TAB(result,*(dpixptr),tabptr);   /* BGR */

                  spixptr += pixel_stride;
                  dpixptr += pixel_stride;
             }
             ccols = cols;
        }
}
```

The third example converts from YCbCr-601, compressed in a 4:2:0 format, into RGB-709 color space. To accommodate the chrominance subsampling, the function accepts pointers to three images, each of which represents one band of the YCbCr image. The format is shown in Figure 7.14.

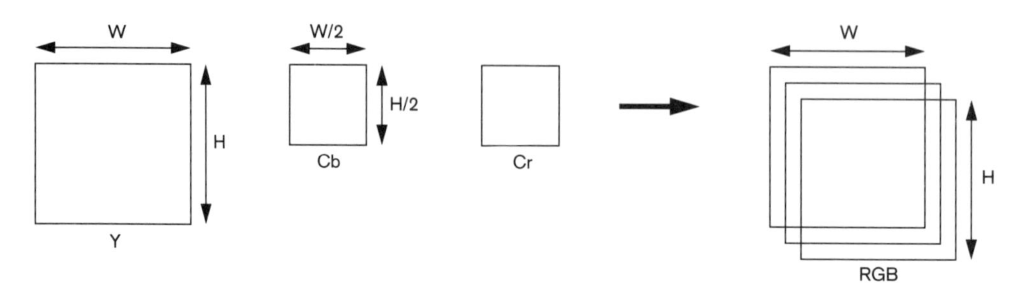

Figure 7.14 Data layout of 4:1:1 YCbCr image

During the color space conversion, *four Y* pixels, *one Cb* pixel, and *one Cr* pixel are used in the computation of *four* RGB triplets, as shown in Figure 7.15. The *Cb* and *Cr* pixels are each replicated two times horizontally and vertically to accomplish this. Therefore, in the function `pci_ycc601_4202rgb709`, the terms involving *Cb* and *Cr* are computed before the four pixels are calculated.

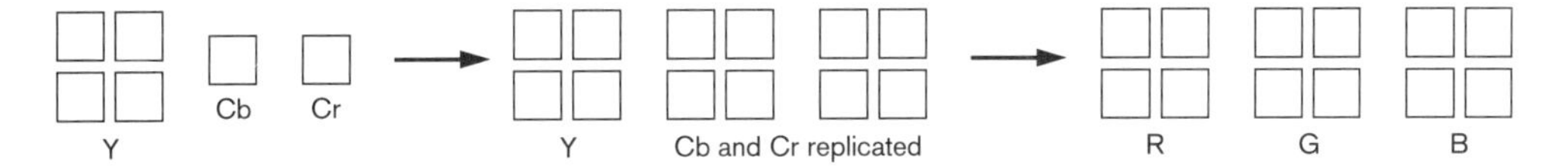

Figure 7.15 Y, Cb, Cr pixels required for computing of four RGB triplets

```
        Function: pci_ycc601_4202rgb709
        Operation:Convert a 4:2:0 YCbCr 601 image into an RGB gamma image
        Input:    y, cb, cr - separate source images
                  the width and height of cb and cr are guaranteed to be
                  half of that of y and dest
        Output:   dest image, laid out in an bgr format
        Conversion Equations:
                  Same as ycc601 to rgb709:
                  r709 = 1.169Y          +  1.6Cr  - 223.5
                  b709 = 1.169Y - .394Cb - .8144Cr + 136
                  g709 = 1.169Y + 2.03Cb           - 278.77
*/

void pci_ycc601_4202rgb709(pci_image *y, pci_image *cb, pci_image *cr,
pci_image *dest)
{
        register unsigned char *ypixptr, *cbpixptr, *crpixptr, *dpixptr;
        register float partial_cr1, partial_cr2, partial_cb1, partial_cb2;
        register float partial_y;
        register float result;
        unsigned char *ypix_sav, *cbpix_sav, *crpix_sav, *dpix_sav;
        unsigned int pixel_stride, rows, ccols;
        unsigned int ylb, cblb, crlb, dlb;
        int cols;

        /* get parameters from the images */
        ypix_sav  = ypixptr  = y->data;
        cbpix_sav = cbpixptr = cb->data;
        crpix_sav = crpixptr = cr->data;
        dpix_sav  = dpixptr  = dest->data;
        dlb  = dest->linebytes;
        ylb  = y->linebytes;
        cblb = cb->linebytes;
        crlb = cr->linebytes;

        /* use size of smaller images to drive the loop */
        rows = cb->height;
        cols = ccols = cb->width;

        while (rows--) {
             while (ccols--) {
```

```
        /* compute the Cb, Cr partials */
        partial_cr1 =  K02* (float)(*crpixptr) + K03;
        partial_cr2 =  K12* (float)(*crpixptr) + K13;
        partial_cb1 =  K11* (float)(*cbpixptr);
        partial_cb2 =  K21* (float)(*cbpixptr) + K23;

        /* compute four output pixels at a time */

        /* ... top left ... */
        partial_y = K00* (float)(*ypixptr);
        result = partial_y + partial_cr1;
        CLIP(result,*(dpixptr+2));
        result = partial_y + partial_cb1 + partial_cr2;
        CLIP(result,*(dpixptr+1));
        result = (partial_y + partial_cb2);
        CLIP(result,*(dpixptr));

        /* ... top right ... */
        partial_y = K00* (float)*(ypixptr+1);
        result = partial_y + partial_cr1;
        CLIP(result,*(dpixptr+5));
        result = partial_y + partial_cb1 + partial_cr2;
        CLIP(result,*(dpixptr+4));
        result = partial_y + partial_cb2;
        CLIP(result,*(dpixptr+3));

        /* ... bottom left ... */
        partial_y = K00* (float)*(ypixptr+ylb);
        result = partial_y + partial_cr1;
        CLIP(result,*(dpixptr+dlb+2));
        result = partial_y + partial_cb1 + partial_cr2;
        CLIP(result,*(dpixptr+dlb+1));
        result = partial_y + partial_cb2;
        CLIP(result,*(dpixptr+dlb));

        /* ... bottom right ... */
        partial_y = K00* (float)*(ypixptr+ylb+1);
        result = partial_y + partial_cr1;
        CLIP(result,*(dpixptr+dlb+5));
        result = partial_y + partial_cb1 + partial_cr2;
        CLIP(result,*(dpixptr+dlb+4));
        result = partial_y + partial_cb2;
        CLIP(result,*(dpixptr+dlb+3));

        /* move to next cbcr pixel... */
        cbpixptr++;
        crpixptr++;
        /* ...and the next 2x2 y and dest pixel-block */
        dpixptr += 6;
        ypixptr += 2;
    }
    ccols = cols;
```

```
            /* update all the pixel pointers */
            cbpixptr = (cbpix_sav += cblb);
            crpixptr = (crpix_sav += crlb);
            dpixptr = (dpix_sav += (dlb<<1));
            ypixptr = (ypix_sav += (ylb<<1));
        }
}
```

YCbCr-709 conversions Since YCbCr-709 is very similar to YCbCr-601, we do not provide additional coding examples. Trivial modifications to the above code (changing the constant values in most cases) will yield YCbCr-709 conversions.

PhotoYCC conversions The following function converts from RGB-709 to PhotoYCC. This, again, is a fairly straightforward conversion.

```
/*
        Function: pci_rgbl2photoycc
        Operation:Convert a linear rgb image to photoycc format
        Input:
        Conversion Equations:
                  Compute rnl, gnl, bnl from r, g, b

                  Then,
                  yphoto  = 181.883(.299rnl + .587gnl + .114bnl)
                  cbphoto = 156 + 111.4 (bnl - yphoto)
                  crphoto = 136 + 135.64 (rnl - yphoto)

                  But, rnl, gnl, and bnl are scaled to 256. So the
                  equations to use in computation are:

                  yphoto  = 181.883(.299rnl + .587gnl + .114bnl)/256.
                          = .2124rnl + .417gnl + .081bnl;
                  cbphoto = 156 + 111.4bnl/256 - 111.4yphoto
                          = 156 + .614bnl - 111.4yphoto
                  crphoto = 136 + 135.64rnl/256 - 135.64yphoto
                          = 136 + .529rnl - 136.64yphoto

        Output:   dest image, laid out in a similar manner to src image
*/

#define K00 (.299/256.)
#define K01 (.587/256.)
#define K02 (.114/256.)
#define K10 156
#define K11 111.4
#define K20 136
#define K21 135.64
```

```
void pci_rgbl2photoycc(pci_image *src, pci_image *dest)
{
        register unsigned char *spixptr, *dpixptr;
        register unsigned char rnl, bnl, gnl;
        register float y_float, result;
        register unsigned char r,g,b;
        register unsigned char *tab;
        unsigned char *spix_sav, *dpix_sav;
        unsigned int pixel_stride, rows, ccols;
        unsigned int slb, dlb;
        int cols;

        /* get parameters from the images */
        spix_sav  = spixptr  = src->data;
        dpix_sav  = dpixptr  = dest->data;
        dlb  = dest->linebytes;
        slb  = src->linebytes;
        rows = src->height;
        cols = ccols = src->width;
        tab = &pci_l2nltab[0];

        while (rows--) {
             while (ccols--) {
                  bnl = *(tab+*spixptr);
                  gnl = *(tab + *(spixptr+1));
                  rnl = *(tab + *(spixptr+2));
                  y_float = K00*(float)rnl + K01*(float)gnl
                       + K02*(float)bnl;
                  CLIP((181.883*y_float),*dpixptr);
                  result = K10 + K11*((float)bnl/256. - y_float);
                  CLIP(result,*(dpixptr+1));
                  result = K20 + K21*((float)gnl/256. - y_float);
                  CLIP(result,*(dpixptr+2));

                  /* move to next pixel */
                  dpixptr += 3;
                  spixptr += 3;
             }
             /* move to next row */
             dpixptr = (dpix_sav += dlb);
             spixptr = (spix_sav += slb);
             ccols = cols;
        }
}
```

CMYK conversions The conversions between RGB and CMYK are fairly straightforward, as follows. Note, however, that color printing products utilize many different variations of these conversions, some of which are proprietary.

```
/*
        Function: pci_rgblinear2cmyk
        Operation:Convert a rgblinear image to cmyk format
        Input:    RGB image, laid out in BGRBGRBGR... format
        Conversion Equations:
                  c = 1 - r;
                  m = 1 - g;
                  y = 1 - b;
                  k = min(c,m,y)
                  c = c - k;
                  m = m - k;
                  y = y - k;
        Output:   CMYK image, laid out in CMYKCMYK... format
*/

#define MAXBYTE 255

void pci_rgbl2cmyk(pci_image *src, pci_image *dest)
{
        register unsigned char c,m,y,k;
        register unsigned char *spixptr, *dpixptr;
        register float y_float;
        register unsigned char r,g,b;
        register unsigned char *tab;
        unsigned char *spix_sav, *dpix_sav;
        unsigned int pixel_stride, rows, ccols;
        unsigned int slb, dlb;
        int cols;

        /* get parameters from the images */
        spix_sav  = spixptr  = src->data;
        dpix_sav  = dpixptr  = dest->data;
        dlb  = dest->linebytes;
        slb  = src->linebytes;
        rows = src->height;
        cols = ccols = src->width;

        while (rows--) {
             while (ccols--) {
                  c = MAXBYTE - *(spixptr+2);
                  m = MAXBYTE - *(spixptr+1);
                  y = MAXBYTE - *spixptr;
                  k = m;
                  if (c<m)
                       k = c;
                  if (y<k)
                       k = y;
                  *dpixptr = c-k;
                  *(dpixptr+1) = m-k;
                  *(dpixptr+2) = y-k;
                  *(dpixptr+3) = k;

                  dpixptr += 4;
                  spixptr += 3;
             }
```

```
                dpixptr = (dpix_sav += dlb);
                spixptr = (spix_sav += slb);
                ccols = cols;
        }
}
```

7.4 The display of color images

So far, we have looked at ways to convert between different color spaces in common use in computer imaging. However, once a color image has been cast into the proper color space, it must still be displayed. In this section, we take a closer look at two important issues involved in the display of color images: *gamma correction* and *color quantization and dithering*.

7.4.1 Gamma correction and display lookup tables

Gamma refers to the inherent nonlinearity between the voltage applied to the electron gun of a CRT and the intensity of the light output into the display by that voltage. A similar nonlinear relationship also exists in the world of photography between the density of an exposed negative and the exposure (that is, the amount of light) the negative received.

In the case of a CRT, the relationship between voltage and intensity follows a power law. Gamma is the exponent in this relationship. The perceived intensity of light is proportional to the $(\text{voltage})^\gamma$. Many computer images, however, are acquired and processed assuming that there is a linear relationship between the pixel values and the intensity of the light. Therefore, *gamma correction* must be applied before the image is displayed to compensate for the nonlinearity.

The value of gamma is dictated by the physics of the CRT and ranges from 1.6 to 2.6. Note that changing the *black-level* (i.e., the brightness setting) of the monitor changes the effective gamma. The CCIR Rec. 709 color space (covered in Section 7.3) uses a gamma value of 2.2, whereas several other devices, such as many Apple Macintosh monitors and color prepress applications, use a gamma of 1.8.

In principle, to apply gamma correction, we must know the monitor's gamma first. Then, the image is passed through a lookup table loaded with a function shown in Figure 7.16, which is the inverse of the gamma function.

The lookup table is generated in software as follows:

```
#define INV_GAMMA 1./2.2
static unsigned char pci_inv_gamma_tab[256];
void pci_gen_inv_gamma_tab()
```

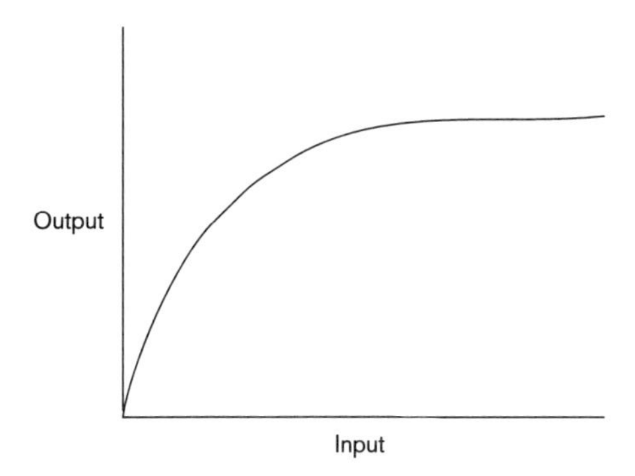

Figure 7.16 A typical gamma correction function

```
{
        int i;
        for (i=0; i<256; i++)
             pci_l2709tab[i] = (unsigned char) (pow(double(i)/256.,
                                                     INV_GAMMA)*255);
        return;
}
```

In practice, gamma correction is often closely tied to the display hardware of the imaging system.

The block diagram of a true-color (24-bit) display system is shown in Figure 7.17. Here, the image is passed through a hardware lookup table before being displayed. There is no additional performance penalty for gamma correction since it comes for free with the hardware.

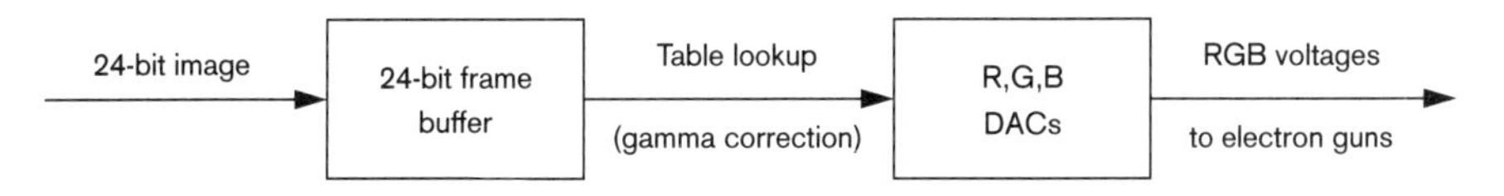

Figure 7.17 Hardware gamma correction in a 24-bit display system

In systems where hardware lookup tables are not available, a software table lookup operation (discussed in Chapter 5) must be performed on the image using the table generated by the above programming example.

7.4.2 Color quantization and dithering

As we saw earlier in the chapter, color is represented using three color components. Limited by the dynamic range of current acquisition technology, most digital color images are represented using 8 bits per color, or 24 bits per RGB pixel. (Some advanced scanners can generate up to 12 bits of data per color component and these are usually represented as 48 bits per RGB pixel, that is, 16 bits per color component, and the following techniques can easily be extended to these cases.)

Displaying 24-bit pixels requires that the display generator be able to display over four million colors. However, a large number of color display cards today can display only 256 colors simultaneously. For this display, color quantization is necessary, and the 24-bit images must be converted to 8 bits before display, using one of several available color quantization techniques.

There are many methods of performing color quantization. In the first order approximation, a *color cube* may be used for *nearest color quantization*. A more advanced technique is *ordered dithering* using a color cube; it reduces some artifacts, such as contouring, created by nearest color quantization. *Error diffusion dithering* is yet more advanced where the errors created by the quantization are minimized.

Color quantization using a color cube The simplest way of performing color quantization is to use a color cube to find the nearest color for display. This is a lookup table operation, with the input of the lookup table being a triplet $\{B,G,R\}$ and the output is between 0 and N, where N is determined by the number of colors that can be displayed. For example, in many 8-bit systems, although 256 colors can be displayed, several of these must be reserved for use by the window system in order to avoid colormap flashing, and N is typically 220–230.

The nearest color operation then finds the triplet in the table closest to the source pixel $\{B,G,R\}$ and uses the corresponding destination color.

For color quantizing 24-bit pixels to 8-bit pixels, a color cube is defined by two arrays: `dimensions[]` and `multipliers[]`. The array `dimensions[]` specifies how many values of each color are present in the input side of the lookup table. For example, if `dimensions[]` is {4,9,6}, then four values of blue, nine values of green, and six values of red are present in the table. The array `multipliers[]` specifies how many times each value is replicated in the table. For example, if `multipliers[]` is {1,4,36}, then the blue values are replicated once, the green values four times, and the red values are replicated 36 times.

We will now construct the 4:9:6 color cube. Four blue values are allowed: These are found by dividing 256 by (4 – 1) (since the first value is always 0). These are 0, 85, 170, and 255. Similarly, the green values are 0, 32, 64, …, and the red values are 0, 51,102, ….

The color cube then looks like Table 7.2.

Table 7.2 Values in a 4:9:6 color cube

Blue	Green	Red	Output
0	0	0	0
85	0	0	1
170	0	0	2
255	0	0	3
0	32	0	4
85	32	0	5
170	32	0	6
255	32	0	7
0	64	0	8
85	64	0	9
170	64	0	10
255	64	0	11
...	...	...	...
0	0	51	36
85	0	51	37
170	0	51	38
255	0	51	39
0	32	51	40
85	32	51	41
170	32	51	42
255	32	51	43
...	...	...	...
0	224	255	208
85	224	255	209
170	224	255	210
255	224	255	211
0	255	255	212
85	255	255	213
170	255	255	214
255	255	255	215

The structure of the color cube should now be clear. For each value of red, all possible values of green and blue are present in the table; for each value of green, all possible values of blue are present in the table. The values in the `multipliers[]` array thus correspond to the total number of distinct combinations of the color or colors to the left of that color in the table. Thus, `multipliers[2]`, for red, is 36, because there are a total of 4 * 9 = 36 combinations of green and blue. `multipliers[1]` for green is 4, because there are four possible values of blue.

The following programming example color quantizes a 24-bit image into an 8-bit image using the 4:9:6 color cube.

```
/*
        Function: pci_nearest_color
        Operation: Convert a 24-bit image into an 8-bit image
                  using color quantization
        Inputs:    src is the source image
                  dimensions and multipliers define the colorcube
                  (see text)
        Output:    dest is the output image
*/

void pci_nearest_color(pci_image *src,pci_image *dest, int *dimensions,
                  int *multipliers)
{
        register unsigned char *spixptr, *dpixptr;
        unsigned int spixel_stride, dpixel_stride, rows, ccols;
        register unsigned int mult_r, mult_g, mult_b;
        register unsigned int pixel, index;
        register unsigned int step_r, step_g, step_b;
        int cols;

        /* get parameters from the images */
        spixptr = src->data;
        dpixptr = dest->data;
        spixel_stride = src->pixel_stride;
        dpixel_stride = dest->pixel_stride;
        ccols = cols = src->width;
        rows = src->height;

        mult_r = *(multipliers+2);
        mult_g = *(multipliers+1);
        mult_b = *(multipliers);

        step_r = 256/(*(dimensions+2)-1);
        step_g = 256/(*(dimensions+1)-1);
        step_b = 256/(*dimensions-1);

        while (rows--) {
             while (ccols--) {
                  /* step down into the appropriate red chunk */
                  pixel = (*(spixptr+2) * 256)/step_r;
                  if ((pixel & 0xff) > 0x80)
                       index = mult_r;
                  index += (pixel>>8) * mult_r;

                  /* now move to the green within the red chunk */
                  pixel = (*(spixptr+1)*256)/step_g;
                  if ((pixel & 0xff) > 0x80)
                       index += mult_g;
                  index += (pixel>>8)*mult_g;
```

```
                    /* finally the blue within the green chunk */
                    pixel = (*(spixptr+1)*256)/step_b;
                    if ((pixel & 0xff) > 0x80)
                         index += mult_b;
                    index += (pixel>>8)*mult_b;

                    *dpixptr = index;

                    /* get the blue, green and red */
                    spixptr += spixel_stride;
                    dpixptr += dpixel_stride;
               }
               ccols = cols;
          }
     }
```

Ordered dither One problem with color quantization using a color cube is that it gives rise to contouring in images due to the relatively coarse quantization. In *ordered dither*, the nearest color lookup is performed. Then, a *dither mask* is used to fine-tune the output value. In effect, this performs averaging, which reduces contouring.

The procedure is now illustrated with an example. A typical dither mask is:

$$\frac{1}{16}\begin{bmatrix} 0 & 8 & 2 & 10 \\ 12 & 4 & 14 & 6 \\ 3 & 11 & 1 & 9 \\ 15 & 7 & 13 & 5 \end{bmatrix}$$

The first step is to replicate this mask over the image. This is shown in Figure 7.18 for a 12×12 image. Now, each pixel in the source image has a dither mask value associated with it.

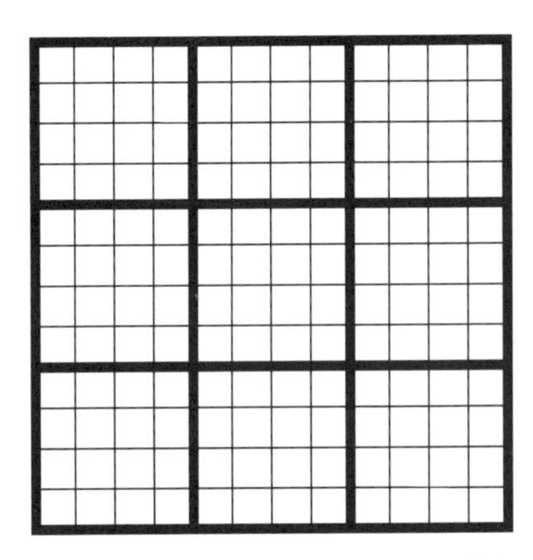

Figure 7.18 Replicating a 4×4 dither mask over a 12×12 image

The second step is to perform the nearest color lookup using the color cube. As we have seen in the example code above, this is equivalent to dividing the individual components of the pixel by *N*, where *N* is (256 / dimensions[] – 1) for that component. For example, if the 4:9:6 color cube is being used, the blue component of the source pixel is divided by 85.

In the third step, the quotient from this division is compared to the dither mask value for that pixel. If the mask value is greater, then the next value in the color cube is used in the dithering.

We illustrate this now with an example (Figure 7.19). Suppose the source pixel at the location (5,5) has B,G,R = (45,62,210). If we used the nearest color method to find the 8-bit value for this pixel, we would obtain the pixel corresponding to the color cube triplet (85, 64, 204), which has the output value of 81.

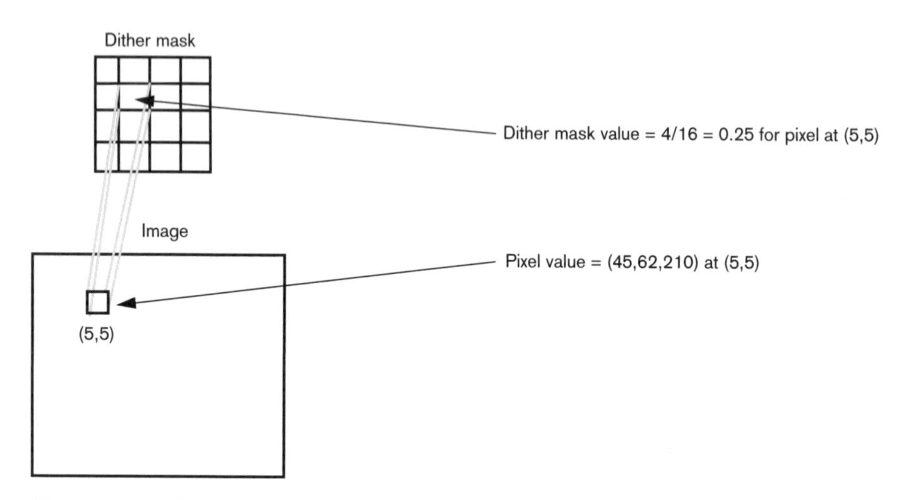

Figure 7.19 Ordered dithering an image

How does ordered dither differ from this? Still using the 4:9:6 color cube, we first divide the 210 by 51 to get 4.12. Similarly, we get 62/32 = 1.9375 and 45/85 = 0 .529. Now, the dither mask value associated with this pixel location is 4/16 = 0.25. Therefore, the color cube entry for red is 51, since 0.12 < 0.25. However, since 0.9375 > 0.25, the color cube entry for green is 64. Similarly, the color cube entry for blue is 85. Thus, the output pixel value corresponds to the entry (85,64,51) and is 45.

Generating the dither mask is a challenging task. The problem is to find matrix values that will evenly distribute the input pixel values in the output space, thus avoiding artifacts and contouring problems. One way to do this is to use a recursive definition of the matrix [8]. The first ordered dither matrix is defined as:

$$M_1 = \frac{1}{256}\begin{bmatrix} 0 & 192 \\ 128 & 64 \end{bmatrix}$$

The 8×8 dither matrix is derived from the 4×4 by adding constants to it:

$$M_8 = \left(\frac{1}{256}\right)\begin{bmatrix} 0 & 192 & 48 & 240 \\ 128 & 64 & 176 & 112 \\ 32 & 224 & 16 & 208 \\ 160 & 96 & 144 & 80 \end{bmatrix}$$

In other words, M_8 is derived from M_4 by the following recursion:

$$M_8 = \begin{bmatrix} M_4 + 0 & M_4 + 48 \\ M_4 + 32 & M_4 + 64 \end{bmatrix}$$

The same recursion can be used to generate M_N, where $N = 2^n$, from $M_{N/2}$. C code for this algorithm can be found in reference 8.

Error-diffusion dither By its very nature, nearest color dither introduces a quantization error when a table lookup is performed to find the destination pixel value. In ordered dither, we use a dither mask to add small perturbations to the location in the table from which the destination pixel is obtained. In *error-diffusion dither*, on the other hand, the idea is to use the error generated by the current source pixel to reduce the quantization errors in quantizing the subsequent pixels.

Associated with error-diffusion dithering is the following kernel:

$$\frac{1}{16}\begin{bmatrix} 0 & 0 & 0 \\ 0 & 0 & 7 \\ 3 & 5 & 1 \end{bmatrix}$$

The operation is performed as follows. The current pixel is dithered using a color cube as before. However, the errors corresponding to the dithering are propagated to the future source pixels before the next source pixel is dithered.

The algorithm is illustrated with the following example, shown in Figure 7.20. Suppose the pixel being processed is the one at the center with the $(B,G,R) = (90,70,210)$. The nearest color triplet that matches this pixel is (85,64,204). Therefore, the error is (90,70,210) – (85,64,204) = (5,6,6).

	(45, 62, 210)	(r1, g1, b1)
(r2, g2, b2)	(r3, g3, b3)	(r4, g4, g4)

0	0	0
0	0	7/16
3/16	5/16	1/16

Figure 7.20 Image window and diffusion kernel

Before the next source pixel is dithered, this error is used to update the source pixels by placing the center of the diffusion kernel at the pixel just dithered. Thus, $r1$ becomes $r1 + (7 / 16) * (6)$, $g1$ becomes $g1 + (7 / 16) * (6)$, and $b1$ becomes $b1 + (7 / 16) * (5)$. Similarly, the pixels ($r2$,$g2$,$b2$), ($r3$,$g3$,$b3$), and ($r4$,$g4$,$b4$) are updated using the weights 3/16, 5/16, and 1/16.

Once the source pixels are updated, we move to the next pixel and repeat these steps. Error-diffusion dithering yields very good results. However, computationally it is quite expensive.

7.5 *Conclusion and further reading*

In this chapter we have considered some of the issues involving color in computer imaging. In particular, we have looked at the quantitative measurement of color using color spaces. We have sorted out several important device-independent and device-dependent color spaces, and have provided coding examples for conversion between them.

We have also looked at the problem of color quantization, useful when the number of colors that can be displayed simultaneously by a display system is less than the number of colors that can be represented by the pixels of the image.

Color science is an involved and mature science, the subtleties of which cannot be addressed within the scope of this book. However, it is hoped that this chapter will provide a starting point for the interested reader. The classic texts on color science are references 1 and 2. An account of color issues related to digital video can be found in reference 7. An overview of various color-related issues related to desktop publishing can be found in reference 5.

7.6 *References*

1 R. W. G. Hunt, *The Reproduction of Color*, 4th edition, New York: Wiley, 1987.

2 G. Wyszecki and W. S. Stiles, *Color Science: Concepts and Methods, Quantitative Data and Formulae*, 2nd edition, New York: Wiley, 1982.

3 T. Young "On the Theory of Light and Colours," *Philosophical Transactions of the Royal Society of London*, Vol. 92, 1802, pp. 20–71.

4 M. Stone, *Device Independent Color Reproduction*, Xerox PARC Tech. Report EDL-92-1, April 1992.

5 *The Desktop Color Book*, San Diego: Verbum Books, 1994.

6 W. K. Pratt, *Digital Image Processing*, 2nd edition, New York: Wiley, 1992.

7 C. Poynton, *A Technical Introduction to Digital Video*, New York: Wiley, 1996.

8 S. Hawley, "Ordered Dither," *Graphics Gems*, A. Glassner, ed., Boston: Academic Press, 1988.

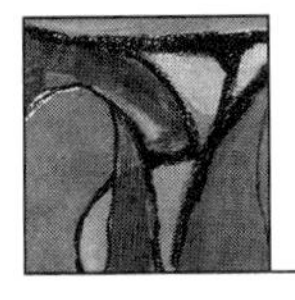

chapter 8

Image geometric operations

8.1 Introduction

Geometric operations on images are primarily concerned with modifying the position—that is, the *address*—of pixels in images using a variety of geometric transformations. Modification of images based on geometry is often called *image warping*. These techniques have found use in a wide range of application areas, including desktop publishing, remote sensing, movies, and other parts of the entertainment industry.

There is a strong aesthetic appeal and fascination in the geometric modification of images. For example, scaling or zooming an image performs an operation that would otherwise require substantial work in the darkroom. For some other types of geometric operations, the changes in the image are often startling and drastic; yet the information content of the image is always accessible to the viewer. For these reasons, geometric operations on images have become popular.

This chapter is about the most widely used image geometric operations. Upon reading it, the user will obtain a basic understanding and relative merit of the available techniques. The engineer in charge of implementing an imaging product will gain an understanding of the fundamental operations needed to implement these techniques. Several coding examples sprinkled throughout the chapter illustrate methods for implementing geometric functions using both straightforward and optimized approaches.

Early techniques used in geometric modification of images were developed for remote sensing applications. For example, the images from space transmitted by satellites often needed to be corrected for distortion. They also needed to be registered so that a large montage of images could be created from many individual pieces. Both correction of distortion and registration require modifying the location of pixels.

Subsequently, simpler geometric modification functions, such as zooming and rotation, became common in hardware image processors (such as the Vicom, described in Chapter 3), as well as software image processing packages. In recent years, with the advent of desktop imaging systems, many of these functions have become popular in desktop publishing, where they are used for numerous purposes from correction of scanning errors to creation of special effects. More sophisticated geometric modification techniques have been popularized by their use in the entertainment industry. Several software products on the market today feature image manipulation capabilities based on geometric modification of images.

There are many types of geometric operations. The ones in common use in imaging products include image scaling, rotation, affine transformations, and transpositions. More sophisticated operations include special-effects filters and *morphing*, a technique that enables the gradual metamorphosis of one object into another using geometric modifications and alpha blending (explored in Chapter 5). In this chapter we will concentrate on the building blocks for these common geometric operations.

There are two fundamental steps that are required in these operations. These are *address computation* and *interpolation*. In Section 8.2, we discuss these steps in detail, providing the theoretical basis on which the rest of the chapter lies. In Section 8.3, we present some general implementation details for performing geometric operations. Armed with the knowledge of the basic steps and implementation choices, we concentrate on specific operations from Section 8.4 onwards. Section 8.4 is about image scaling; because of the importance of scaling, we look at several different implementation techniques. In Section 8.5 we look at image rotation. Affine transformation, a powerful linear transformation, is covered in Section 8.6. Image transposition is the topic of Section 8.7. In Section 8.8, we examine some special-effects filters.

8.2 Steps in the implementation of geometric operations

The geometric operations described in this chapter involve two images: a source image $f(x,y)$ and a destination image $g(x',y')$. The destination image is created by applying a geometric operation to the source image. Note that for the purposes of this chapter, x corresponds to the column and y to the row of the image.

The pixel $S(x,y)$ at location (x,y) of the source is mapped to the pixel $D(x',y')$ at location (x',y') of the destination according to the geometric transformation defined by *X*[] and *Y*[], as follows:

$$\begin{aligned} x' &= X(x, y) \\ y' &= Y(x, y) \end{aligned} \tag{8.1}$$

The mappings *X*[] and *Y*[] are illustrated in Figure 8.1 for an example that scales an image. A pixel in the source image, (x_1,y_1), is mapped to destination coordinates (x'_1,y'_1) using the forward mappings *X*[] and *Y*[]. A destination point (x'_2,y'_2) is mapped to the source point (x_2,y_2) using the inverse mapping X^{-1}[] and Y^{-1}[].

As we indicated earlier, there are two fundamental steps involved. These are *address computation* and *interpolation*.

8.2.1 Address computation

For some of the geometric operations discussed in this chapter, such as scaling and rotation about the point (0,0), address computation is performed using a geometric transformation of the form shown in Equation 8.2.

$$\begin{bmatrix} x' \\ y' \end{bmatrix} = \begin{bmatrix} a & b \\ c & d \end{bmatrix} \begin{bmatrix} x \\ y \end{bmatrix} \tag{8.2}$$

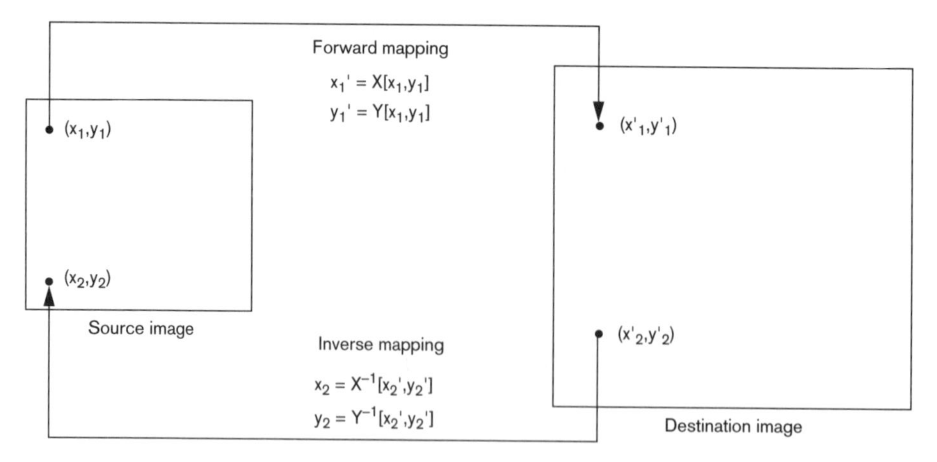

Figure 8.1 Forward and inverse mapping

This is, in fact, a special case of a more general transformation, which we use wherever the operation requires translations in the image coordinates, such as rotations about arbitrary points, affine transformations, and translations:

$$\begin{bmatrix} x' \\ y' \\ 1 \end{bmatrix} = \begin{bmatrix} a & b & c \\ d & e & f \\ g & h & i \end{bmatrix} \begin{bmatrix} x \\ y \\ 1 \end{bmatrix} \tag{8.3}$$

The 1 is added to the two-dimensional coordinates to simplify mathematical notations and derivations. This notation is known as *homogeneous coordinates*, where an $(n + 1)$-dimensional vector is used to represent an n-dimensional coordinate system.

Which notation should we follow in this chapter? Our goal is to keep things as simple as possible but not simpler; therefore, we will avoid using the homogeneous coordinate system unless we are dealing with operations, such as rotate and affine transformations, where its use simplifies the mathematics considerably.

In practice, implementations based on forward mapping have some problems associated with them. For example, rounding can result in skipped pixels. After rounding, the forward mapping can be a one-to-many mapping, which is difficult to implement correctly in practice. For these reasons, inverse mapping is used most often. To perform inverse mapping, the transformation matrices shown in the previous equations must be inverted. In this chapter, we will be concerned exclusively with inverse mappings.

8.2.2 Interpolation

The use of inverse mapping creates a new problem. Consider the case shown in Figure 8.2. The integer coordinates (x',y') of the destination image are inverse mapped to a point (x,y) in the source image. However, the point (x,y) may not be an integer value. In this case, how do we decide what pixel value, corresponding to the source image at (x,y), is copied to the destination?

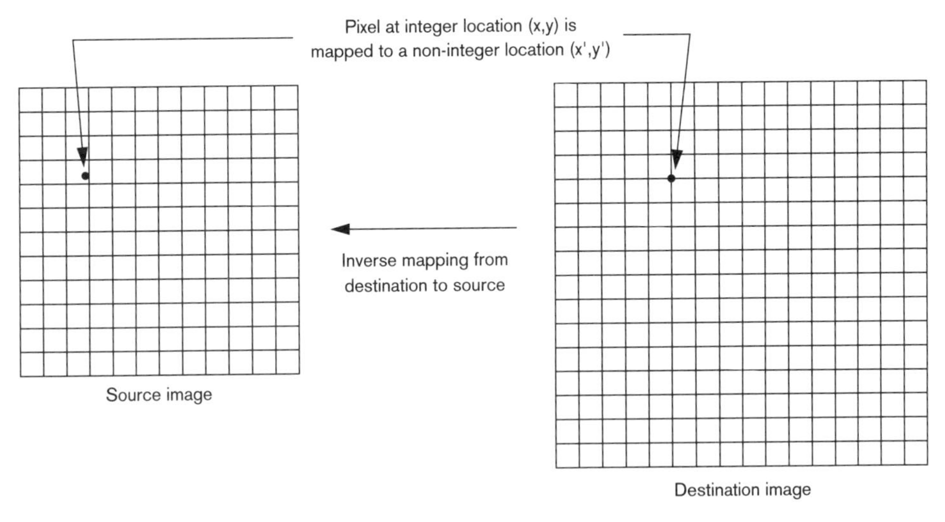

Figure 8.2 Inverse mapping an integer coordinate (x',y') to noninteger (x,y)

Intuitively, it is clear that the source image value at (x,y) can be *interpolated* from the nearby pixels. Interpolation is an important step in geometric operations because it greatly affects the quality of the output. To gain a firm understanding of interpolation, we need to look at the creation of digital images and the underlying principles.

The digital image, a two-dimensional array of pixels, is a sampled and quantized version of a two-dimensional continuous-valued function. As we saw in Chapter 2, the Sampling Theorem guarantees that the continuous-valued function can be reconstructed perfectly, in principle, if the analog function is sampled at a frequency greater than the Nyquist frequency, i.e., twice the maximum frequency present in the analog signal.

In order to find the value of the source image at (x',y'), we would like to replace the discrete pixels of the source with the original two-dimensional analog function. Then, the value of this function at (x,y), $f(x,y)$, would be the source pixel to copy into (x',y'). This is called *ideal interpolation* or *ideal resampling*.

How would one implement ideal interpolation? To answer this, we need to examine the reconstruction of a sampled signal. For the sake of clarity, let us consider a hypothetical one-dimensional band-limited signal. The input signal $f(x)$ and its spectrum $F(u)$ are shown in Figure 8.3.

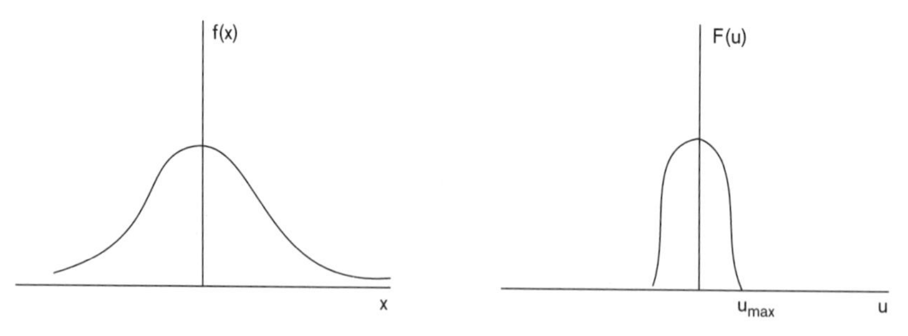

Figure 8.3 A hypothetical signal f(x) and its spectrum F(u)

The signal, when sampled at a rate above the Nyquist frequency of $2u_{max}$, yields a sampled function $f(x_k)$, as shown in Figure 8.4. Sampling the function $f(x)$ is equivalent to multiplying it by a comb function. This means that the spectrum of the sampled signal is obtained by convolving the spectrum of the original signal by the spectrum of a comb function, which is also a comb function. The resulting spectrum $G(u)$ is also shown in Figure 8.4.

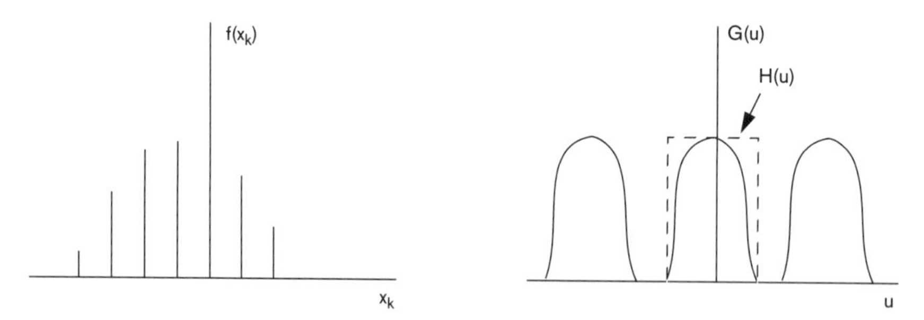

Figure 8.4 A sampled signal and its spectrum

To exactly reconstruct the signal $f(x)$, we need to extract $F(u)$ from $G(u)$. This is accomplished by multiplying $G(u)$ by a box function $H(u)$, as shown in Figure 8.4. The signal $h(x)$ corresponding to $H(u)$ is $\text{sinc}(x) = \sin(x) / x$. Therefore, the analog signal is exactly reconstructed if the sampled signal is convolved with the sinc function.

$$f(x) = \sum_{k=-\infty}^{\infty} \text{sinc}(x - x_{kf}) f(x_k) \tag{8.4}$$

This is called *ideal reconstruction*. Although our discussion above uses a one-dimensional signal, the ideas are easily extended to two-dimensional signals using *separability*. Using the same ideas that we encountered in separable convolution in Chapter 6, we can perform ideal reconstruction on two-dimensional signals by convolving separately in the horizontal and vertical directions. As we will see, many of the geometric algorithms implemented in this chapter exploit separability of the transformation.

In practice, interpolation through ideal reconstruction is impossible to realize, because the sinc function is of infinite duration. Therefore, we approximate the ideal interpolation kernel, the sinc function, with other filters of finite duration, usually obtained by truncating infinite duration functions with "window" functions.

The choice of window functions and interpolation kernels is a topic of much research and debate. The subject has been covered extensively in references 1 and 2. For all practical purposes, this choice depends on two criteria: speed of implementation and quality of the result.

Let us assume for the moment that we have decided on a particular interpolation kernel, $h(x)$. Given this kernel, and a one-dimensional sampled signal $f(x_k)$, what are the steps needed to interpolate? The technique used to interpolate between samples of $f(x_k)$ is shown in Figure 8.5. (As before, it is easily generalized to two dimensions by repeating the process along the rows and then along the columns of the image.)

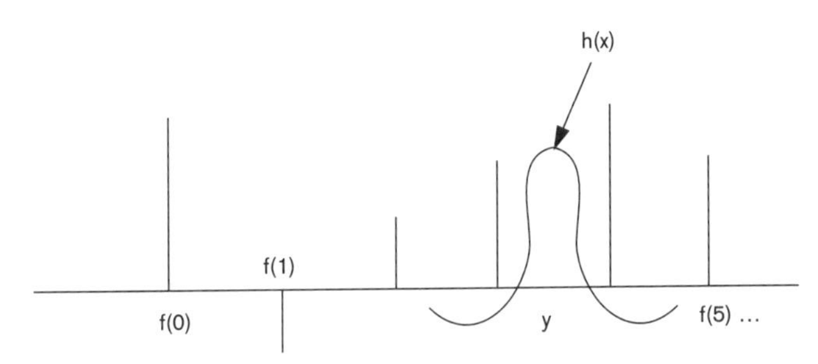

Figure 8.5 Interpolating between values of a sampled function using an interpolation kernel

In Figure 8.5, the interpolating kernel $h(x)$ is placed with its center at the point y where the value of the function $f(x_k)$ is to be interpolated. Then the interpolated value at y is found by the following convolution:

$$f(y) = \sum_{k=-\infty}^{\infty} f(x_k)h(y - x_k) \tag{8.5}$$

In practice, $h(x)$ is always of finite duration. The source pixels that span the duration of $h(x)$ make up the *region of support* of the convolution.

Practical interpolation methods Now that we understand the basics of the interpolation, we can return to our two-dimensional images. The three most commonly used interpolation techniques in images are *nearest neighbor interpolation*, *bilinear interpolation*, and *polynomial interpolation using convolution*.

The simplest and most commonly used form of interpolation is nearest-neighbor interpolation, often called *pixel replication*. The source pixel closest to the inverse mapped point (x,y) whose value is to be approximated, is copied to the destination (Figure 8.6).

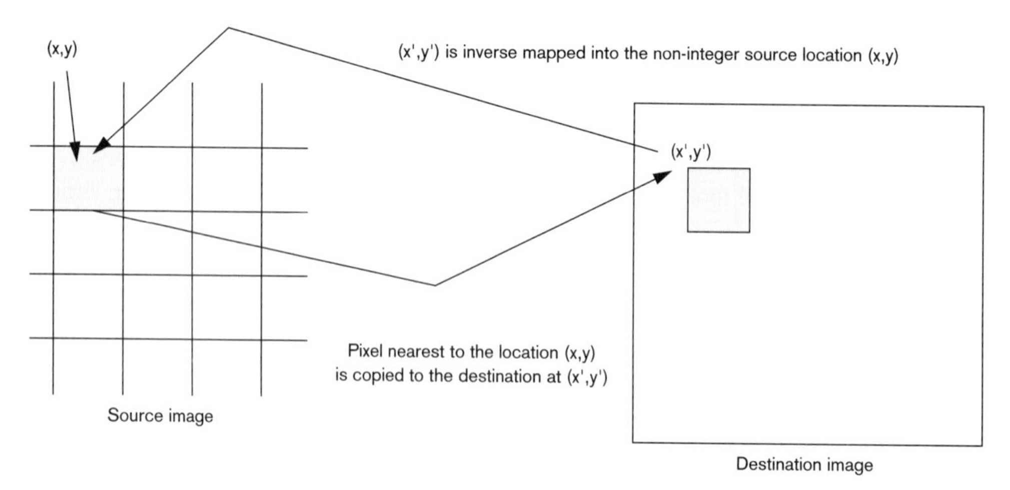

Figure 8.6 Nearest neighbor interpolation

In one dimension, the nearest neighbor algorithm can be stated as:

$$f(x) = f(n), n-\frac{1}{2} \leq x \leq n+\frac{1}{2} \tag{8.6}$$

and the extension to two dimensions is as follows (m and n are integers):

$$f(x, y) = f(m, n), m-\frac{1}{2} \leq x \leq m+\frac{1}{2}, n-\frac{1}{2} \leq y \leq n+\frac{1}{2} \tag{8.7}$$

The interpolation kernel is shown in Figure 8.7a for the one-dimensional case. Nearest neighbor interpolation is simple and fast to implement; however, the quality of the resultant image suffers from jaggies and blockiness. This occurs because the interpolation kernel does not do any averaging of the pixel values being interpolated.

The second most commonly used interpolation is *bilinear interpolation*. This is also separable; therefore, we first consider the one-dimensional case here. The idea is to fit a straight line between the two sample points on either side of the point to be interpolated. Then we use this line to find the value $f(x)$ for any x between the points. For two sample points at the coordinates x_0 and x_1 with values $f(x_0)$ and $f(x_1)$, the equation of the line passing through them is (Figure 8.8):

$$\frac{f(x)-f(x_0)}{x-x0} = \frac{f(x_1)-f(x_0)}{x1-x0} \tag{8.8}$$

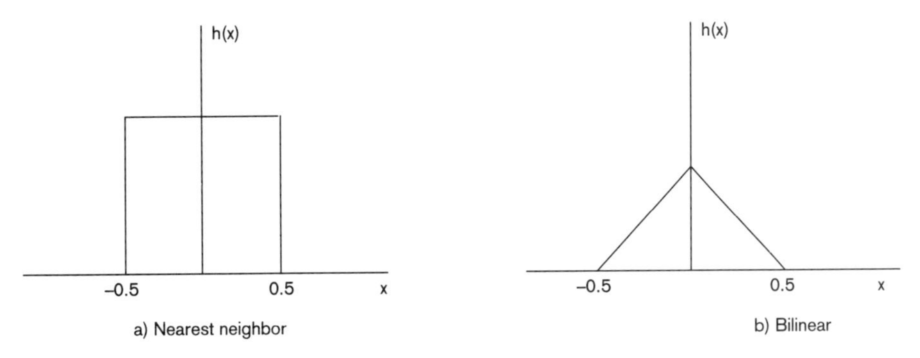

Figure 8.7 Interpolation kernels for nearest neighbor and bilinear interpolations

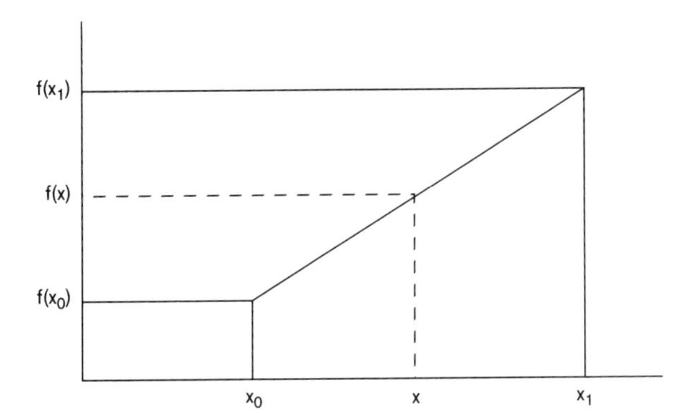

Figure 8.8 Interpolating using a straight line

which yields

$$f(x) = \frac{x - x_0}{f(x_1) - f(x_0)}(f(x_1) - f(x_0)) + f(x_0) \tag{8.9}$$

For our images, which are sampled on a uniform grid, $(x_1 - x_0) = 1$. If we let $p = (x - x_0)$, then Equation 8.9 reduces to:

$$f(x) = (1 - p)f(x_0) + pf(x_1) \tag{8.10}$$

Extension of bilinear interpolation to two dimensions is shown in Figure 8.9. The interpolation is first performed along the rows and then along the columns. Interpolation along the two rows yields:

$$\begin{aligned} a &= pf(x_0 + 1, y_0) + (1 - p)f(x_0, y_0) \\ b &= pf(x_0 + 1, y_0 + 1) + (1 - p)f(x_0, y_0 + 1) \end{aligned} \tag{8.11}$$

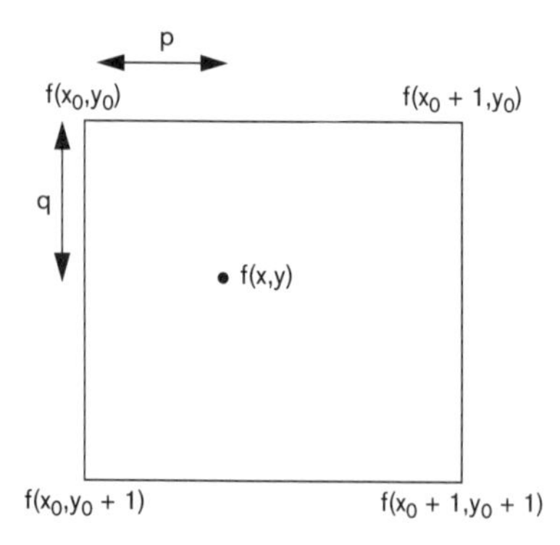

Figure 8.9 Bilinear interpolation in two dimensions

Finally,

$$f(x, y) = qb + (1 - q)a \tag{8.12}$$

The one-dimensional interpolation kernel for bilinear interpolation is shown in Equation 8.13 and graphically in Figure 8.7b:

$$\begin{aligned} h(x) &= 1 - |x|,\ 0 \le |x| \le 1 \\ h(x) &= 0 \text{ otherwise} \end{aligned} \tag{8.13}$$

Bilinear interpolation often represents a sensible trade-off between the quality of the image and speed of implementation. The quality of the result, while being better and smoother than nearest neighbor, is not quite as good as, for example, a more general polynomial interpolation. However, due to the relative simplicity of the computations, high speeds can often be achieved.

Beyond nearest neighbor and bilinear interpolation, more advanced interpolation kernels can be designed to be close to the sinc function. Usually, the design criteria are closely tied to visual quality of the interpolated image as well as the frequency response characteristics of the window function used to create the kernel. Several examples are shown in reference 2. One such example is the interpolation kernel, which, in one dimension, is given by

$$\begin{aligned} h(x) &= (a+2)|x|^3 - (a+3)|x|^2 + 1,\ 0 \le |x| < 1 \\ h(x) &= a|x|^3 - 5a|x|^2 + 8a|x| - 4a,\ 1 \le |x| \le 2 \\ h(x) &= 0 \text{ otherwise} \end{aligned} \tag{8.14}$$

The value of a can be tuned for different effects. For example, $a = -1$ results in $h(x)$ having the same slope as the sinc function at $x = 1$. Setting $a = -1/2$ allows the interpolation function to approximate the original unsampled image to as high a degree as possible in the sense of a power series expansion.

Interpolating with a general kernel is a powerful technique that yields high-quality results. However, the computational cost can be quite prohibitive.

As we will see in the coding examples in Section 8.4, given an arbitrary interpolation kernel, an efficient implementation technique involves precomputing and quantizing the interpolation kernel values into a table, which saves us from having to compute expensive polynomial functions within an inner loop.

8.3 Some general implementation details

The steps shown in Section 8.2 answer the question, "What are the steps that must be applied to each pixel in the image to implement a geometric operation?" The next logical question to ask is, "How do we deal with all the pixels of the image?" This and other questions about nitty-gritty details of implementation are addressed in this section. It is essential to have this big picture clear in the mind before diving into the code examples in Section 8.4.

8.3.1 One-pass versus multipass implementations

The implementor of a geometric operation must decide during the initial phases of the design whether to use a one-pass or a multipass implementation.

In a one-pass implementation, the geometric operation is completed in one pass through the image. The nearest neighbor scale program shown in Section 8.4 is a one-pass implementation.

The pseudocode for a commonly used one-pass implementation using inverse mapping is as follows:

```
for (all x' in destination){
        for (all y' in destination) {
             inverse map the location (x',y') into (x,y) in source
             interpolate to get source(x,y) if necessary
             copy source(x,y) into destination(x',y')
        }
}
```

In image scaling, one-pass methods work very well. However, in more complex transformations, one-pass implementations, while leading to straightforward code, can also yield suboptimal implementations due to lost locality of reference. This happens because consecutive pixel accesses may touch pixels from different parts of the image. This is discussed further in Section 8.5.2.

In multipass implementations, the geometric operation is completed in more than one pass through the image. For example, a two-pass implementation makes two passes through the image: The first pass computes transformations for one axis and stores the result in an intermediate image. The second pass takes this intermediate image and computes the transformation along the other axis. The two-pass rotate program in Section 8.5 is an example of a two-pass implementation.

A two-pass implementation using inverse mapping requires three images: the source image, the destination image, and an intermediate image for storing the results of the first pass. Pseudocode for such an implementation is as follows:

```
for (all x' in intermediate image){
            inverse map the location (x',y) into (x,y) in source
            interpolate to get source(x,y) if necessary
            copy source(x,y) into intermediate(x',y)
       }
}
for (all y' in destination image){
            inverse map the location (x',y') into (x',y) in intermediate
            interpolate to get intermediate(x',y) if necessary
            copy intermediate(x',y) into destination (x',y')
       }
}
```

Two-pass methods are useful in transformations such as rotate and affine, where they offer performance advantages (due to their scan line access of pixels) compared to one-pass methods.

Some algorithms require more than two passes through the image—most notably, Paeth's three-pass algorithm for rotation [3] which avoids multiplications at the expense of the three passes. Clearly, this kind of an algorithm must have strong advantages to offset possible inefficiencies caused by the multiple passes.

8.3.2 *Table-driven implementations*

In table-driven implementations, information about the transformation is precomputed and stored in tables. This is done for situations where computing transformations on the fly is expensive. Therefore, these computations can be done once and the results stored in a table. For example, in implementing special-effects filters, computation of the destination (x',y') for every source (x,y) may be done once and stored in two two-dimensional tables. A fast special-effects filter can then be implemented by looking up the table values. An implementation of this is discussed in Section 8.7.

The pseudocode for this type of implementation is as follows:

```
for (all x' in destination){
      for (all y' in destination) {
           compute (x,y) from xtable[x'][y'] and ytable[x'][y']
           copy source(x,y) into destination(x',y')
      }
}
```

Another example of a table-driven implementation is interpolation using a general filter, as discussed in Section 8.4.3. As indicated earlier, storing the sampled and quantized values of the interpolation kernels yields significant computation savings.

Clearly, table-driven implementations can be embedded in either one-pass or two-pass implementations. For transformations that require expensive mathematical calculations, this method yields efficient results.

8.3.3 Clipping

During the implementation of geometric imaging operations, we must make sure that pixels that lie outside the boundaries of the image are never touched by the program. This is called *clipping*.

Consider the example of rotating an image shown in Figure 8.10. The pixels in the unshaded part of the destination map to locations outside the source image. If we are performing inverse mapping, we need to verify that the source coordinates obtained from the inverse mapping are within the bounds of the source image. During a forward mapping, we need to verify that every destination pixel location obtained using forward mapping is within the destination image.

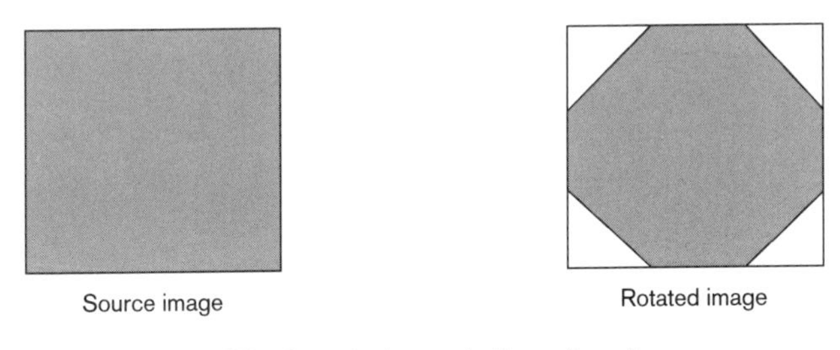

Figure 8.10 Clipping during rotation of an image

8.3.4 Boundary conditions

In Chapter 6, we saw that during one- or two-dimensional discrete convolution, incomplete overlap occurs at the edges of the source image. This requires the specification of edge conditions to use while performing convolution.

The same issue arises during the implementation of geometric imaging operations, because interpolation is a convolution. One of the four edge conditions discussed in Chapter 6—zero-fill, don't write, extend the edges, or reflect the edges—can be specified as the edge condition around the edges of the images.

Failure to specify the boundary conditions completely results in an incomplete software specification and product.

Having looked at the building blocks and the strategies for constructing geometric operations, we are now ready to look at the implementation of individual geometric operations.

8.4 Image scaling

Image scaling is one of the most important functions performed in many imaging products. It is useful for a variety of applications, serving to magnify and minify image data for viewing and storing. Because of this importance, we will consider several methods of implementing image scaling using a variety of interpolation methods and implementation strategies.

The transformation required to scale an image is given by:

$$\begin{bmatrix} x' \\ y' \end{bmatrix} = \begin{bmatrix} s & 0 \\ 0 & t \end{bmatrix} \begin{bmatrix} x \\ y \end{bmatrix} \tag{8.15}$$

where s and t are the scaling factors in the x and y directions, respectively. The corresponding inverse transformation is:

$$\begin{bmatrix} x \\ y \end{bmatrix} = \begin{bmatrix} \frac{1}{s} & 0 \\ 0 & \frac{1}{t} \end{bmatrix} \begin{bmatrix} x' \\ y' \end{bmatrix} \tag{8.16}$$

Inverse mapping is commonly used, and interpolation is needed after the transformation. The three most common methods of interpolation are nearest neighbor, bilinear, and general filtering (with an arbitrary interpolation kernel such as a bicubic).

The following coding examples demonstrate efficient image scaling using nearest neighbor, bilinear, and general interpolation. Examples of these types of scaling operations are shown in Figures 8.11 through 8.14.

All of the implementations are one-pass. Inverse mapping is used in each case. The code loops over all the pixels of the destination image, mapping back to the source image. In general, the algorithm follows this general structure:

```
for (each row in the destination)
     for (each column in the destination) {
          inverse map (row, column) to the source
          interpolate as necessary
          copy interpolated pixel to the destination
     }
```

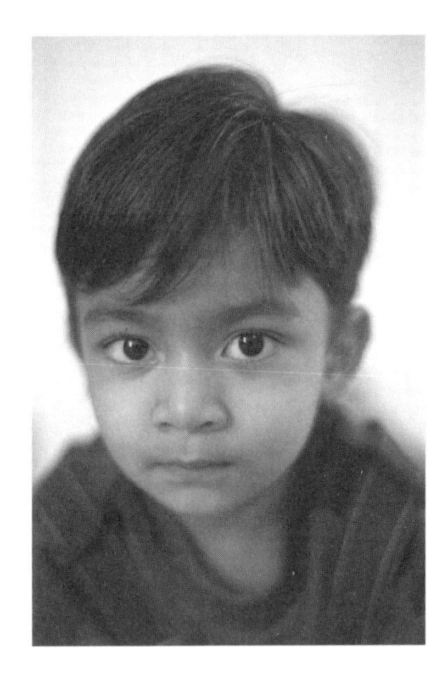

Figure 8.11 Ihsan original

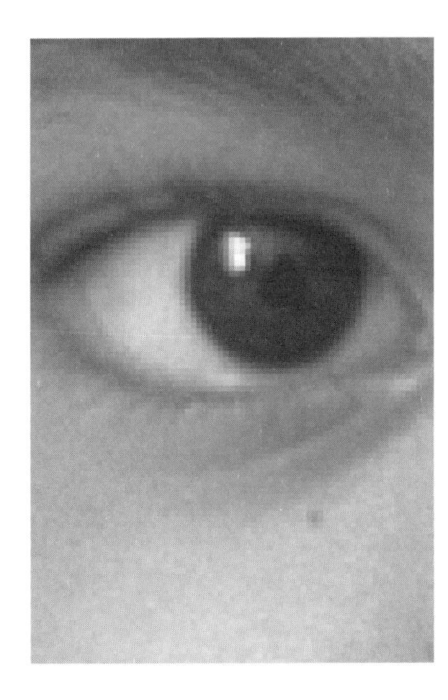

Figure 8.12 Ihsan, 7× zoom, nearest neighbor

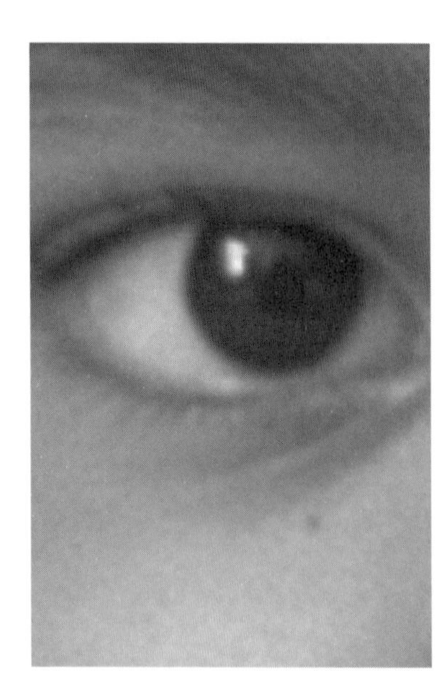

Figure 8.13 Ihsan, 7× zoom, bilinear

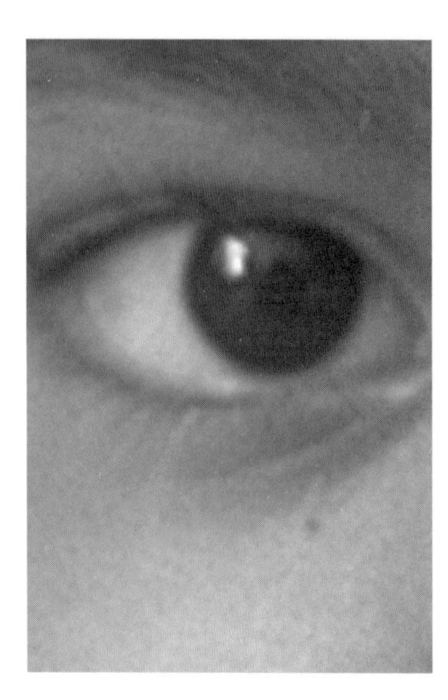

Figure 8.14 Ihsan, 7× zoom, bicubic

8.4.1 Nearest neighbor scale

Two examples of nearest neighbor scaling are shown. The first is close to a brute-force implementation but minimizes the number of multiplications. The second eliminates multiplications in the inner loop using a Bresenham-like algorithm.

In the first example, which can be considered a naive implementation, we take advantage of the following. Given an arbitrary pixel at the destination, Equation 8.15 indicates that we need to inverse map both the row and the column coordinate. However, if we fix the row in the destination image, we know that all pixels for that row come from the same row in the source image. Therefore, we compute the source row once for each destination row, thus eliminating one multiplication from the inner loop. Note, however, that this implementation should never be used in a real product because of the (expensive) division in the inner loop.

```
/*
        Function: pci_scale_nn
        Operation:Scale a source image into a destination image
                  scale factors are defined by image sizes
        Inputs:   src is the source image
        Output:   dest is the output image
*/

void
pci_scale_nn (pci_image *src, pci_image *dest)
{
 register int drows, dcols, srows, scols;
 register unsigned char *spixptr, *dpixptr, *dpix, *spix;
 register unsigned int slb, dlb;
 register int i, j;
        /* get parameters from the images */
        dpix = dpixptr = dest->data;
        spixptr = src->data;
        srows = src->height;
        scols = src->width;
        drows = dest->height;
        dcols = dest->width;
        slb = src->linebytes;
        dlb = dest->linebytes;
        for (i=0; i<drows; i++) {
             /* find the 1st pixel of the source row for this dest row */
             spix = spixptr + slb*((srows*i)/drows);
             for (j=0; j<dcols; j++) {
                  *dpixptr = *(spix + (scols*j)/dcols);
                  dpixptr++;
             }
             dpixptr = (dpix += dlb);
        }
}
```

In our second example, we introduce a variation of the Bresenham line-drawing algorithm [4] to track which source pixel to copy to the destination. Three new variables, `limit`, `frac`, and `counter` are introduced. `counter` tells us what is the exact pixel location of the inverse map. It is incremented by `frac` every time a new destination pixel is created. When `counter` exceeds `limit`, we need to read in a new source pixel.

Note that in the following code, we only apply this principle to row interpolation for the sake of clarity. It could equally well be applied to column interpolation; however, the gain is not great and it would make the code more confusing. Also, note that the code works only for magnification (not for minification).

```
/*
        Function: pci_scale_nn_bres
        Operation:Scale a source image into a destination image
                  Scale factors are defined by image sizes
                  Use Bresenham-like stepping to go through the rows
        Inputs:   src is the source image
        Output:   dest is the output image
*/

void
pci_scale_nn_bres (pci_image *src, pci_image *dest)
{
        register int drows, dcols, srows, scols;
        register unsigned char *spixptr, *dpixptr, *dpix, *spix;
        register unsigned int slb, dlb;
        register int i,j;
        register int limit, frac, counter; /* used in stepping */
        /* get parameters from the images */
        dpix = dpixptr = dest->data;
        spixptr = src->data;
        srows = src->height;
        scols = src->width;
        drows = dest->height;
        dcols = dest->width;
        slb = src->linebytes;
        dlb = dest->linebytes;
        /* check for magnification only */
        if ((dcols < scols) || (drows < srows))
             return;
        /* initialize the counters */
        limit = dcols;
        frac = scols % dcols;
        counter = dcols/2;
        for (i=0; i<drows; i++) {
             /* find the 1st pixel in the source row for this dest row */
             spix = spixptr + slb*((srows*i)/drows);
             while (dcols--) {
                  *dpixptr = *spix;
                  counter += frac;
                  if (counter > limit) {
                       counter -= limit;
                       spix++;
                  }
```

```
                dpixptr++;
            }
            dpixptr = (dpix += dlb);
            counter = dcols/2;
            dcols = dest->width;
        }
}
```

The algorithm is illustrated in Figure 8.15 for the case of a one-dimensional scale from a five-pixel row to a 16-pixel row. For this case, `limit` = 16, `frac` = 5, and `counter` is initialized to 8 (so that source pixels going over the 1/2 border map to the next destination pixel).

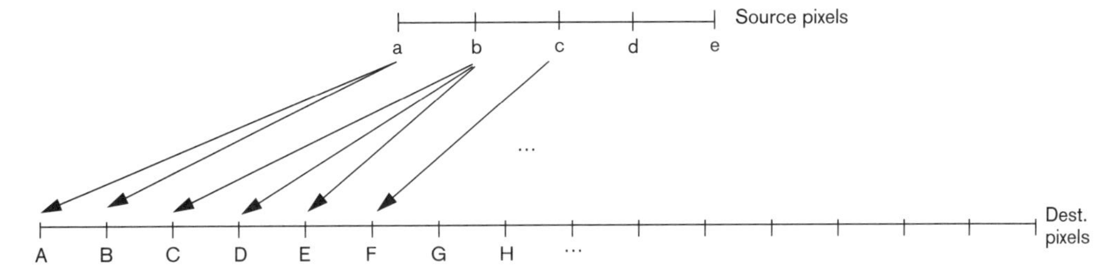

Figure 8.15 Scaling a row using the Bresenham algorithm: mapping of source pixels to destination pixels

The progression of the algorithm is shown in Table 8.1 for the first six pixels in the destination. The first column represents the loop counter, counting up to the number of pixels in the destination row. The next two columns show the pixel being copied to the destination pixel and the current value of counter. Once the source pixel is copied, the location in the source must be updated: This is done by updating the counter. If the pointer to the source pixel needs updating, this is done in column 5. The last column is a sanity-check column, which shows the true inverse mapped location and the source pixel nearest to that location.

Table 8.1 Progression of a Bresenham-like scaling algorithm

Loop counter	Pixel movement	Counter	Updated counter	Updated source pixel pointer	Inverse mapped location
0	a → A	8	13	a	0.0000 (A)
1	a → B	13	18 → 2	b	0.3125 (A)
2	b → C	2	7	b	0.6250 (B)
3	b → D	7	12	b	0.9375 (B)
4	b → E	12	17 → 1	c	1.2500 (B)
5	c → F	1	6	c	1.5625 (C)

Performance of the second implementation is significantly better than the brute-force algorithm shown in the earlier example, primarily because multiplications and divisions have been eliminated from the inner loop. For example, on a SPARCstation 10SX, which has a SuperSPARC processor with an integer multiply instruction, it runs over four times faster.

8.4.2 Bilinear scale

We now look at scaling an image using bilinear interpolation. The first implementation, once again a naive implementation, uses a straightforward bilinear interpolation for every pixel in the destination.

```
/*
        Function: pci_scale_bl
        Operation:Scale a source image into a destination image using
                  bilinear interpolation.
                  Scale factors are defined by image sizes
                  The quantities p and q are as defined in Sec. 8.2.2
        Input:    src is the source image
        Output:   dest is the output image
*/

void
pci_scale_bl (pci_image *src, pci_image *dest)
{
        register int drows, dcols, srows, scols;
        register unsigned char *spixptr, *dpixptr, *spix1, *spix2, *dpix;
        register unsigned int slb, dlb;
        register int i, j;
        float p,q,p1,q1;            /* as shown in Figure 8.9 */
        float xfac, yfac;           /* scale factors */
        float xfac_inv, yfac_inv;   /* inverse scale factors */
        float src_y_f, src_x_f;     /* inv mapped point in src (float) */
        int src_y_i, src_x_i;       /* inv mapped point in src (int) */
        register unsigned char a,b,c,d; /* src pixels used in interp. */
        float result;               /* interpolated pixel value */

        /* get parameters from the images */
        dpix = dpixptr = dest->data;
        spixptr = src->data;
        srows = src->height;
        scols = src->width;
        drows = dest->height;
        dcols = dest->width;
        slb = src->linebytes;
        dlb = dest->linebytes;

        /* compute scale factors and inverse scale factors */
        xfac = (float) dcols/ (float) scols;
        yfac = (float) drows/ (float) srows;
```

```
        xfac_inv = 1.0/xfac;
        yfac_inv = 1.0/yfac;

        /* perform the zoom by inverse mapping from dest to source */
        for (i=0; i<drows-1; i++) {
             src_y_f = i * yfac_inv;
             src_y_i = (int) src_y_f;
             q = src_y_f - (float) src_y_i;
             q1 = 1.0 - q;
             spix1 = spixptr + src_y_i*slb;
             spix2 = spixptr + (src_y_i+1)*slb;
             for (j=0; j<dcols-1; j++) {
             src_x_f = j * xfac_inv;
             src_x_i = (int) src_x_f;
             p = src_x_f - (float) src_x_i;
             p1 = 1.0 - p;

             /* get the four pixels */
             a = *(spix1 + src_x_i);
             b = *(spix1 + src_x_i + 1);
             c = *(spix2 + src_x_i);
             d = *(spix2 + src_x_i + 1);

             /* compute the interpolated pixel value */
             result = (((float)a*p1 + (float)b*p)*q1 +
                   ((float)c*p1 +(float)d*p)*q);

             *dpix++ = (unsigned char) result;
             }
        dpix = (dpixptr += dlb);
        }
}
```

The second example implementation of bilinear scale optimizes based on the fact that the quantities p and $(1 - p)$, as shown in Figure 8.9, do not change from row to row. Therefore, we compute these once for all the pixels in a row, and then store them in a buffer. The buffer is an array of the structure:

```
typedef struct {
        int x_i;
        float p;
        float p1;
} row_values;
```

where `x_i` is the inverse mapped value of the source column in integer, and `p1` is $1 - p$.

```
/*
        Function: pci_scale_bl2
        Operation:Scale a source image into a destination image
                  using bilinear interpolation.
                  Scale factors are defined by image sizes.
                  Uses faster algorithm.
                  The quantities p and q are as defined in Sec. 8.2.2.
```

```
        Input:    src is the source image
        Output:   dest is the output image
*/

typedef struct {
        int x_i;                    /* inv mapped source column (int)*/
        float p;                    /* the quantity p from Figure 8.9 */
        float p1;                   /* 1-p */
} row_values;
void
pci_scale_bl2(pci_image *src, pci_image *dest)
{
        register int drows, dcols, srows, scols;
        register unsigned char *spixptr, *dpixptr, *spix1, *spix2, *dpix;
        register unsigned int slb, dlb;
        register int i, j;
        float p,q,p1,q1;            /* as shown in Figure 8.9 */
        float xfac, yfac;           /* scale factors */
        float xfac_inv, yfac_inv;   /* inverse scale factors */
        float src_y_f, src_x_f;     /* inv mapped point in src (float) */
        int src_y_i, src_x_i;       /* inv mapped point in src (int) */
        register unsigned char a,b,c,d; /* src pixels used in interp.*/
        row_values *r, *rptr;       /* buffer for storing p and p1 */
        float x_f;                  /* inv mapped source column (float) */
        float result;               /* interpolated pixel value */
        /* get parameters from the images */
        dpix = dpixptr = dest->data;
        spixptr = src->data;
        srows = src->height;
        scols = src->width;
        drows = dest->height;
        dcols = dest->width;
        slb = src->linebytes;
        dlb = dest->linebytes;
        /* compute scale factors and inverse scale factors */
        xfac = (float) dcols/ (float) scols;
        yfac = (float) drows/ (float) srows;
        xfac_inv = 1.0/xfac;
        yfac_inv = 1.0/yfac;
        /* allocate buffer for storing the values of m and m1 */
        rptr = r = (row_values *) malloc (sizeof(row_values) * dcols);
        /* fill up the buffer once; to be used for each row */
        for (i=0; i<dcols; i++) {
             x_f = i*xfac_inv;
             rptr->x_i = (int)x_f;
             rptr->p = x_f - (float) rptr->x_i;
             rptr->p1 = 1.0 - rptr->p;
             rptr++;
        }

        /* perform the zoom by inverse mapping from dest to source */
        for (i=0; i<drows-1; i++) {
             src_y_f = i * yfac_inv;
             src_y_i = (int) src_y_f;
```

```
                q = src_y_f - (float) src_y_i;
                q1 = 1.0 - q;
                spix1 = spixptr + src_y_i*slb;
                spix2 = spixptr + (src_y_i+1)*slb;
                rptr = r;
                for (j=0; j<dcols-1; j++,rptr++) {
                     src_x_i = rptr->x_i;
                     p = rptr->p;
                     p1 = rptr->p1;

                     /* get the four pixels */
                     a = *(spix1 + src_x_i);
                     b = *(spix1 + src_x_i + 1);
                     c = *(spix2 + src_x_i);
                     d = *(spix2 + src_x_i + 1);
                     /* compute the interpolated pixel value */
                     result = (((float)a*p1 + (float)b*p)*q1
                            + ((float)c*p1 +(float)d*p)*q);

                     *dpix++ = (unsigned char) result;
                }
                dpix = (dpixptr += dlb);
          }
          free(r);
}
```

Before leaving bilinear interpolation, we note that bilinear scaling with integer scale factors, especially for scale factors of 2 and 0.5, can be made to run significantly faster than arbitrary scaled bilinear zooms. For the case of a 2× bilinear zoom, as shown in Figure 8.16, the destination pixels $l = (a + b) >> 1$, $n = (c + d) >> 1$, and $m = (l + n) >> 1$, where a, b, c, and d are the source pixels. Therefore, all multiplications can be avoided. In addition, because we know exactly which pixels to load for performing the interpolation, we can optimize by moving the largest possible chunks of pixels that can be moved by a given instruction.

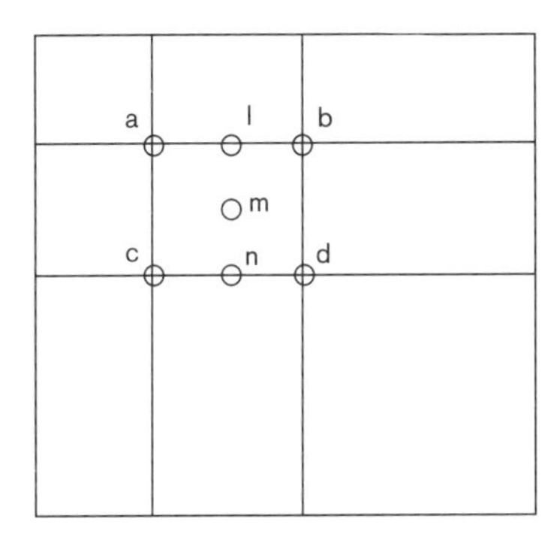

Figure 8.16 2× bilinear zoom

8.4.3 General filtered scale

While nearest neighbor and bilinear scaling functions are both popular and reasonably easy to implement, they do not always yield the highest quality in the scaled image. When quality is of utmost importance, general filtered scaling should be used. In this scheme, an arbitrary interpolation kernel, such as a cubic function, is specified, and the image is scaled using this interpolation kernel. While computationally more expensive than either nearest neighbor or bilinear scaling, this method gives us a smoother looking scaled image.

Given an interpolation kernel $h(x)$, a two-pass implementation scales the image along each row followed by scaling along each column (or vice versa). An example of scaling along rows using a general interpolation kernel is shown in Figure 8.17.

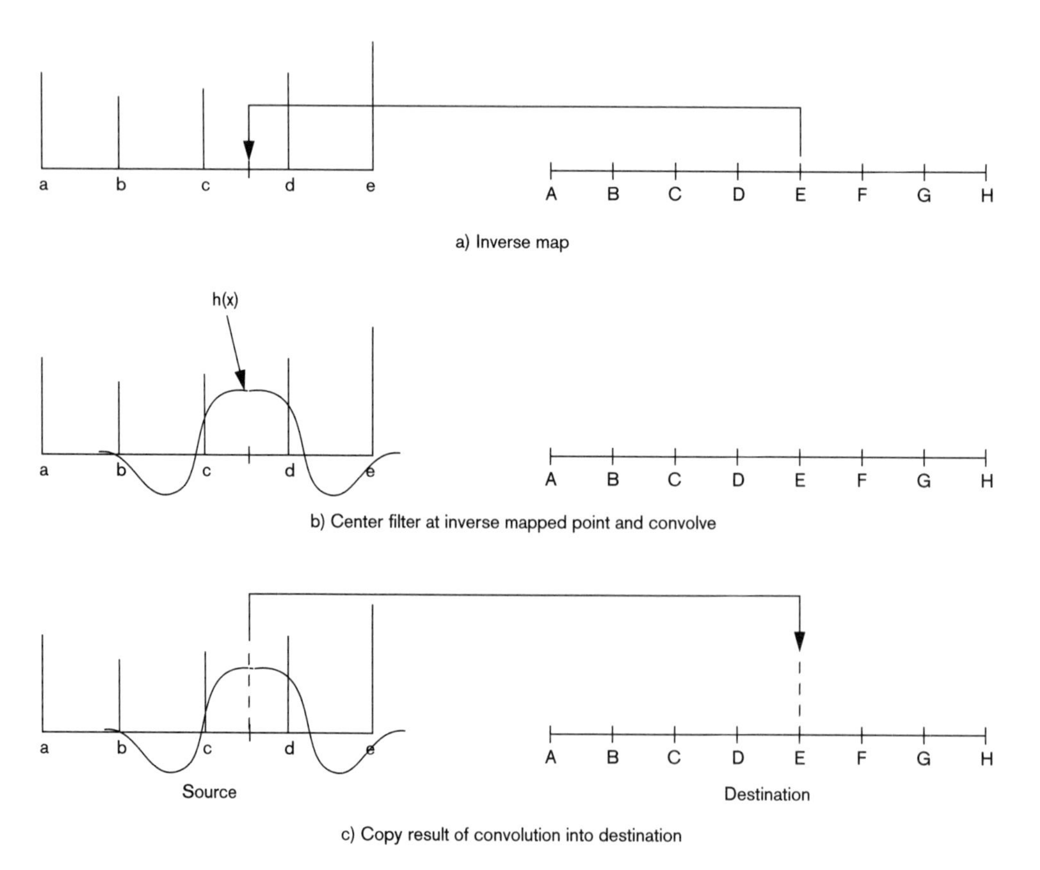

Figure 8.17 Scaling using general filtered interpolation

When scaling along rows, the x coordinate of each destination pixel must be inverse mapped to the source. Then, $h(x)$ must be centered on the exact x location in the source, and the values of $h(x)$ coinciding with the source input pixels in the support region must be computed. The convolution between the values of $h(x)$ and the source pixels is computed, and the result of the convolution is copied to the destination pixel location.

From the intermediate image thus created, a similar set of steps is performed in the y direction to complete the scaling operation.

The computation of $h(x)$ at several values for each destination pixel is an expensive task. An alternative suggested by Ward and Cok [5] uses table lookup. The distance between two adjacent pixels is divided into a finite number of subpixel positions, $(x_0, \ldots, x_n)$. For each subpixel position, the contribution of the filter over the set of pixels in the region of support is computed and stored in a table.

For example, in Figure 8.18, the sampled values of the kernel for its subpixel location (0.4) for each of the four pixels, $h(-0.4)$, $h(-1 -0.4)$, $h(1 -0.4)$, and $h(2 -0.4)$, in the support region are stored. A similar set of values is stored for positioning the kernel in the other subpixel locations.

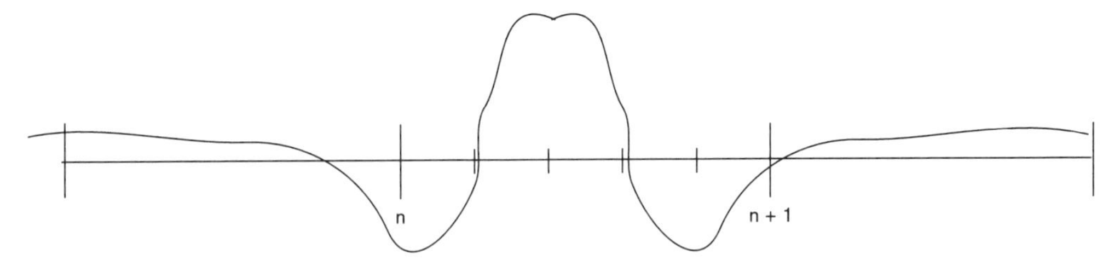

Figure 8.18 An interpolation filter with a region of support of four pixels

When the inverse mapping takes place, the subpixel position closest to the actual x position in the source is found, and the kernel is assumed to be centered there. The convolution is then performed as we have described before.

The following code sample shows an implementation of a general interpolated zoom using this technique. The region of support has been predefined to be five for the sake of code clarity. However, the program can easily be modified to make this a variable.

The main routine, `pci_scale_bc()`, calls a sequence of routines:

- `pci_alloc_image()` allocated the intermediate image
- `pci_scale_bc_htab()` computes table for horizontal scale
- `pci_scale_bc_h()` scales the image horizontally
- `pci_scale_bc_vtab()` computes the table for vertical scale
- `pci_scale_bc_v()` scales the image vertically

The structure pci_ftab{} has two elements: the offset of the first source pixel, starting from the beginning of the source row (or column) used in the computation of the current destination pixel, and an array containing the five values of the interpolation kernel where it overlaps with the source pixels in the region of support.

The interpolation kernel chosen in this case is the cubic function of Equation 8.14. Any other suitable kernel could be used as well.

The main routine is as follows:

```
/*
        Function: pci_scale_bc
        Operation:This function performs a general interpolated zoom on the
             source pixel to get the destination image.
             Zooming is done in two passes. In the first pass, the image
             is zoomed in the horizontal direction and placed in a buffer.
             In the second pass, the buffered image is zoomed in the
             vertical direction.
             Code below considers a bicubic zoom, i.e., one that uses
             cubic interpolation in the horizontal and vertical directions.
             As before, source and destination image sizes determine
             scale factor.
        Input:    src is the source image
        Output:   dest is the destination image
*/

/* #defines related to this scaling algorithm */
#define SPACC   5                  /* SubPixelACCuracy = 5 */
#define SUPPORT 5                  /* region of support for filter */
#define SCA     4096               /* filter coeffs scaled to 12 bits */
#define SHIFT   12
#define HALF    (1<<11)
#define EDGE   2              /* two edge pixels at boundary - no write */
void
pci_scale_bc (pci_image *src, pci_image *dest)
{
        register int drows, dcols, srows, scols;
        register unsigned int slb, dlb;
        register int i, j;
        float result;
        row_values *r, *rptr;
        pci_ftab *htabptr, *vtabptr;
        int filter[SPACC*SUPPORT];
        pci_image inter;

        /* get parameters from the images */
        srows = src->height;
        scols = src->width;
        drows = dest->height;
        dcols = dest->width;

        /* allocate intermed image: width(dest)X height(src)X1-banded */
        pci_alloc_image(&inter, srows, dcols, 1);
```

```
        /* allocate memory for the tables */
        htabptr = (pci_ftab *) malloc(sizeof(pci_ftab) * dcols);
        vtabptr = (pci_ftab *) malloc(sizeof(pci_ftab) * drows);

        /* compute table for one row */
        pci_scale_bc_htab(src,dest,htabptr);

        /* perform the horizontal scale */
        pci_scale_bc_h(src,&inter,htabptr);

        /* compute table for one column */
        pci_scale_bc_vtab(&inter,dest,vtabptr);

        /* perform the vertical zoom */
        pci_scale_bc_v(&inter,dest,vtabptr);
}
```

The routine `pci_scale_bc_htab()` computes the horizontal filter table for interpolation as follows:

```
/*
        Function: pci_scale_bc_htab
        Operation:Create filter table for horizontal interpolation
        Input:    src is the source image
                  dest is the destination image
        Output:   htable is the horizontal table
*/

void
pci_scale_bc_htab(pci_image *src, pci_image *dest, pci_ftab *htable)
{
        int i;
        int scols, dcols;
        unsigned char *sptr;
        float xfac, inv_xfac;
        float fsx;              /* dest col mapped back to src */
        int sx;                 /* integer col in src */
        float frac;             /* fractional part in src */
        float sub;              /* subpixel position of filter */
        sptr = src->data;
        scols = src->width;
        dcols = dest->width;
        inv_xfac = (float)scols/(float)dcols;

        /* fill up the table */
        for (i=0; i<dcols; i++) {
             fsx = i*inv_xfac;
             frac = fsx - (int) fsx;
             /* find the closest subpixel position for frac */
             sub = (float) ((int)(frac*SPACC + 0.5)/(float)SPACC);
             htable->src_off = fsx;
             htable->coeff[0] = (int) (pci_cubic(-sub-2)*SCA);
             htable->coeff[1] = (int) (pci_cubic(-sub-1)*SCA);
```

```
            htable->coeff[2] = (int) (pci_cubic(sub)*SCA);
            htable->coeff[3] = (int) (pci_cubic(1-sub)*SCA);
            htable->coeff[4] = (int) (pci_cubic(2-sub)*SCA);
            htable++;
        }
}
```

Similarly, the routine `pci_scale_bc_vtab()` computes the vertical filter table for interpolation as follows:

```
/*
        Function: pci_scale_bc_vtab
        Operation:Create filter table for vertical interpolation
        Input:    src is the source image
                  dest is the destination image
        Output:   vtable is the vertical table
*/

void
pci_scale_bc_vtab(pci_image *src, pci_image *dest, pci_ftab *vtable)
{
        int i;
        int srows, drows;
        unsigned char *sptr;
        float yfac, inv_yfac;
        float fsy;                  /* dest row mapped back to src */
        int sy;                     /* integer row in src */
        float frac;                 /* fractional part in src */
        float sub;                  /* subpixel position of filter */
        sptr = src->data;
        srows = src->height;
        drows = dest->height;
        inv_yfac = (float) srows/(float)drows;

        /* fill up the table */
        for (i=0; i<drows; i++) {
            fsy = i*inv_yfac;
            frac = fsy - (int) fsy;

            /* find the closest subpixel position for frac */
            sub = (float) ((int)(frac*SPACC + 0.5)/(float)SPACC);
            vtable->src_off = fsy;
            vtable->coeff[0] = (int) (pci_cubic(-sub-2)*SCA);
            vtable->coeff[1] = (int) (pci_cubic(-sub-1)*SCA);
            vtable->coeff[2] = (int) (pci_cubic(sub)*SCA);
            vtable->coeff[3] = (int) (pci_cubic(1-sub)*SCA);
            vtable->coeff[4] = (int) (pci_cubic(2-sub)*SCA);
            vtable++;
        }
}
```

Both pci_scale_bc_vtab() and pci_scale_bc_htab() utilize the routine pci_cubic() for generating the filter coefficients. This routine can be modified to use any other kind of filter.

```
/*
        Function:  pci_cubic
        Operation: Returns a floating-point value for bicubic function
*/

#define A -1.
float pci_cubic(float x)
{
        if (x<0)
                x = -x;
        if (x>=2)
                return (0);
        else if (x>=1)
                return (A*x*x*x -5*A*x*x +8*A*x -4*A);
        else
                return ((A+2)*x*x*x -(A+3)*x*x +1);
}
```

The routines pci_scale_bc_h() and pci_scale_bc_v() perform the actual scaling in horizontal and vertical directions:

```
/*
        Function: pci_scale_bc_h
        Operation:Scale horizontally using a table of filters
        Input:    src is the source image
                  inter is the destination (typically an intermediate) image
        Output:   htable is the horizontal table
*/

void
pci_scale_bc_h(pci_image *src, pci_image *inter, pci_ftab *htable)
{
        int srows,icols;
        int j,k;
        unsigned char *s, *sptr, *i, *iptr;
        int result;
        int slb, ilb;
        pci_ftab *hptr;
        s = src->data;
        i = inter->data;
        srows = src->height;
        icols = inter->width;
        ilb = inter->linebytes;
        slb = src->linebytes;
        for (j=0; j<srows; j++) {
             hptr = htable+EDGE; /* no-write condition at edge */
             iptr = i;
             for (k=2; k<icols; k++) {
```

```
                    /* note that src_off corresponds to the offset
                    of the first src pixel to the left of the current
                    subpixel position (i.e., dest mapped back to src) */
                    sptr = s + hptr->src_off;
                    /* compute convolution using info from htable */
                    result =  (int) sptr[-2] * hptr->coeff[0]
                         + (int) sptr[-1] * hptr->coeff[1]
                         + (int) sptr[0]  * hptr->coeff[2]
                         + (int) sptr[1]  * hptr->coeff[3]
                         + (int) sptr[2]  * hptr->coeff[4];

                    /* convert to 8 bits */
                    result = (result + HALF)>>SHIFT;
                    if (result < 0) result = 0;
                    if (result > MAXBYTE) result = MAXBYTE;
                    *iptr = result;
                    iptr++;
                    hptr++;
               }
               s += slb;
               i += ilb;
          }
}

/*
          Function: pci_scale_bc_v
          Operation:Scale vertically using a table of filters
          Input:    inter is the source (typically an intermediate) image
                    dest is the destination image
          Output:   vtable is the vertical table
*/

void
pci_scale_bc_v(pci_image *inter, pci_image *dest, pci_ftab *vtable)
{
          int icols, drows;
          int j,k;
          unsigned char *d, *dptr, *i, *iptr;
          int result;
          int dlb, ilb;
          pci_ftab *vptr;

          /* copy image information into local area */
          d = dest->data;
          i = inter->data;
          icols = inter->width;
          drows = dest->height;
          ilb = inter->linebytes;
          dlb = dest->linebytes;

          /* do all the columns */
          for (j=0; j<icols; j++) {
               vptr = vtable + EDGE;  /* no-write condition at edge */
               dptr = d;
```

```
            /* go over all the pixels */
            for (k=2; k<drows; k++) {
                /* note that src_off corresponds to the offset
                of the first src pixel to the left of the current
                subpixel position (i.e., dest mapped back to src) */
                iptr = i + vptr->src_off*ilb;

                /* compute convolution using info from vtable */
                result =  (int) iptr[-2*ilb] * vptr->coeff[0]
                    + (int) iptr[-1*ilb] * vptr->coeff[1]
                    + (int) iptr[0] * vptr->coeff[2]
                    + (int) iptr[ilb] * vptr->coeff[3]
                    + (int) iptr[2*ilb] * vptr->coeff[4];

                /* convert to 8 bits */
                result = (result + HALF)>>SHIFT;
                if (result < 0) result = 0;
                if (result > MAXBYTE) result = MAXBYTE;

                *dptr = result;

                dptr += dlb;
                vptr++;
            }
            d++;
            i++;
        }
}
```

The above example freely uses floating-point arithmetic and multiplications to perform inverse mapping and interpolation. However, both of these are computationally expensive operations. The Bresenham technique we saw earlier can be applied once again to reduce floating-point operations and multiplications. An example is given by Wolberg [1].

This concludes our study of image scaling. We now proceed to other image geometric operations, particularly image rotation.

8.5 Image rotation

Image rotation is of interest in many applications, such as image registration and correction for scanner mistakes. After looking at the transformations necessary to perform useful rotations, we will look at two methods of implementation. The first is a straightforward implementation of the transformation in one pass through the image. The second implementation method uses a two-pass scan line algorithm.

The transformation for image rotation by a counterclockwise angle θ about the origin (0,0) of the image is as follows. Note that we use homogeneous coordinates here because of the simplicity of notation in the following derivation.

$$\begin{bmatrix} x' \\ y' \\ 1 \end{bmatrix} = \begin{bmatrix} \cos\theta & -\sin\theta & 0 \\ \sin\theta & \cos\theta & 0 \\ 0 & 0 & 1 \end{bmatrix} \begin{bmatrix} x \\ y \\ 1 \end{bmatrix} \tag{8.17}$$

For rotating about a point (a,b), the image must be translated by (a,b), rotated, and then translated by $(-a,-b)$. The corresponding transformations are:

$$\begin{bmatrix} 1 & 0 & a \\ 0 & 1 & b \\ 0 & 0 & 1 \end{bmatrix} \begin{bmatrix} \cos\theta & -\sin\theta & 0 \\ \sin\theta & \cos\theta & 0 \\ 0 & 0 & 1 \end{bmatrix} \begin{bmatrix} 1 & 0 & -a \\ 0 & 1 & -b \\ 0 & 0 & 1 \end{bmatrix} = \begin{bmatrix} \cos\theta & -\sin\theta & -a\cos\theta + b\sin\theta + a \\ \sin\theta & \cos\theta & -a\sin\theta - b\cos\theta + b \\ 0 & 0 & 1 \end{bmatrix} \tag{8.18}$$

When implementing using an inverse mapping, the corresponding inverse transformation is:

$$\begin{bmatrix} \cos\theta & \sin\theta & -a\cos\theta - b\sin\theta + a \\ -\sin\theta & \cos\theta & a\sin\theta - b\cos\theta + b \\ 0 & 0 & 1 \end{bmatrix} \tag{8.19}$$

In image rotation, we have to explicitly deal with clipping. Therefore, in the following coding examples, we have included a test to perform clipping.

8.5.1 One-pass rotation

The following code performs a one-pass rotation using the transformations shown in Equations 8.17 through 8.19.

```
/*
        Function:  pci_rotate
        Operation: Rotates an image about (a,b) using
                   nearest neighbor interpolation.
                   The inverse transformation used is
                   given in Section 8.5 Eq. 8.19
        Inputs:    src is the source image
                   degrees is the angle to rotate
                   (xorig, yorig) is the origin of rotation
        Output:    dest is the rotated image
*/

void
pci_rotate (pci_image *src, pci_image *dest, int xorig, int yorig, float
degrees)
{
        double cosine, sine, radians;
        unsigned char *dpix, *spix, *dpixptr, *spixptr;
        int rows, cols, ccols;
        float xoff, yoff;
```

```
        register float cosf, sinf;
        register float src_xf, src_yf;
        int i,j;
        register int src_x, src_y;
        register int xoffset, yoffset;
        int slb, dlb;

        /* get parameters from the images */
        rows = src->height;
        cols = src->width;
        dpix = dpixptr = dest->data;
        spix = spixptr = src->data;
        slb = src->linebytes;
        dlb = dest->linebytes;

        /* convert angle into radians */
        radians = M_PI*(double)degrees/180.0;

        /* get the cosine and sine */
        cosine = cos(radians);
        sine = sin(radians);
        cosf = (float)cosine;
        sinf = (float)sine;

        /* precompute transformation parameters */
        xoff = -(float)xorig*cosf - (float)yorig*sinf + (float)xorig;
        yoff = (float)xorig*sinf - (float)yorig*cosf + (float)yorig;
        xoffset = (int)xoff;
        yoffset = (int)yoff;

        /* now inverse map the destination into the source */
        for (i=0; i<rows; i++) {
             for (j=0; j<cols; j++) {
                  src_xf = cosf * j + sinf * i + xoffset;
                  src_yf = (-sinf)*j + cosf*i + yoffset;
                  ROUNDDOWN(src_x,src_xf)
                  ROUNDDOWN(src_y,src_yf)
                  /* clip the source pixel */
                  if ((src_x>=0) && (src_x<cols) &&
                       (src_y>=0) && (src_y<rows))
                        *dpixptr = *(spix + src_y*slb + src_x);
                        dpixptr++;
             }
             dpixptr = (dpix += dlb);
        }
}
```

8.5.2 *Multipass rotation*

One disadvantage of the above implementation of image rotation is that scan line ordering of source pixels is not preserved. This has negative implications for performance. Because of the loss of locality of reference, caching advantages are lost in software implementations. In addition, hardware implementations often require simple addressing schemes requiring access to pixels in scan line order.

A solution to this was provided by a two-pass algorithm by Catmull and Smith [6]. The idea is as follows: We know that the rotation transformation is given by:

$$\begin{bmatrix} x' \\ y' \\ 1 \end{bmatrix} = \begin{bmatrix} \cos\theta & -\sin\theta & 0 \\ \sin\theta & \cos\theta & 0 \\ 0 & 0 & 1 \end{bmatrix} \begin{bmatrix} x \\ y \\ 1 \end{bmatrix} \tag{8.20}$$

However, instead of computing both x' and y' in one pass, we compute the image with coordinates (x',y) first, so that

$$\begin{aligned} x' &= x\cos\theta - y\sin\theta \\ y' &= y \end{aligned} \tag{8.21}$$

This is the intermediate image, where only the x coordinates of the pixels have changed. In the second pass, the (x', y') image is computed from the intermediate image. Note, however, that we cannot use $y' = x\sin\theta + y\sin\theta$ because x has changed. The new value of x is

$$x = \frac{x' + y\sin\theta}{\cos\theta} \tag{8.22}$$

Therefore, y' is computed in the second pass (when x' does not change) as:

$$y' = x\sin\theta + y\cos\theta = \frac{x'\sin\theta + y}{\cos\theta} \tag{8.23}$$

In the following coding example, we show how to perform a two-pass rotation using this algorithm. Note that the example takes into account the center of rotation, which makes the equations a little more involved.

```
/*
        Function: pci_rotate_2pass

        Operation:
        Rotate an image about (a, b) using the Catmull/Smith two-pass
        algorithm. Use nearest neighbor interpolation.
        First we derive the forward mapping.
        Let (x,y) be the source pixel and (x',y') be the destination
        pixel coordinates.
        The forward transform is equivalent to translating the image
        by (a,b), doing the rotation, and translating it by (-a, -b).

        Forward = 1 0 a   * c -s 1    *  1 0 -a where c = cos and s = sin
                  0 1 b     s  c 1       0 1 -b
                  0 0 1     0  0 1       0 0  1

        Then, x'   = Forward   * x
              y'                 y
```

```
        Multiplying the matrix we get:

        Forward = c   -s   A   where A = -ac + bs + a
                  s    c   B   where B = -as - bc + b
                  0    0   1
        Now, the two-pass method. In the first pass, we keep the y
        coordinate constant and vary only the x coordinate. This gives
        us an intermediate image (x', y) where

             x' = cx - sy + A
             y' = y

        In the second pass we want to compute y'. However, we cannot
        use the equation y = sx + cy + B directly because x has changed
        Using the new x  = (x' + y - A)/c, we get

             y' = (x's + x - As + Bc)/c

        Now we really want to implement this using an inverse map
        because the forward map can yield holes in the image. Then,
        the inverse map for the first pass is:

             x = (x' + ys - A)/c
             y = y;

        The inverse map for the second pass is:

             x = x'
             y = y'c - xs + As - Bc

        These are the equations implemented in the code.

        Inputs:   src is the source image
                  degrees is the angle to rotate
                  (xorig, yorig) is the origin of rotation

        Outputs:  dest is the destination image
                  inter is the intermediate image
*/

void
pci_rotate_2pass(pci_image *src, pci_image *inter, pci_image *dest, int
xorig, int yorig, float degrees)
{
        register int rows, cols;
        register unsigned char *spixptr, *ipixptr, *dpixptr;
        register unsigned char *dpix, *ipix, *spix;
        register unsigned int slb, ilb, dlb;
        register int i, j;
        double cosine, sine, radians, cosinv;
        double A, B;
        float sinf, cosf;
        register float src_xf, src_yf;
        register int src_x, src_y;
        float K;
```

```
/* get parameters from the images */
dpix = dpixptr = dest->data;
ipix = ipixptr = inter->data;
spix = spixptr = src->data;
rows = src->height;
cols = src->width;
slb = src->linebytes;
ilb = inter->linebytes;
dlb = dest->linebytes;

/* convert angle into radians and get cosine and sine */
radians = M_PI*(double)degrees/180.0;
cosine = cos(radians);
cosinv = 1./cosine;
sine = sin(radians);
cosf = (float)cosine;
sinf = (float)sine;

/* compute A, B */
A = -(float)xorig*cosf + (float)yorig*sinf + (float)xorig;
B = -(float)xorig*(sinf) - (float)yorig*cosf + (float) yorig;
/*
* .... first pass ....
* for every point (x',y) in the intermediate image, we want to
* fetch the pixel at ((x' + ys - A)/c, y) in the source image
* source pixel's x can be written as x'*cosinv + K where K is
* (ys-A)*cosinv = constant for each row. Compute K once for each row
*/

for (i=0; i<rows; i++) {
     /* precompute things that do not change in one row */
     spixptr = spix + i*slb;
     ipixptr = ipix + i*ilb;
     K = ((float)i*sinf - (float)A) * (float)cosinv;

     /* now go over the columns for this row */
     for (j=0; j<cols; j++) {
          src_xf = j*cosinv + K;
          ROUNDDOWN(src_x, src_xf)
          /* clip the source pixel */
          if ((src_x >= 0) && (src_x < cols))
               *ipixptr = *(spixptr + src_x);
          ipixptr++;
     }
}

/*
 * .... second pass ....
 * for every point (x',y') in the destination image, we want to
 * fetch the pixel at (x',y'c-x's+As-Bc) in the intermediate
 * image
 */
```

```
        for (j=0; j<cols; j++) {
            /* precompute things that do not change in one col */
            ipixptr = ipix + j;
            dpixptr = dpix + j;
            K = -(float)j*sinf + A*sinf - B*cosf;

            /* now go over the rows for this column */
            for (i=0; i<rows; i++) {
                src_yf = i*cosf + K;
                ROUNDDOWN(src_y,src_yf)

                /* clip the source pixel */
                if ((src_y >= 0) && (src_y < rows))
                    *dpixptr = *(ipixptr + src_y*ilb);
                dpixptr += dlb;
            }
        }
}
```

Examples of two-pass rotation of an image are shown in Figures 8.19 through 8.21.

Figure 8.19 Museum original

Figure 8.20 Result of pass one

Figure 8.21 Result of pass two

Another algorithm for multipass rotation was developed by Paeth [3]. It is a three-pass algorithm based on skewing the image; scaling never has to take place. The idea is to decompose the rotation matrix as follows:

$$\begin{bmatrix} \cos\theta & -\sin\theta \\ \sin\theta & \cos\theta \end{bmatrix} = \begin{bmatrix} 1 & -\tan\frac{\theta}{2} \\ 0 & 1 \end{bmatrix} \begin{bmatrix} 1 & 0 \\ \sin\theta & 1 \end{bmatrix} \begin{bmatrix} 1 & -\tan\frac{\theta}{2} \\ 0 & -1 \end{bmatrix} \tag{8.24}$$

While this is a three-pass algorithm, the main advantage of this algorithm is that it avoids scaling (and thus expensive multiplication operations).

8.6 *Affine transformation*

Both the scaling and rotation operations studied in Sections 8.4 and 8.5 are special cases of a more general transformation called the affine transformation, whose equation is as follows:

$$\begin{bmatrix} x' \\ y' \\ 1 \end{bmatrix} = \begin{bmatrix} a & b & c \\ d & e & f \\ 0 & 0 & 1 \end{bmatrix} \begin{bmatrix} x \\ y \\ 1 \end{bmatrix} \tag{8.25}$$

The transformation combines translation, rotation, scaling, and shearing into one operation. The values of the constants needed for the different operations are shown in Table 8.2.

Table 8.2 Values of coefficients in affine matrix

Function	Constant values
scale	a = xscale, e = yscale, rest = 0
rotate	a = cosθ, b = –sinθ, d = cosθ, e = sinθ, rest = 0
translate	a = 1, c = xtranslate, e = 1, f = ytranslate, rest = 0
shear	a = 1, b = xshear, d = yshear, e = 1, rest = 0

The main advantage of the affine transformation is its generality. It can be used for a combinations of the individual functions noted above by concatenating the appropriate affine matrices.

An example of an affine transformation is shown in Figure 8.22 (boat original) and Figure 8.23, which is the result of affine transformation using the following affine matrix:

$$M = \begin{bmatrix} 2.0 & 1.0 & 0.0 \\ 1.0 & 2.0 & 0.0 \end{bmatrix}$$

Figure 8.22 Boat original

Figure 8.23 Affine transformation

8.7 Image transposition

In image transposition, the goal is to reflect the image along a given axis. There are four flavors of transposing an image, determined by the axis. These axes are shown in Figure 8.24.

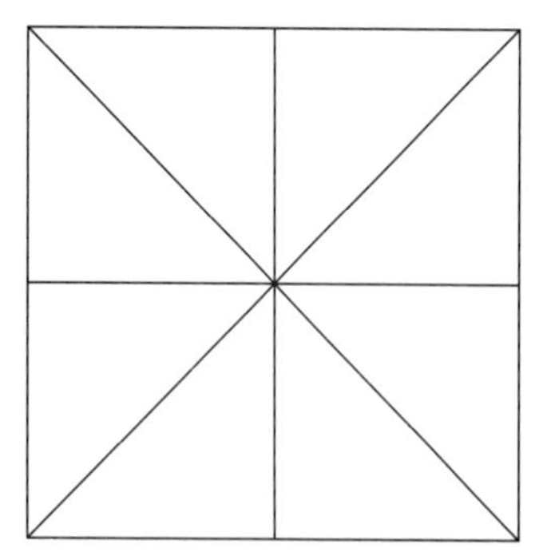

Figure 8.24 Axes for transposing an image

Assuming an image of width M and height N, reflection along the horizontal axis is given by:

$$\begin{aligned} x' &= x \\ y' &= N - y \end{aligned} \tag{8.26}$$

Reflection along the vertical axis is given by:

$$\begin{aligned} x' &= M - x \\ y' &= y \end{aligned} \tag{8.27}$$

Reflection along the main diagonal is given by:

$$\begin{aligned} x' &= y \\ y' &= x \end{aligned} \tag{8.28}$$

while reflection along the antidiagonal it is given by:

$$\begin{aligned} x' &= M - x \\ y' &= N - y \end{aligned} \tag{8.29}$$

Implementation of image transposition is straightforward; therefore, we will proceed to the next section on special-effects filters.

8.8 Special-effects filters

Special-effects filters are implemented by manipulating the (x,y) coordinates of the image in innovative and imaginative ways. Examples, shown in reference 7, include manipulating pixel coordinates in both Cartesian coordinates and polar coordinates. As an example, if the source pixel location (x,y) is often converted into polar coordinates (r,θ) for radial effects, such as:

$$\begin{aligned} r' &= \sqrt{rR} \\ a' &= a \end{aligned} \tag{8.30}$$

The C code for this example is shown below.

```
/*
        Function: pci_sfx
        Operation:Transforms an image using a radial transformation:
                  r' = sqrt(r*R);
                  a' = a;
                  Let (mx, my) be the center of the source image.
                  Then r = sqrt((x-mx)**2+(y-my)**2) and
                       a = atan((y-my)/(x-mx))
                  For a given (r,a) pair,
                       x = rcosa
                       y = rsina
                  The implementation is like this:
                  for all (y in destination)
                       for all (x in destination) {
                            find (r', a')
                            find (r,a)
                            invert into (x,y)
                            copy src(x,y) into dest(x,y)
                  }
        Inputs:   src is the source image
                  R controls the amount of warping
        Output:   dest is the warped image
*/

void
pci_sfx (pci_image *src, pci_image *dest, double R)
{
        register int rows, cols;
        register unsigned char *spixptr, *dpixptr;
        register unsigned char *dpix, *spix;
        register unsigned int slb, dlb;
        register int i,j;
        register double r,a;
        register float src_cf, src_rf;
        register int src_c, src_r;
        register double locR;
        register int midrow, midcol;
```

```
        /* get parameters from the images */
        dpix = dpixptr = dest->data;
        spix = spixptr = src->data;
        rows = src->height;
        cols = src->width;
        slb = src->linebytes;
        dlb = dest->linebytes;
        locR = R;
        midrow = rows/2;
        midcol = cols/2;

        /* do the whole image */
        for (i=0; i<rows; i++) {
             for (j=0; j<cols; j++) {
                  if ((i!=midrow) || (j!= midcol))
                       a=atan2((double)(midrow-i),
                            (double)(j-midcol));

                  else
                       a=0.0;
                  r = hypot((double)(j-midcol),
                       (double)(midrow-i));
                  /* find (r,a) */
                  r = r*r/locR;     /* a does not change */

                  /* invert into (x,y) */
                  src_cf = (float) ((double)midcol + r*cos(a));
                  src_rf = (float) (-r*sin(a) +
                       (double) midrow);

                  /* prepare for copying src(x,y) into dest(x,y) */
                  ROUNDDOWN(src_c, src_cf)
                  ROUNDDOWN(src_r,src_rf)

                  /* clip the source pixel */
                  if ((src_c >= 0) && (src_c < cols) &&
                      (src_r >= 0) && (src_r < rows))
                       *dpixptr
                            =*(spixptr+src_c+src_r*slb);
                  dpixptr++;
             }
             dpixptr = (dpix += dlb);
        }
}
```

This code is inefficient, since there are several computations involving doubles and math functions inside the inner loop. One way to speed up the process is to compute the (x,y) locations of the source pixel corresponding to every destination pixel. These locations are stored in two tables: `xtab[][]` and `ytab[][]`. These tables can then be used whenever the transformation needs to be performed. This method is particularly useful when interactive speeds are needed and the same transformation needs to be performed repeatedly (on different images or parts of the same image).

The code for performing this transformation using tables is shown below. The function `pci_sfx_gentab()` computes the tables `xtab` and `ytab` (which are tables of shorts, since the x and y coordinates can exceed 256). The function `pci_sfx_tab()` then computes the corresponding transformation.

Note that for radially symmetric transformations, the size of `xtab[]` and `ytab[][]` can be substantially reduced, since only one octant of information is needed.

```
/*
        Function: pci_sfx_gentab
        Operation: Generate tables for special-effects warping
        Input:     src is the source image
                   R controls the amount of warp
        Outputs:   xtab and ytab, which are tables containing the warp
*/

void
pci_sfx_gentab(pci_image *src, pci_image *xtab, pci_image *ytab, float R)
{
        register int rows, cols;
        register short *xptr, *yptr;
        register unsigned int slb, xlb, ylb;
        register int i, j;
        register double r,a;
        register float src_cf, src_rf;
        register int src_c, src_r;
        register double locR;
        register int midrow, midcol;

        /* get parameters from the images */
        rows = src->height;
        cols = src->width;
        slb = src->linebytes;
        locR = (double) R;
        midrow = rows/2;
        midcol = cols/2;
        xptr = (short *)xtab->data;
        yptr = (short *)ytab->data;
        xlb = xtab->linebytes;
        ylb = ytab->linebytes;
        xlb /= 2;
        ylb /= 2;

        /* generate the tables; loop over dest image coordinates */
        for (i=0; i<rows; i++) {
             for (j=0; j<cols; j++) {
                  if ((i!=midrow)||(j!=midcol))
                       a = atan2((double)(midrow-i),(double)(j-midcol));
                  else
                       a = 0.0;
                  r = hypot((double)(j-midcol),(double)(midrow-i));
                  r = r*r/locR;
```

```
                /* invert into (x,y) */
                src_cf = (float) ((double)midcol + r*cos(a));
                src_rf = (float) (-r*sin(a) + (double) midrow);

                /* find the corresponding integer coordinate in dest */
                ROUNDDOWN(src_c, src_cf)
                ROUNDDOWN(src_r, src_rf)

                /* clip the row and col value in dest */
                if (src_c <0)
                     src_c = 0;
                if (src_c >cols)
                     src_c = cols;
                if (src_r <0)
                     src_r = 0;
                if (src_r >rows)
                     src_r = rows;

                /* xtab[i][j] contains x displacement */
                *(xptr + i*xlb + j) = (signed short) src_c;

                /* ytab[i][j] contains y displacement */
                *(yptr + i*ylb + j) = (signed short) src_r;
           }
      }
}

/*
        Function: pci_sfx2
        Operation: This function performs the same operation as the function
                  pci_sfx. However, the x and y coordinates are now placed
                  in two tables xtab[][] and ytab[][]. That is,
                       dest[r][c] = src[xtab[r][c]][ytab[r][c]]
        Inputs:   src is the source image
                  xtab and ytab are the two-dimensional tables
        Output:   dest is the destination image
*/

void
pci_sfx2(pci_image *src, pci_image *dest, pci_image *xtab, pci_image *ytab)
{
        register int rows, cols;
        register unsigned char *spixptr, *dpixptr;
        register unsigned char *dpix;
        register short *xpix, *ypix;
        register unsigned int xlb, ylb, slb, dlb;
        register int i, j;
        register int src_c, src_r;
        register int midrow, midcol;

        /* get parameters from the images */
        dpix = dpixptr = dest->data;
        spixptr = src->data;
        xpix = (short *)xtab->data;
```

```
        ypix = (short *)ytab->data;
        rows = src->height;
        cols = src->width;
        xlb = xtab->linebytes/2;
        ylb = ytab->linebytes/2;
        dlb = dest->linebytes;
        slb = src->linebytes;
        midrow = rows/2;
        midcol = cols/2;

        /* do the whole image */
        for (i=0; i<rows; i++) {
             for (j=0; j<cols; j++) {
                  src_c = (int) *(xpix + i*xlb +j);
                  src_r = (int) *(ypix + i*ylb +j);
                  *dpixptr = *(spixptr + src_c + src_r*slb);
                  dpixptr++;
             }
             dpixptr = dpix += dlb;
        }
}
```

Example images of special effects are shown in Figures 8.25 and 8.26.

Figure 8.25 Museum original

Figure 8.26 Transformation applied

8.9 Conclusion and further reading

In this chapter we have covered a variety of image geometric operations and examined their implementation. We have looked at several ways to implement image scaling, including nearest neighbor, bilinear, and bicubic interpolations. We have also looked at rotation and affine transformation algorithms. We have seen image transposition and special-effects algorithms. For all of these operations, we has examined efficient ways of implementation.

To probe further, reference 1 forms the basis for much of the material discussed in this section. A set of interesting geometric warping functions can be found in reference 7.

8.10 References

1 G. Wolberg, *Digital Image Warping*, Los Alamitos, CA: IEEE Computer Society Press, 1990.

2 W. K. Pratt, *Digital Image Processing*, 2nd edition, New York: Wiley, 1988.

3 A. Paeth, "A Fast Algorithm for General Raster Rotation," *Graphics Gems*, San Diego, CA: Academic Press, 1990.

4 J. Bresenham, "Algorithm for Computer Control of Digital Plotters," *IBM System Journal*, Vol. 4, No. 1, 1965, pp. 25–30.

5 J. Ward and D. R. Cok, "Resampling Algorithms for Image Resizing and Rotation," *Proceedings of the SPIE Digital Image Processing Application*, Vol 1075, 1989, pp. 260–269.

6 E. Catmull, and A. R. Smith, "3-D Transformations of Images in Scanline Order," *Computer Graphics*, (SIGGRAPH 1980 Proceedings), Vol. 14, No. 3, July 1980, pp. 279–285.

7 G. Holtzmann, *Beyond Photography*, Englewood Cliffs, NJ: Prentice Hall, 1985.

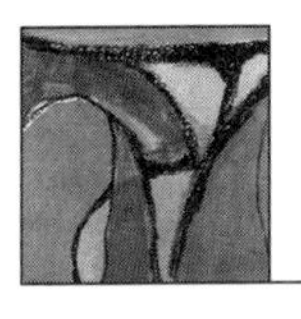

chapter 9

Image data compression

9.1 Introduction

Image data compression is a remarkably prolific area of activity, where the goal is to compress images by reducing the number of bits required to represent them. This is accomplished by eliminating or reducing the redundancies that are inherent in all natural images. Image data compression is useful in both still imaging and digital video applications. As computer imaging and video become widespread computer application areas, the need for image data compression increases. The arrival of numerous image compression products in both hardware and software, and their acceptance by customers, attests to the success of this technology.

Because of the large amounts of memory needed to represent digital images, compression was an area of early research effort in image processing. The need for compression is most acutely felt in the storage and transmission of images. Applications in image storage include archiving of medical images, movies, books, and general-purpose image databases, such as one a photographer might use. In image transmission, applications include digital television, video teleconferencing, satellite picture transmission, facsimile transmission, and video education.

Additionally, with the rapid growth of the Internet, transmission of both still and moving images over a network with finite (and often narrow) bandwidth has taken on an even more important role.

Success of image compression has been accelerated by the establishment of several standards for the compression of binary, gray-scale, color, and moving images. Interoperability between brand names—virtually guaranteed with conformance to these standards—has popularized compression tools among users. Many of these products can compress images by factors as large as 100 with little or no degradation to the image.

This chapter is about the techniques used in image compression, and the construction of systems based on these techniques. Section 9.1 presents motivating examples and discusses the concept of redundancy. An example application of image data compression is also shown. Section 9.2 describes the building blocks used in the construction of these systems. Section 9.3 is a closer look at the implementation of image compression systems based on the industry standard CCITT (Comité Consultatif International de Téléphonique et Télégraphique) Group 3, JPEG (Joint Photographic Experts Group), and MPEG (Motion Picture Experts Group) schemata.

9.1.1 Definitions and motivating examples

In most imaging operations, we are accustomed to the pixel as the fundamental unit of a digital image. During compression, however, the image is converted into a stream of bits (which is converted back into the image during decompression). Therefore, the word *pixel* may not be applied meaningfully to the compressed bitstream.

The goal of image data compression is to reduce the number of bits needed to represent the image. This is done by reducing inherent redundancies in images. Compression is useful in storing, retrieving, and transmitting images.

If a compression system can compress an image of A bytes down to B bytes, then the *compression ratio* is said to be A:B. To be useful, a compression system must have a decompression system associated with it. Some of the desirable qualities in a compression/decompression system are high compression ratio, good quality of decompressed image, and computationally inexpensive compression and decompression.

In the literature, as well as the rest of this chapter, the term *image coding* is often used in place of image data compression. The terms *encoder* and *decoder* are often used interchangeably for compression system and decompression system, respectively. An encoder-decoder pair is sometimes called a *codec*.

Lossless compression is said to have been performed if the original image is exactly recovered after the compress-decompress cycle. The compression is called *lossy* if the original image cannot be exactly recovered. To achieve lossless compression, predictive techniques (Section 9.2.3) are used, since transform compression methods (when coupled with quantization) introduce losses to the original image. In this chapter we examine both lossless and lossy compression techniques.

Two examples, illustrated in Figure 9.1, show the need for image compression.

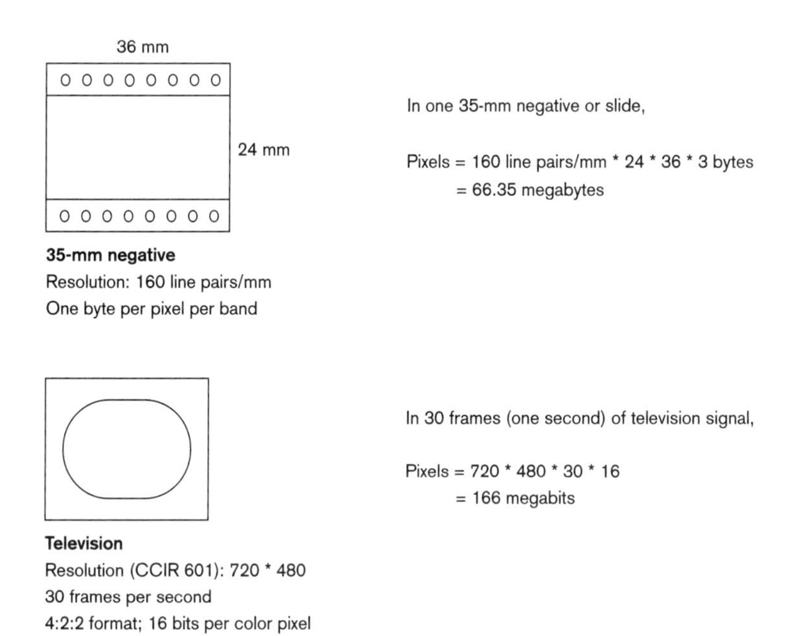

Figure 9.1 Storage and bandwidth requirements of uncompressed images

In *image storage*, compression reduces the amount of storage media needed. Consider digitally archiving a 35-mm color negative, whose size is actually 24 mm × 36 mm. At the typical film resolution of 160 line pairs per mm, requiring 160 * 160 pixels per square mm, the number of pixels in one frame is 160 * 160 * 24 * 36 = 22,118,400 pixels. Assigning one byte for each of the red, green, and blue components requires 66.35 megabytes of storage. The baseline JPEG compressor—using chrominance subsampling, described in Chapter 7—can compress this storage to as little as 3 megabytes while maintaining reasonably good visual fidelity.

In *image transmission*, compression reduces the bandwidth requirement on the communication channel. Transmission of television-quality digital images requires large bandwidths. Consider the CCIR 601 standard video format, which requires 720 * 480 pixels per frame. Using 4:2:2 chrominance subsampling, an average of 16 bits is needed for each color pixel. Therefore, real-time (i.e., 30 frames per second) transmission requires 720 * 480 * 30 * 16 = 166 Mbits per second. With an efficient compression system such as MPEG 2 this rate can be reduced to 10 Mbits per second without substantial loss of fidelity.

These examples illustrate the practical needs for image compression techniques. In Figure 9.2 we show a videoconferencing system as an example of the use of image data compression techniques that are considered in this chapter. In this system, two persons who may be far away from one another are able to meet face-to-face. The conferencing system consists of a camera, a video monitor, a compression system, and a decompression system. Each participant's image is compressed, sent over the channel, and decompressed for display.

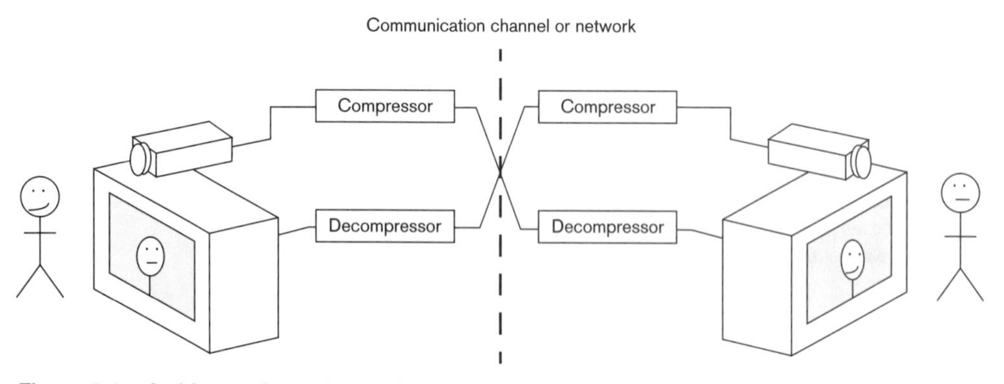

Figure 9.2 A videoconferencing system

To be useful, this system also needs sound. Therefore, microphones and speakers will be needed in addition to the video equipment for a complete conferencing system. In addition, we need to synchronize the audio portions of the conference with the video portions. The videoconferencing standards H.320 and H.323 allow this kind of synchronization for a complete "system" solution.

We now examine the reasons why image data compression techniques work.

9.1.2 *Redundancy in images*

Compression techniques work because most natural images have large amounts of redundancies. Useful images are seldom made of random pixels. Rather, they are two-dimensional representations of scenes from the real world. Thus, most of the pixel values in these images are related in some fashion.This correlation gives rise to redundancies in the pixels. Image compression techniques reduce or eliminate these redundancies. There are various ways of looking at redundancy; the types of redundancies we are concerned with are *spatial* redundancy, *spectral* redundancy, and *temporal* redundancy.

Consider a gray-scale image of a face, where one byte is sufficient to represent the dynamic range of the image. The pixels on the forehead, for example, are likely to be close together in value because the forehead is usually a smooth area, evenly reflecting light. Therefore, it is unnecessary to use one byte to represent each pixel. Instead, a reference pixel can be represented with one byte, and every other pixel in the forehead can be represented by a few bits (smaller than a byte) containing the *difference* with the reference pixel. This is an example of *spatial* redundancy.

Another type of redundancy occurs in the frequency domain representation of an image, which we discussed in Chapter 1. Consider again the image of a face. Compared to the area occupied by the image, there are few places of large *rates of change* in intensity (with respect to distances in the image). That is, the proportion of high frequency is small compared to middle-frequency and low frequency. Therefore, in a frequency domain representation of the image, fewer bits can be used to assign high-frequency representation. Alternatively, we can discard those frequency components whose contributions are negligibly small. This is an example of the reduction of *spectral* redundancy to yield compression.

Temporal redundancy occurs in a sequence of images representing a moving picture. In successive frames of the same scene, moving objects undergo motion, which can be described parametrically. For example, translational motion can be described by a displacement vector. Thus, in compressing a sequence of images, areas in motion—modeled as rectangular blocks in the image—can be represented using the initial block and a set of displacement vectors representing the motion of the block through a corresponding set of images.

The above examples provide intuitive reasons why compression techniques work so well in computer imaging practice. In the next section we describe the individual compression techniques that can be used in the construction of image compression systems.

9.2 *Building blocks for image compression*

Over the past three decades, numerous techniques have been developed for image data compression. In practice, most image compression systems must utilize more than one compression technique. This is because one technique is usually not sufficient to achieve a high compression ratio while maintaining acceptable picture quality after decompression. We can think of the compression system as being constructed out of *building blocks,* each of which is an individual compression technique. The rest of this section describes some important building blocks available to the designer of the image compression system.

Classical image compression building blocks include *variable-length coding*, *run-length coding*, *transform compression*, *predictive compression*, *vector quantization*, *motion estimation*, and *subband coding techniques*. Usually an image compression system utilizes transform or predictive compression to decorrelate the pixels, which are then quantized and variable-length coded for further compression. If sequences of images, depicting motion, are to be compressed, then motion estimation is used in conjunction with the above techniques.

In variable-length coding, the statistical properties of the image are exploited to encode it. For example, values that are more likely to occur are given shorter code words so that the average code length is minimized.

Run-length coding, particularly useful in the compression of binary images, reduces image redundancies by replacing a consecutive set of pixels of the same value with a number representing the count of those pixels.

In transform compression, a transform (usually the Discrete Cosine Transform) of the image is taken. The coefficients computed by the transform correspond to relative proportions of two-dimensional spatial frequency components present in the image. If some frequencies are present in negligible quantities, the corresponding coefficients can be discarded, resulting in compression.

Predictive coding exploits the spatial correlation of image pixels. Most images are composed of continuous objects, and neighboring pixels within an object are usually highly correlated. During coding, this correlation is used to predict the next pixel from the current pixel, and the difference (error) between the prediction and the actual pixel is coded. The same predictor is used during decoding to reconstruct the pixel using the error. The storage or bandwidth required to represent the error is substantially less than the requirement for representing the pixel itself.

Vector quantization is another way to achieve compression. Instead of encoding individual pixel values, vectors of pixels or transform coefficients are encoded using this technique. This method works well for images belonging to the same general class, for example, a series of faces. Another advantage of vector quantization is efficient decoding.

Motion estimation compresses time-varying images by predicting the linear displacement of blocks of the image, and then using the motion vector to represent those subblocks.

In subband coding, the frequency spectrum of the image is divided into several bands and the image is decomposed into a series of subbands, one for each frequency band. Each subband can then be individually quantized and coded for maximal compression.

We summarize some of the trade-offs of these building blocks in Table 9.1.

Table 9.1 **Comparison of compression building blocks**

Building block	Suitable types of image	Relative expense	Typical compression ratio	Artifacts
Variable-length coding	All kinds of images	Inexpensive	2:1	None (lossless)
Run-length coding	Best for binary image	Inexpensive	5:1	None (lossless)
Transform compression	All kinds of images	Moderate	10:1 (JPEG)	Blockiness
Predictive compression	All kinds of images	Inexpensive	2:1	None (lossless)
Vector quantization	Images with similar statistics	Expensive (coding); inexpensive (decoding)	?	Blockiness
Motion estimation	Motion picture	Expensive (coding); inexpensive (decoding)	50:1 (MPEG)	Motion artifacts
Subband coding	All types of images	Moderate	20:1	Contours

In the following sections, we present more details on these building blocks.

9.2.1 Variable-length coding

In variable-length coding, the statistical properties of the image are used to reduce redundancy. Code words are generated for the pixel values, with smaller-length words for those that occur more frequently. The amount of compression that can be achieved by this kind of coder has an upper bound dictated by a statistical measure of the image called the entropy.

Two algorithms for generating variable-length codes are Huffman coding and arithmetic coding. Of these, the Huffman coder is the more commonly used. Arithmetic coding is not covered in this book and the reader should see reference 1 for a detailed discussion.

Variable-length (VL) coders are general-purpose devices that can be used for coding signals other than images. In general, a VL coder maps a set of *source symbols*, S, into a set of *code words*, C, and each code word is composed of symbols from a code alphabet, D (which consists of the bits 0 and 1 for all the cases of interest to us). In the case of images, the most obvious source symbol S is a pixel; however, other quantities representing the image, such as transform coefficients, are effective candidates for variable-length encoding. In general, a codebook is used by the coder to look up the code words for each input source symbol. Figure 9.3 shows such a coder.

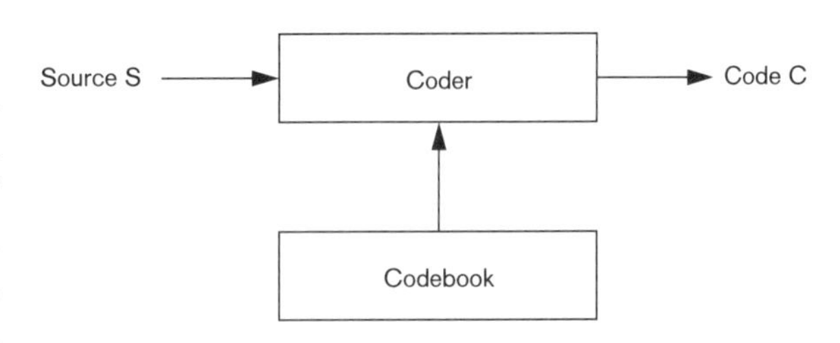

Figure 9.3 A coder

An important measure of the VL code's usefulness, given a set of source symbols having the same statistical properties, is the average length of a code word. Thus, for example, if a set of images having the same statistical characteristics is to be compressed using a VL code, then we would like to know what will be the average number of bits—called the *rate* of the code—needed to represent a pixel.

It turns out that there is a theoretical upper boundary to the best rate that can be obtained by a VL coder. According to the Source Coding Theorem [2], the *entropy* of a source is the best possible rate for a variable-length code used to encode that source.

Consider a source with M symbols with probabilities $p_0, p_1, \ldots, p_{M-1}$, and a variable-length coder, which assigns the code words $C_0, C_1, \ldots, C_{M-1}$ of length $l_0, l_1, \ldots, l_{M-1}$ bits, to these symbols. Then the average length of a code word is

$$L = \sum_{k=0}^{M-1} l_k p_k \tag{9.1}$$

whereas the entropy of the source is defined as

$$H = -\sum_{k=0}^{M-1} p_k \log_2 p_k \tag{9.2}$$

The Source Coding Theorem tells us that no matter how clever the design of the coder, its average code length will be greater than the entropy of the source.

$$L > H \tag{9.3}$$

Huffman coding Huffman coding, invented by D. A. Huffman in 1952, is an ingenious algorithm for generating variable-length codes [3]. The resulting code minimizes the average code word length, L, subject to the Source Coding Theorem.

While Huffman coding is a general-purpose algorithm, which works for a coder using an alphabet D of any size, we will be concerned here with the special case where D is 2. That is, our Huffman codes will be composed of 0s and 1s.

In order to generate a Huffman code, we need to know the probability of occurrence of each source symbol. A *Huffman tree* is constructed from these probabilities. Once the tree is constructed, code words for each source symbol can be assigned from it.

During tree building, the source symbols are arranged in descending order of their probabilities. The lowest two probabilities are added, and the new set of probabilities is once again arranged in descending order. The two lowest probabilities of the new tree are merged again, and this process is repeated until only one probability remains: 1.0. This point is called the root of the tree.

During code assignment, starting from the root of the tree, 1s and 0s are assigned to each branch until the source symbol is reached.

We demonstrate the algorithm with examples of tree building and code assignment. Consider an image with the pixel values $\{r_0, r_1, \ldots, r_6\}$, with probabilities {0.4, 0.2, 0.15, 0.10, 0.08, 0.04, 0.03}. The first step in building the Huffman code for this set of values is to build the

Huffman tree. On the left side of the tree, the symbols and their probabilities are written in descending order, as shown in Figure 9.4.

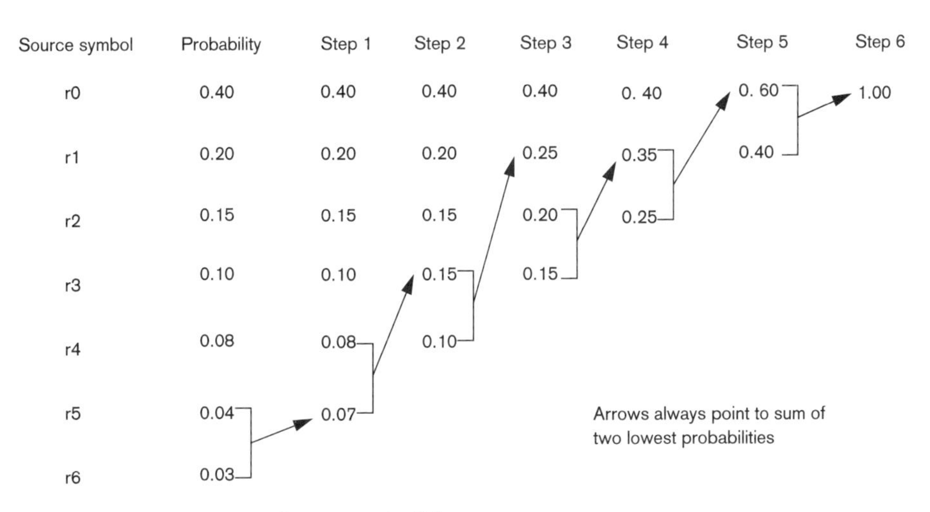

Figure 9.4 Huffman coding—tree building

In Step 1, the two lowest probabilities, 0.04 and 0.03, are added to get 0.07, and the new probabilities are arranged in descending order. In Step 2, 0.08 and 0.07 are added to 0.15, and the probabilities rearranged once again. This procedure is repeated until the probabilities have merged to one. This completes building the Huffman tree.

During code assignment, one starts at the root of the tree. At each node, the two children are assigned 0 and 1 (or 1 and 0—the order does not matter). This is repeated for all the nodes in the tree. The code for a symbol can now be read by concatenating all the bits that are encountered while traversing the tree from its root to the leaf corresponding to that symbol. Thus, for example, in Figure 9.5, the code for r_4 is 1011.

For this example, the entropy of the code, which is the best average bit rate that can be achieved, is

$$H = \sum_{k=0}^{7} p_k \log_2 p_k = 2.36 \tag{9.4}$$

while the average bit rate of the Huffman code generated in Figure 9.5 is

$$L = \sum_{k=0}^{7} l_k p_k = 2.42 \tag{9.5}$$

where l_k is the length of the kth code word.

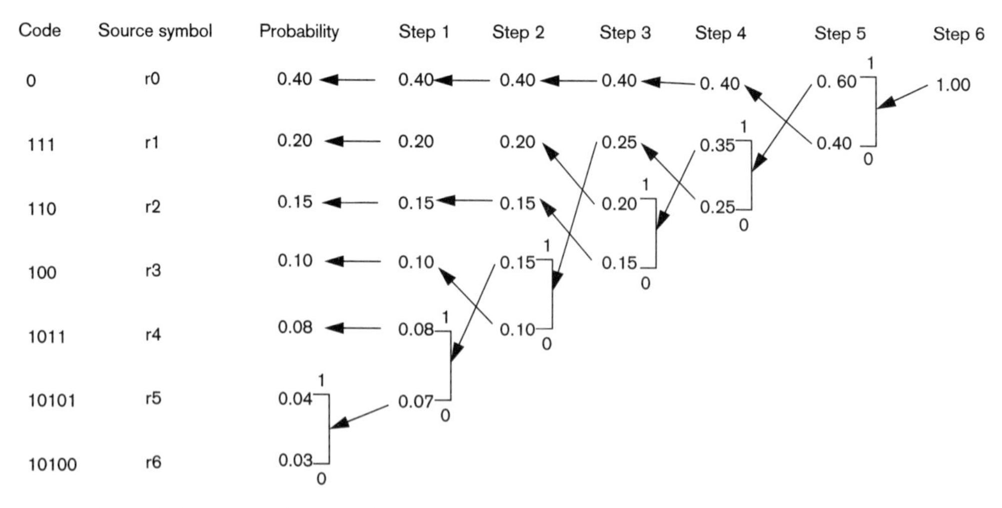

Figure 9.5 Huffman coding—code assignment

Huffman coding implementation Software implementation of Huffman coding is logically done in three phases. As we indicated earlier, the probabilities of the source symbols are needed to build the code. During the first phase, source probabilities are approximated by calculating the histogram of the source symbols. For example, if the source symbols to be coded are transform coefficients, then those coefficients must first be computed. Using these histogram approximations, the Huffman tree is built in the second phase. In the third phase, code assignment takes place.

In most real-image compression applications, the first two phases are executed once using a set of sample images. These samples are chosen so that their statistical properties are close to the target images that are to be coded. The third phase, coding, is then executed repeatedly for all the images that are to be coded. An example implementation of all three phases can be found in reference 4.

During decoding, the bitstream must first be parsed to extract the code words. Then, a table lookup approach is usually employed to decode the code words. Notice that information specifying the Huffman table must either be implicitly understood by the decoder, or it must be embedded in the compressed data. As we will see in our study of compression standards, Huffman coding is used most effectively in compressing transform coefficients.

Modified Huffman coding One disadvantage of Huffman coding for a large set of pixel values (8-bit pixels have 256 possible values) is that long code words are generated for the smallest probabilities. This is common in image compression where decorrelation leads to a large number of source symbols having small probabilities. Including the Huffman codes for every one of these symbols results in a large Huffman table.

An alternative is provided by Modified Huffman coding [5], where the source symbol value N is expressed in terms of a base M as $N = q * M + r$. If the value is $N - 1$ or less, then Huffman coding is used. If it is N or greater, then two tables are used, one for q (called the Makeup code) and the other for r (called the Terminating code).

Modified Huffman coding has been used effectively in the CCITT Groups 3 and 4 Recommendation used in facsimile transmissions. A more detailed discussion can be found in Section 9.4.2.

9.2.2 Run-length coding

Like variable-length coding, run-length coding also exploits the statistical properties of the image to reduce redundancy. However, run-length coding is different from variable-length coding because the code words generated are of a fixed length.

Run-length coding can be applied to gray-scale and binary images. In gray-scale images, the run-length code for a row of pixels of a constant value is a set of numbers $\{g_i, l_i\}$, where g_i is the pixel value and l_i is the number of consecutive pixels with value g_i. This is shown in Figure 9.6a.

In binary images, all pixels have the value 0 or 1. Therefore, the run-length coding need only keep track of one kind of pixel. The code is then a set of numbers $\{a_i, l_i\}$, where a_i is the location of the run of pixels and l_i is its length. This is shown in Figure 9.6b, where the pixels valued 1 are run-length encoded.

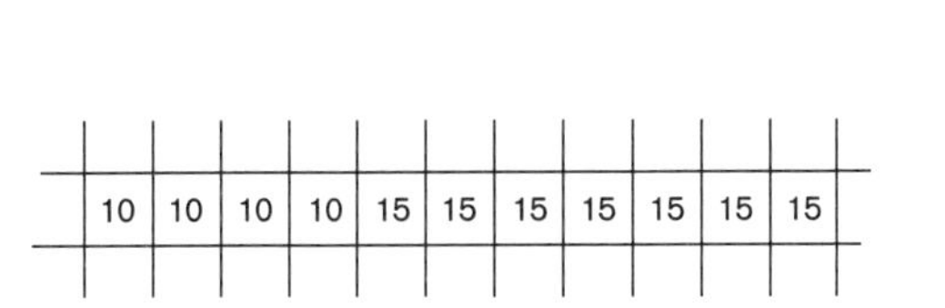

a) Gray-scale run-length coding. The code for this span of pixels is 10/4 15/7.

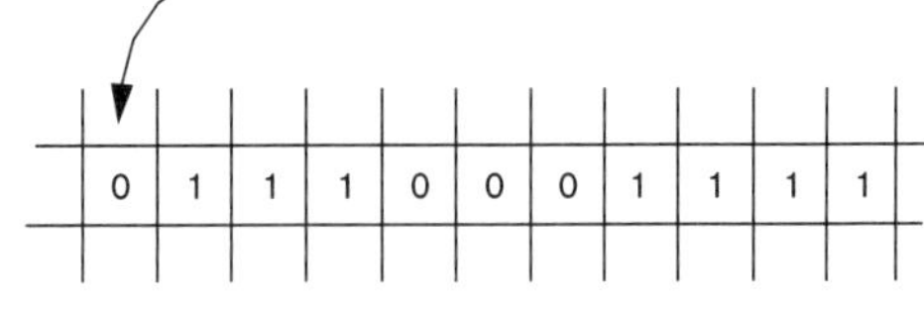

b) Binary image run-length coding. The code for this span of pixels is 1/3 7/4.

Figure 9.6 Run-length coding for images

Run-length encoding is a simple scheme that works well for binary images. Considerable work must be done before it can be made useful for gray-scale images, primarily because pixel sequences of exactly the same value do not occur frequently in real images. Binary run-length encoding can be applied to gray-scale imagery by decomposing the gray-scale image into a set of (binary) bit-plane images, each representing a bit slice of the gray-scale image.

9.2.3 *Transform coding*

Transform coding is a powerful tool often used in image compression systems in conjunction with quantization and variable-length coding. Transform coding assists the compression process by compacting the energy of the image into a smaller area. The image is typically subdivided into $N \times N$ blocks. Then, a mathematical operation called a *unitary transform* is applied to each block. The result of the operation is another block representing the transform coefficients.

A unitary transform of the signal *decorrelates* it by expressing it in terms of a set of orthogonal basis signals. While a detailed study of unitary transforms is beyond the scope of this book, we note here that they have several nice properties, including *energy compaction* and *energy preservation*. Energy compaction means that most of the energy of the signal being transformed is concentrated in the first few terms of the transformed signal. Energy preservation means that both the original and the transformed signal have the same energy.

To understand why transform coding works, let us first consider applying a unitary transform, such as the Discrete Fourier Transform, to a one-dimensional sampled signal. The Discrete Fourier Transform is a way to express the signal as a weighted sum of sinusoids, each of a different frequency. The transform coefficients are simply the weights. If the sampled signal is a low-frequency signal, then the coefficients corresponding to the low-frequency sinusoids will be large, and those corresponding to the high-frequency sinusoids will be small.

Similarly, the transform of an image (which is a two-dimensional signal) is a way to write that image as a weighted sum of *basis images*. The use of transforms effectively *decorrelates* the pixels by expressing them in terms of these orthogonal basis images. Most images have large portions of homogeneous areas where pixel values change slowly. A smaller number of pixels at the edges of these areas change faster. Homogeneous regions correspond to low frequency while edges correspond to high-frequency content of the image. Thus, most images have a large proportion of low-frequency content. Therefore, the transform concentrates the energy of the image in the coefficients corresponding to low frequencies, that is, on the top left part of the transform. A suitable method can then be applied to discount the high-frequency coefficients without causing substantial image degradation.

Note, however, that taking a transform of an image does not, by itself, provide compression. The unimportant transform coefficients must somehow be eliminated, and the important ones must be variable-length encoded to achieve compression. This is shown in Figure 9.7.

Having looked at the basics of transform coding, we are now ready to look at design issues and implementations.

Design issues in transform compression Before implementing a transform coding system, several practical issues need to be addressed. The choice of the transform, size of the blocks, and how to choose the right coefficients, are some of these design decisions.

The choice of transform is largely driven by its energy compaction property and ease of implementation. For all intents and purposes, the Discrete Cosine Transform (DCT) is the overwhelming choice due to its *energy compaction property*. That is, the DCT is able to pack the

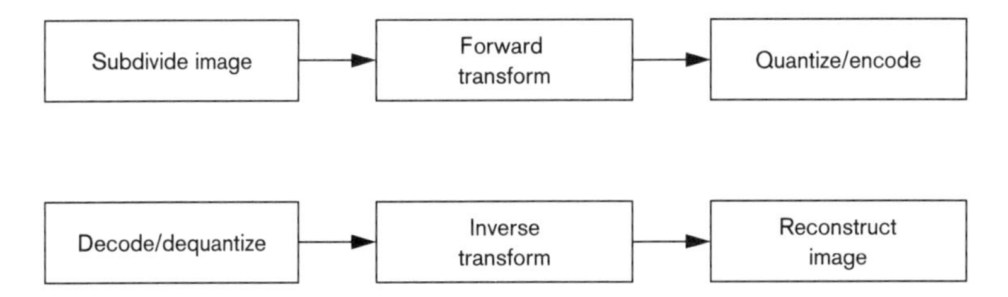

Figure 9.7 Transform compression

image's energy into the lowest transform coefficients in a very efficient manner. (There is only one other transform, the Karhunen-Loeve transform [5], which has better energy compaction, but the DCT comes very close to it and is much less expensive to compute.) The DCT is also a *separable* transform, which means that the column and row transforms can be computed independently, yielding significant computational advantages. The computation savings are similar to the separable convolution we saw in Chapter 6 and the separable image scaling we saw in Chapter 8.

The choice of the block size is a trade-off between computational complexity, memory requirements, and compression ratio. The sweet spot is a block size of 8 × 8, which, while providing good energy compaction, does not create excessive demands on hardware and software implementations.

In one dimension, the DCT of a sequence $f(i)$, $i = 0, \ldots, 7$, yields another sequence $F(u)$ according to Equation 9.6. Here, $C(x) = \frac{1}{\sqrt{2}}$ for $x = 0$ and $C(x) = 1$ for nonzero x.

$$F(u) = \frac{1}{2}C(u)\sum_{i=0}^{7} f(i)\cos\left[(2i+1)\frac{u\pi}{16}\right] \tag{9.6}$$

In two dimensions, the DCT of a block $f(i,j)$, $i, j = 0, \ldots, 7$, is computed as

$$F(u,v) = \frac{1}{4}C(u)C(v)\sum_{i=0}^{7}\sum_{j=0}^{7} f(i,j)\cos\left[(2i+1)u\frac{\pi}{16}\right]\cos\left[(2j+1)v\frac{\pi}{16}\right] \tag{9.7}$$

to yield another block $F(u,v)$, where $u, v = 0, \ldots, 7$.

Once the transform and block size are chosen, another issue is the strategy to be used when eliminating smaller transform coefficients. As we have seen earlier, transform coefficients tend to be concentrated on the top left corner, corresponding to the low frequencies. Only the coefficients on that zone of the transform matrix could be chosen for quantization and encoding. This is called *zonal selection*. The main problem with zonal selection is that nonzero transform coefficients outside the zone are ignored once the zone is chosen. This problem is solved using *threshold masking*, where a threshold is applied to the transform coefficients. Those values that fall above the threshold are retained; those below the threshold are discarded. The JPEG image compression standard discussed later has opted for a variation of threshold masking, allowing the user to choose the threshold values in the quantization matrix.

Implementation of the discrete cosine transform (DCT) From the programmer's point of view, there are two options for implementation of the DCT that depend on the hardware to be used. If the target hardware is dedicated for image compression, then it probably has a built-in DCT. In this case, the programmer's task becomes simple: Call the subroutine (or set the registers) to execute the DCT. However, if the target hardware lacks this feature (for example, in the case of a general-purpose computer), then the DCT must be implemented in software. The rest of this section is intended to help the programmer with the second option.

A straightforward implementation of Equation 9.7 requires 64 multiplication and addition operations per transform coefficient. This number can immediately be lowered by using separability. Using Equation 9.6, a two-dimensional DCT is executed as a sequence of eight one-dimensional DCTs over the rows of the input block to create an intermediate block, followed by eight one-dimensional DCTs over the columns of the intermediate block to create the final result. This is the approach taken in the first example program in this section.

For one approach to "fast" DCT algorithms, note that Equation 9.6 reveals that taking the one-dimensional DCT consists of premultiplying the input vector of pixels by the DCT matrix $\boldsymbol{C}$, whose elements are given by

$$\boldsymbol{C} = \left[\cos\left((2c+1)r\frac{\pi}{16}\right)\right] \tag{9.8}$$

where $0 \leq r, c \leq 7$ for the 8×1 DCT.

The fast DCT algorithms involve decomposition of the DCT matrix $\boldsymbol{C}$ into the product of a series of sparse matrices. These sparse matrices have zeros in most places. Computation of the DCT then requires a reduced number of multiplications and additions.

The computational burden can be lowered further in the case of transform compression by noting that the step immediately following DCT computation is quantization, which is implemented by a pointwise division of the DCT coefficients by a quantization array. This implies that if each coefficient at the output of the forward transform block in Figure 9.7 were off by a scale factor, it is acceptable, provided this scale factor gets absorbed in the quantization matrix. This is called the *scaled DCT* (SDCT). The second program in this section illustrates the implementation of the SDCT using a decomposition by Feig [6].

While all of the above comments apply equally to the DCT and the IDCT, Westover and Macmillan [7] have taken advantage of some specific properties commonly encountered in the IDCT step. In the last example of this section, their "splat" algorithm is discussed and example pseudocode is provided.

The efficient implementation of the DCT is a well-researched topic, and there are several other algorithms that we have not addressed here. The reader should see reference 1 for a survey of these algorithms.

There is a tendency in the literature to evaluate the effectiveness of a fast DCT algorithm by counting the number multiplications and additions. However, for implementation on a general-purpose computer, we note that there are several caveats in this kind of an approach. Several of these algorithms require significant data movement in the intermediate steps. This can add substantially to the overhead, or prevent efficient parallelization of the algorithms for implementation on SIMD architectures such as the SX. In addition, the general-purpose microprocessor usually has a finite number of registers, and the storage requirements of the intermediate values in the fast algorithm may exceed this number. As a result, the programmer must be cautious when judging a "fast" algorithm solely on the basis of the number of multiplications and additions.

A final caveat is that not all implementations of the DCT yield correct precision, especially in floating point. Some applications, such as the H.261 standard used in videoconferencing, require exact precision in the DCT coefficients, and significant work must be done to achieve this precision, particularly if the computations are being done in integer arithmetic.

DCT example 1: Separability In the example program, the intermediate block, `inter[][]` is computed by taking the $1-d$ DCT of each row of the input block `pixels[][]` according to:

$$inter[i][v] = \frac{1}{2}C(v)\sum_{j=0}^{7} pixels[i][j]\cos\left[(2j+1)v\frac{\pi}{16}\right] \tag{9.9}$$

Next, the $1-d$ DCT along each column of `inter[][]` is computed to yield `dct[][]`.

$$dct[u][v] = \frac{1}{2}C(u)\sum_{i=0}^{7} inter[i][v]\cos\left[(2i+1)u\frac{\pi}{16}\right] \tag{9.10}$$

Note that in order to preserve precision of the result, `inter[][]` is declared as a float.

```
#include <math.h>
#include <stdio.h>

#define C(x)  ((x==0)?.7071:1.)

/*
        Function: pci_dct
        Operation:Find the 8x8 DCT of an array using separable DCT
                  First, find 1 - d DCT along rows, storing the result
                  in inter[][]. Then, 1 - d DCT along columns of inter[][]
                  is found.
        Input:    pixels is the 8x8 input array
        Output:   dct is the 8x8 output array
*/

pci_dct(int pixels[8][8], int dct[8][8])
{
        int inr, inc;           /* rows and columns of input image */
        int intr, intc;         /* rows and columns of intermediate image */
```

```
        int outr, outc;         /* rows and columns of dct */
        double f_val;                /* cumulative sum */
        float inter[8][8];           /* stores intermediate result */
        int i,j;

        /* find 1-d dct along rows */
        for (intr=0; intr<8; intr++)
             for (intc=0; intc<8; intc++) {
                  for (i=0,f_val=0; i<8; i++) {
                       f_val += (double)(pixels[intr][i]) *
                       cos((double)(2*i+1)*(double)intc*M_PI/16);
                  }
                  inter[intr][intc] =  (C(intc)*f_val*0.5);
             }

        /* find 1-d dct along columns */
        for (outc=0; outc<8; outc++)
             for (outr=0; outr<8; outr++) {
                  for (i=0,f_val=0; i<8; i++) {
                       f_val += (double)inter[i][outc] *
                       cos((double)(2*i+1)*(double)outr*M_PI/16);
                  }
                  dct[outr][outc] = (int) (C(outr)*f_val*0.5);
             }
}
```

DCT example 2: The scaled DCT As we mentioned above, the scaled DCT is capable of absorbing the quantization and dequantization steps within the DCT computation. One example of a scaled DCT is due to Feig, who has shown that the DCT matrix $\boldsymbol{C}$ can be decomposed according to

$$\boldsymbol{C} = \boldsymbol{D}\boldsymbol{A}_3\boldsymbol{A}_2\boldsymbol{A}_1\boldsymbol{A}_0 \tag{9.11}$$

where the matrices $\boldsymbol{A}_0, \boldsymbol{A}_1, \boldsymbol{A}_2, \boldsymbol{A}_3$, and $\boldsymbol{D}$ are defined as:

$$\boldsymbol{A}_0 = \begin{bmatrix} 1 & 0 & 0 & 0 & 0 & 0 & 0 & 1 \\ 0 & 1 & 0 & 0 & 0 & 0 & 1 & 0 \\ 0 & 0 & 0 & 1 & 1 & 0 & 0 & 0 \\ 0 & 0 & 1 & 0 & 0 & 1 & 0 & 0 \\ 1 & 0 & 0 & 0 & 0 & 0 & 0 & -1 \\ 0 & 1 & 0 & 0 & 0 & 0 & -1 & 0 \\ 0 & 0 & 0 & 1 & -1 & 0 & 0 & 0 \\ 0 & 0 & -1 & 0 & 0 & 1 & 0 & 0 \end{bmatrix} \qquad \boldsymbol{A}_1 = \begin{bmatrix} 1 & 0 & 1 & 0 & 0 & 0 & 0 & 0 \\ 0 & 1 & 0 & 1 & 0 & 0 & 0 & 0 \\ 1 & 0 & -1 & 0 & 0 & 0 & 0 & 0 \\ 0 & 1 & 0 & -1 & 0 & 0 & 0 & 0 \\ 0 & 0 & 0 & 0 & 1 & 0 & 0 & 0 \\ 0 & 0 & 0 & 0 & 0 & 1 & 0 & 1 \\ 0 & 0 & 0 & 0 & 0 & 0 & 1 & 0 \\ 0 & 0 & 0 & 0 & 0 & 1 & 0 & -1 \end{bmatrix}$$

$$A_2 = \begin{bmatrix} 1 & 0 & 0 & 0 & 0 & 0 & 0 & 0 \\ 0 & 1 & 0 & 0 & 0 & 0 & 0 & 0 \\ 0 & 0 & 1 & 0 & 0 & 0 & 0 & 0 \\ 0 & 0 & 0 & 1 & 0 & 0 & 0 & 0 \\ 0 & 0 & 0 & 0 & 1 & 0 & 0 & c_4 \\ 0 & 0 & 0 & 0 & 0 & c_4 & 1 & 0 \\ 0 & 0 & 0 & 0 & 0 & -c_4 & 1 & 0 \\ 0 & 0 & 0 & 0 & 1 & 0 & 0 & -c_4 \end{bmatrix} \qquad A_3 = \begin{bmatrix} 1 & 1 & 0 & 0 & 0 & 0 & 0 & 0 \\ 0 & 0 & 0 & 0 & 1 & t_1 & 0 & 0 \\ 0 & 0 & 1 & t_2 & 0 & 0 & 0 & 0 \\ 0 & 0 & 0 & 0 & 0 & 0 & -1 & t_5 \\ 1 & -1 & 0 & 0 & 0 & 0 & 0 & 0 \\ 0 & 0 & 0 & 0 & 0 & 0 & t_5 & 1 \\ 0 & 0 & t_2 & -1 & 0 & 0 & 0 & 0 \\ 0 & 0 & 0 & 0 & 0 & 1 & 0 & -1 \end{bmatrix}$$

$$D = \begin{bmatrix} c_4 & 0 & 0 & 0 & 0 & 0 & 0 & 0 \\ 0 & c_1 & 0 & 0 & 0 & 0 & 0 & 0 \\ 0 & 0 & c_2 & 0 & 0 & 0 & 0 & 0 \\ 0 & 0 & 0 & c_5 & 0 & 0 & 0 & 0 \\ 0 & 0 & 0 & 0 & c_4 & 0 & 0 & 0 \\ 0 & 0 & 0 & 0 & 0 & c_5 & 0 & 0 \\ 0 & 0 & 0 & 0 & 0 & 0 & c_2 & 0 \\ 0 & 0 & 0 & 0 & 0 & 0 & 0 & c_1 \end{bmatrix}$$

where $c_j = \cos(\pi j/16)$ and $t_j = \tan(\pi j/16)$.

The following code shows the implementation of a one-dimensional DCT in floating point using this algorithm. Note that the DCT coefficients computed by this program are scaled by a factor of 8 from the "true" DCT coefficients.

```
/*
        Function: pci_fast_dct
        Operation:Find the 8x1 DCT of an array using Feig's algorithm
        Input:    in is the 8x1 input array
                  delta is distance to next relevant element
                  delta = 1 for row DCT; =8 for col DCT
        Output:   out is the 8x1 output array
*/

/* unscaled cosine coefficients */
#define C0      1.0
#define C1      .98079
#define C2      .92388
#define C3      .83147
#define C4      .70711
#define C5      .55557
#define C6      .38268
#define C7      .19509

/* unscaled tangent coefficients */
#define T1      .19891
#define T2      .41421
```

```
#define T5      1.49661
void
pci_fast_dct(int *in, int *out, int delta)
{
        float i0, i1, i2, i3, i4, i5, i6, i7;
        float a0, a1, a2, a3, a4, a5, a6, a7;

        /* get the input vector */
        i0 = (float) *in;
        in += delta;
        i1 = (float) *in;
        in += delta;
        i2 = (float) *in;
        in += delta;
        i3 = (float) *in;
        in += delta;
        i4 = (float) *in;
        in += delta;
        i5 = (float) *in;
        in += delta;
        i6 = (float) *in;
        in += delta;
        i7 = (float) *in;

        /* compute result of premultiplying by matrix A0 */
        /* result is the vector [a0, a1, a2, a3, a4, a5, a6, a7] */
        /* where a0...a7 are computed as below */
        a0 = i0 + i7;
        a1 = i1 + i6;
        a2 = i3 + i4;
        a3 = i2 + i5;
        a4 = i0 - i7;
        a5 = i1 - i6;
        a6 = i3 - i4;
        a7 = i5 - i2;

        /* compute the result of premultiplying by matrix A1 */
        /* result is the vector [i0, i1, i2, i3, a4, i5, a6, i7] */
        /* where i0, i1, i2, i3, i5, and i7 are computed as below */
        i0 = a0 + a2;
        i1 = a1 + a3;
        i2 = a0 - a2;
        i3 = a1 - a3;
        i5 = a5 + a7;
        i7 = a5 - a7;

        /* compute the result of premultiplying by matrix A2 */
        /* result is the vector [i0, i1, i2, i3, a4, a5, a6, a7] */
        /* where a4, a5, a6, a7 are computed as below */
        a7 = a4 - C4*i7;
        a4 = a4 + C4*i7;
        a5 = C4*i5 + a6;
        a6 = -C4*i5 + a6;
```

```
        /* compute the result of premultiplying by matrix A3 */
        /* result is the vector [a0, a1, a2, a3, i4, i5, i6, i7] */
        /* where a0...a7 are computed as below */
        a0 = i0 + i1;
        a1 = a4 + T1*a5;
        a2 = i2 + T2*i3;
        a3 = -a6 + T5*a7;
        i4 = i0 - i1;
        i5 = T5*a6 + a7;
        i6 = T2*i2 - i3;
        i7 = T1*a4 - a5;

        /* compute the result of premultiplying by matrix D and store */
        *out = (int) (a0);
        out += delta;
        *out = (int) (a1);
        out += delta;
        *out = (int) (a2);
        out += delta;
        *out = (int) (a3);
        out += delta;
        *out = (int) (i4);
        out += delta;
        *out = (int) (i5);
        out += delta;
        *out = (int) (i6);
        out += delta;
        *out = (int) (i7);
}
```

To compute a two-dimensional DCT using the one-dimensional DCT program above, the following code may be used.

```
main ()
{
        int i,j;
        int f_in[8][8], f_out[8][8];/* for fast algo */

        /* initialize the input block */
        ...

        /* do the row transforms */
        pci_fast_dct(&f_in[0][0], &f_out[0][0], 1);
        pci_fast_dct(&f_in[1][0], &f_out[1][0], 1);
        pci_fast_dct(&f_in[2][0], &f_out[2][0], 1);
        pci_fast_dct(&f_in[3][0], &f_out[3][0], 1);
        pci_fast_dct(&f_in[4][0], &f_out[4][0], 1);
        pci_fast_dct(&f_in[5][0], &f_out[5][0], 1);
        pci_fast_dct(&f_in[6][0], &f_out[6][0], 1);
        pci_fast_dct(&f_in[7][0], &f_out[7][0], 1);

        /* do the column transforms */
```

```
            pci_fast_dct(&f_out[0][0], &f_in[0][0], 8);
            pci_fast_dct(&f_out[0][1], &f_in[0][1], 8);
            pci_fast_dct(&f_out[0][2], &f_in[0][2], 8);
            pci_fast_dct(&f_out[0][3], &f_in[0][3], 8);
            pci_fast_dct(&f_out[0][4], &f_in[0][4], 8);
            pci_fast_dct(&f_out[0][5], &f_in[0][5], 8);
            pci_fast_dct(&f_out[0][6], &f_in[0][6], 8);
            pci_fast_dct(&f_out[0][7], &f_in[0][7], 8);
}
```

The inverse DCT is computed using

$$f(i, j) = \frac{1}{4}\sum_{u=0}^{7}\sum_{v=0}^{7} C(u)C(v)F(u, v)\cos\left[(2i+1)u\frac{\pi}{16}\right]\cos\left[(2j+1)v\frac{\pi}{16}\right] \tag{9.12}$$

and can be computed using the same method as the forward DCT.

DCT example 3: The splat algorithm for IDCT As we have seen, quantization forces many DCT coefficients to zero during compression. During decompression, the blocks of transform coefficients, whose inverse DCT are to be computed, are thus sparse. That is, for most of the blocks, the nonzero elements are few and concentrated on the top left corner. Taking advantage of this sparseness, another approach for computing the inverse DCT was proposed by Westover and Macmillan [7]. The splat approach, as shown in Figure 9.8, yields fast decoding by storing all the basis images of the DCT in tables. During decode, the tables are looked up and accumulated for all the DCT coefficients.

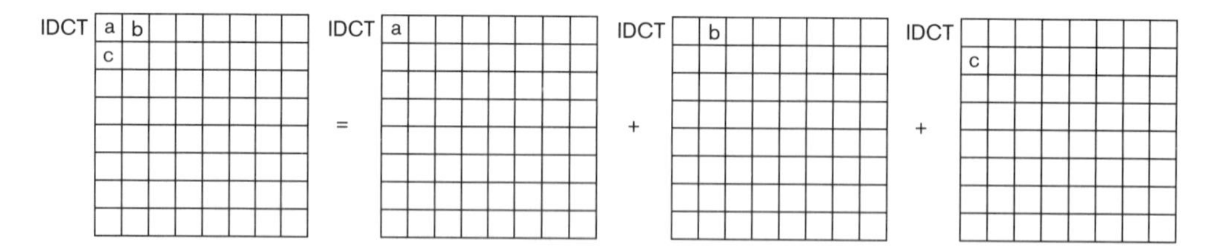

Figure 9.8 An illustration of the splat algorithm for computing the IDCT. The empty boxes are populated with zeros.

Splat's advantage is that relatively few (5–6) nonzero DCT coefficients need to be splatted to compute the IDCT of an 8×8 block. During decoding, splat IDCT computation for these coefficients takes 320–384 additions and no multiplications, compared with 1024 multiplications and 1024 additions necessary for a straightforward IDCT over the 8×8 block.

Pseudocode for implementing IDCT on a JPEG block using splat follows.

```
while (get_next_symbol != EOB) {
      if (symbol != 0)
      dct_table = dct_table + splat_table[symbol]
}
```

Westover and Macmillan have also shown that symmetries inherent in the DCT basis images can be exploited further to take advantage of the cache structure of the processor.

9.2.4 *Predictive coding*

The transform compression methods described previously reduce spectral redundancy. As we saw earlier in Section 9.1.2, another kind of redundancy, spatial redundancy, arises because in most images a pixel and its immediate neighbors are likely to belong to the same physical object or region. Their values are thus likely to be close. Predictive coding, which reduces spatial redundancy in images, exploits this fact.

During compression, the value of the current pixel is *predicted* using one or more of the "previous" pixels. (For most practical applications, a previous pixel is one that is encountered prior to the current pixel while scanning the image row by row, starting at the top left.) The error between the current pixel and its predicted value is transmitted or stored.

Implementation of predictive coding An example of predictive coding is shown in Figure 9.9, where the pixel $f(3)$ is being encoded. The predicted value, $f'(3)$, of the pixel $f(3)$ is computed using the past pixel values $\{f(0), f(1), f(2)\}$. Then the error $e(3) = f(3) - f'(3)$ is computed, possibly quantized, and encoded using a variable-length coding scheme.

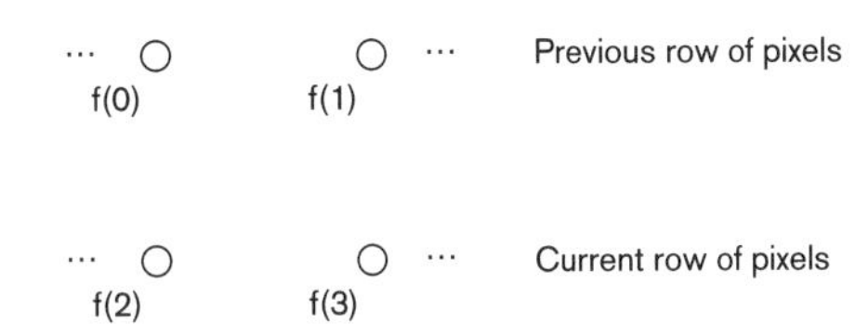

Figure 9.9 Pixels in predictive coding

During decoding, the same predictor as the compressor predicts the current pixel value and adds it to the error to reconstruct the original pixel.

A typical formula used in forming this prediction is [8]

$$f'(3) = \sum_{k=0}^{2} \alpha(k) f(k) \tag{9.13}$$

The coefficients $\alpha(k)$ are determined such that the error e has minimum variance. This requires imposing a statistical model on the image pixel values and finding the autocorrelation matrix of the image [9]. However, it is often difficult or impossible to compute the autocorrelation for every image that is being encoded, and a set of constant values is used for the $\alpha(k)$. Some typical predictors are:

$$\begin{aligned} f'(3) &= -0.81f(0) + 0.90f(1) + 0.90f(2) \\ f'(3) &= -0.50f(0) + 0.75f(1) + 0.75f(2) \end{aligned} \tag{9.14}$$

Other example values of $\alpha(k)$ can be found in Section 9.3.3 in the discussion of lossless JPEG.

This scheme is formalized in Differential Pulse Code Modulation (DPCM), the most common predictive technique in use. Compression using DPCM is shown in Figure 9.10. Using the optional quantizer increases compression efficiency at the cost of lossy compression.

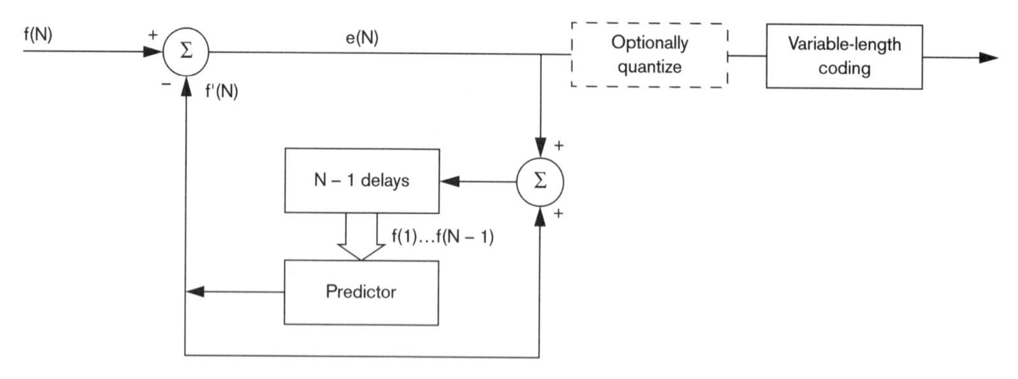

Figure 9.10 DPCM compression

During decompression (Figure 9.11), the same predictor is used to predict the values of the pixels. The error signal is then added to the signal to reproduce the original signal.

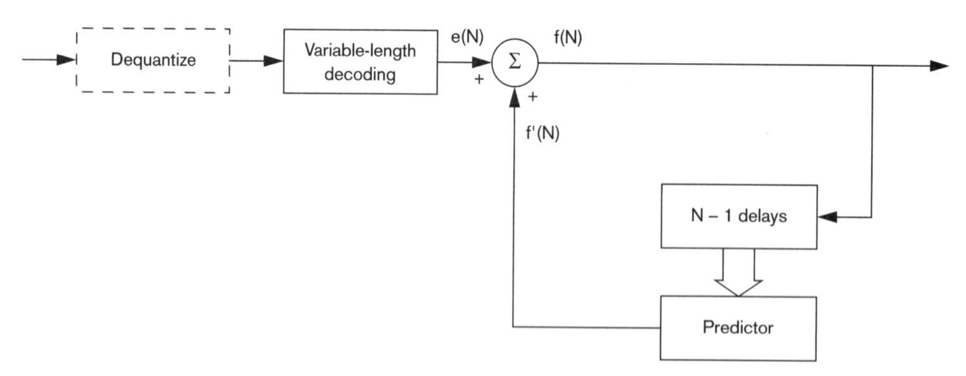

Figure 9.11 DPCM decompression

Note that if the optional quantizer is not used, then we have lossless compression. Predictive compression is commonly used in situations where lossless compression is necessary (e.g., medical imaging). The most common example of a predictive compression system is the lossless JPEG decompressor to be discussed in Section 9.3.

9.2.5 *Motion estimation*

So far, all of the compression building blocks that we have examined remove redundancies while implicitly assuming a still image. However, as we saw in Section 9.1.2, sequences of images representing moving pictures have another type of redundancy—temporal redundancy. Motion estimation is used to reduce temporal redundancy in time-varying images. The idea is simple: If objects in the image are undergoing translational motion (Figure 9.12), their future position can be predicted by a displacement vector.

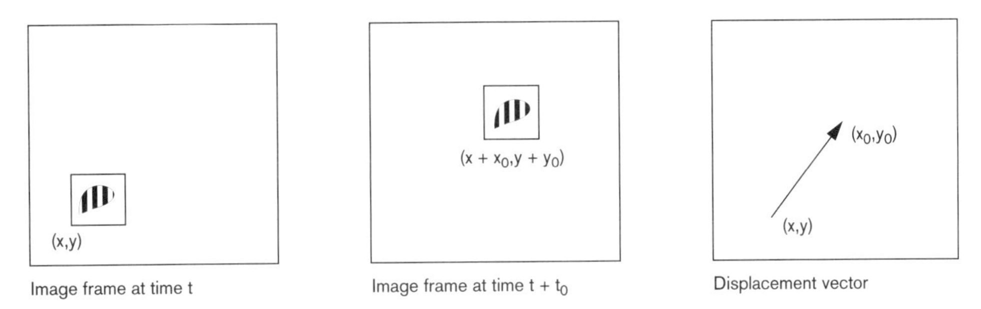

Figure 9.12 Object undergoing translational motion

In practice, a block-matching technique is used [10]. The block of size $M \times M$ (typically 16×16) at location (x,y) at time t is assumed to have moved to location $(x + x_0,y + y_0)$ at time $(t + t_0)$. During encoding, the problem is to estimate (x_0,y_0). During decoding, the block is reconstructed at time $(t + t_0)$ by simply translating the block at time t by (x_0,y_0).

The "best" displacement vector (x_0,y_0) is one that minimizes an error criterion, for example,

$$E(x_0, y_0) = \sum_{\text{all pixels of block}} F(f(x, y, t) - f(x + x_0, y + y_0, t + t_0)) \quad (9.15)$$

where $f(x,y,t)$ is the pixel value at (x,y) at time t, and $F()$ is an appropriate distance measure, such as absolute value or square.

Implementation of motion estimation In practice, the area over which the motion vector is searched, must be limited in order to make the computation feasible. For example, an area of $\pm d$ pixels in the x and y directions can be used. An exhaustive search requires computing $E(x_0,y_0)$ over $(2d + 1)^2$ blocks. As d increases, this becomes prohibitively expensive.

A number of alternatives to exhaustive search have been proposed. In *logarithmic search* [10], shown in Figure 9.13, starting at 0 displacement, the search is carried out in all directions using an initial step size (typically 4 or 5). The location yielding the smallest errors is used as the center of the search, this time with a reduced step size. This process is repeated until the minimum is found. The algorithm thus homes in on the solution.

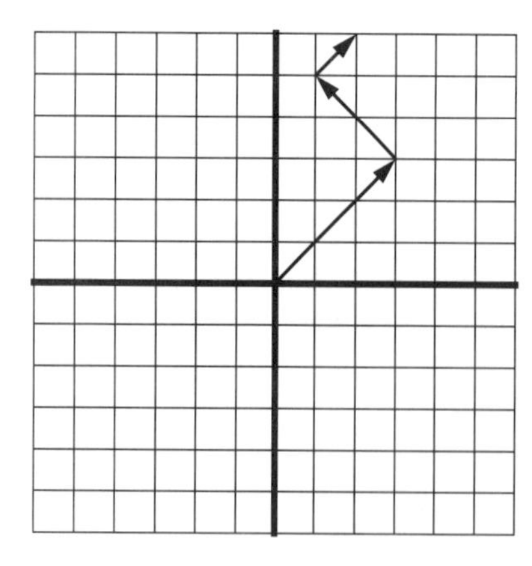

Figure 9.13 Logarithmic search for motion estimation

When motion estimation must be made on more than one frame from a single frame, a technique called *telescopic search* is sometimes used in conjunction with logarithmic search. After the first estimation is done, the base of the block is adjusted to the new displacement vector before motion estimation for the next frame.

Also popular is *hierarchical search* using a pyramid of images [11]. In this technique, the block is subsampled by an integer factor, and the motion vector in the subsampled block is found. Using this vector as the starting point, the block is upsampled and the new vector is found. This process is repeated until the full size of the block is reached. In order to avoid aliasing effects, the image must be filtered before subsampling. An added advantage of this method is that camera noise, which can mislead the search for the motion vector, is substantially reduced during the subsampling/filtering process.

Several hardware solutions for motion estimation are in the market today. These are devices that perform encoding for MPEG or H.320 standards. In addition, the Visual Instruction Set Extension of the UltraSPARC processor has a dedicated instruction, `pdist`, for computing motion estimation. This is discussed in detail Chapter 3.

9.2.6 Vector quantization

Vector quantization is an encoding scheme that attempts to overcome the fundamental limits imposed by the Source Coding Theorem on the compression ratio when encoding individual symbols. Instead, vectors of symbols are coded, as shown in Figure 9.14.

The image is usually subdivided into a series of vectors (for example, by dividing it into blocks whose rows are concatenated into vectors). For each input vector X, the corresponding vector $\overline{X}_m$ is found from a set of vectors $\{X_1, \ldots, X_N\}$, called the codebook, such that the error between X and $\overline{X}_m$ is less than the error between X and all other X_i, $i = 1, \ldots, N$. The index m is then used to represent the vector X. This is the compression. During decompression, a table lookup is performed to obtain X from the index m.

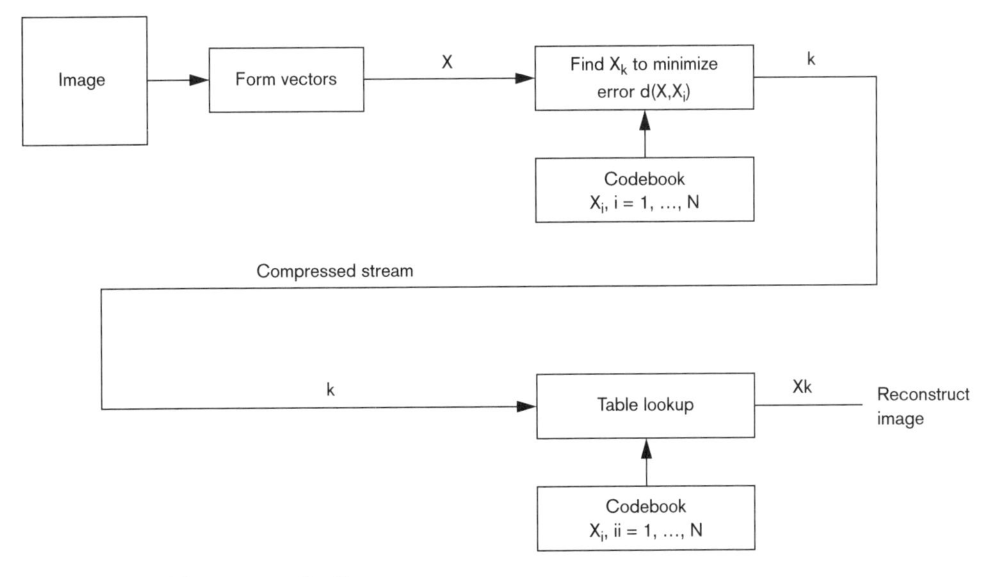

Figure 9.14 Vector quantization

Vector quantization is attractive because it enables quick decompression, which is simply a lookup operation. During transform-based decoding, a straightforward implementation of the IDCT requires 16 multiplies, 16 additions, and 16 memory accesses for each pixel. In contrast, for vector quantization decoding, each pixel requires one memory access only.

However, compression using vector quantization is expensive, because for each vector of pixels being encoded, the entire set of codebook vectors must be searched to find the best match. Methods such as tree-searched search can reduce this time by up to factors of eight or more.

Implementation of vector quantization In implementing vector quantization a codebook must first be generated using a set of representative vectors called training vectors. Encoding and decoding are straightforward once the codebook is known.

The Linde-Buzo-Gray (LBG) algorithm [12], described below, is commonly used to generate the codebook. It is an iterative algorithm that converges on the best possible candidates for the codebook.

1. To start, we need a set of sample vectors called *training vectors*. We also need an initial codebook $\overline{X}_{i,1}$, $i = 1, \ldots, N$. We also need to define a distortion measure d (this could be, for example, the mean squared error between two vectors). Finally, we need to define a fractional distortion change threshold e such that if the fractional error between subsequent iterations is less than e, then the algorithm has converged.

2. Before beginning iteration, set the iteration counter k to 1 and set the initial distortion D_0 to be very high.

3 Encode the training vectors by mapping each vector to its nearest code vector. Compute the distortion D_k caused by this mapping. If the fractional change in distortion is acceptable, that is,

$$\frac{D_{k-1} - D_k}{D_{k-1}} \leq e \tag{9.16}$$

then stop, else go to Step 4.

4 Update the codebook as follows. Replace each code vector $\overline{X}_{i,k}$ by a new $\overline{X}_{i,k+1}$ that minimizes the error. In the common case where the error is measured as Mean Squared Error, each component of the updated vector is computed by averaging the corresponding components of the training vectors that were associated with that particular codebook vector.

The algorithm converges to a local minimum—therefore, in order to avoid spurious results, the choice of initial codebook is extremely important.

There are many variations of vector quantization. These variations include construction of tree-structured codebooks so that the time spent in searching during encoding is minimized.

9.2.7 Subband coding

Subband coding of images is a relatively new compression technique. The main idea behind subband coding is to split up the frequency band of the signal or image using a set of bandpass filters into several different subbands. Each subband, which contains a portion of the bandwidth of the original signal or image, can then be quantized and encoded separately. Like transform compression, compression is achieved in subband coding during quantization and encoding.

Subband coding was originally used to compress speech signals [13]. Then, subband coding was carried over into image compression [9], where a set of subband images is created from the original image, again using a set of bandpass filters. It was found that clever encoding of the subbands can lead to substantially improved compression results.

Implementation of subband coding A simplified, two-band subband image compression scheme is shown in Figure 9.15.

The filters used in creating the subbands are called *analysis* filters. Once the subbands are created, they can be subsampled without aliasing since their frequency bandwidth has been reduced by the bandpass filters. Following the subsampling, the bands are optionally quantized and encoded. The compressed bitstream created in this manner can then be transmitted or stored like other compressed bitstreams.

The decompression scheme corresponding to the two-band compression scheme is shown in Figure 9.16.

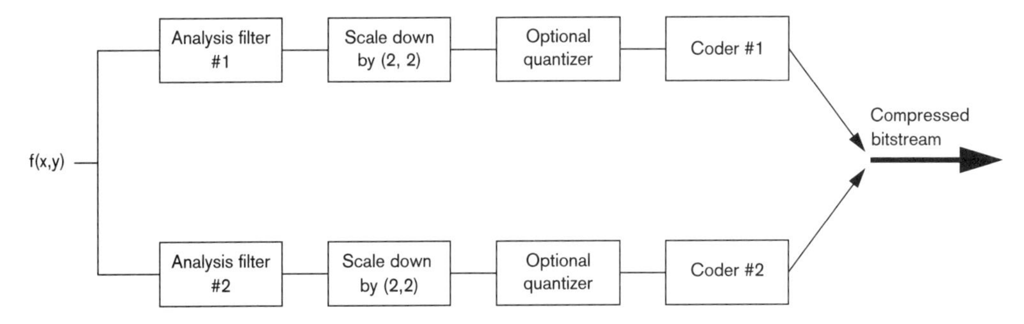

Figure 9.15 Subband coding

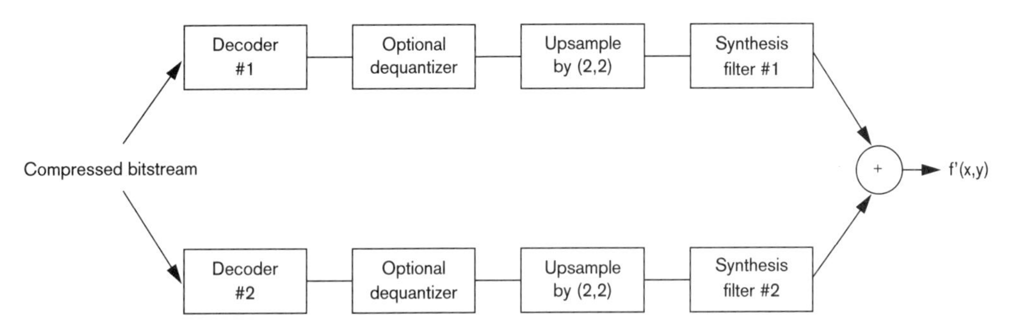

Figure 9.16 Two-band subband decompression

During decompression, a decoder must be applied to decode each subband. After dequantization, which is only needed if the optional quantizers were used during compression, the image is upsampled. *Synthesis* filters are then used to reconstruct the original image.

For n-banded ($n > 2$) subband coding, the outputs of the scale-down blocks in Figure 9.15 are input into a second tier of filters. An example is shown in Figure 9.17.

The design of a subband image coding scheme consists of the design of the analysis/synthesis filters and the choice of the coding and quantization methods used.

Choice of filters There are several requirements on the filters used in subband coding. The bandpass filters need to divide the frequency band of the signal perfectly, that is, without overlaps. These ideal filters do not exist in reality, and implementable bandpass filters create bands whose extremities overlap. This may lead to aliasing problems, as discussed in Chapter 2.

A class of filters called Quadrature Mirror Filters (QMFs) is shown to cancel the aliasing artifacts. QMF filters come in pairs, where one can be derived easily from the other. For example, a QMF pair could be used as the analysis filters #1 and #2 in Figure 9.15. In one dimension, a QMF filter pair $h(n)$ and $g(n)$, of length L, satisfies the relationships $g(n) = (-1)^n h(n)$ and $g(n) = h(L - 1 - n)$.

For images, we need two-dimensional filters. These can be constructed by using separable one-dimensional filters.

An example of a four-band subband coder using separable QMF filters is shown in Figure 9.17. Note that to achieve compression we would still need to encode the decomposed subbands of the figure. Commonly used QMF filters include the Johnston filters, which are described in reference 14.

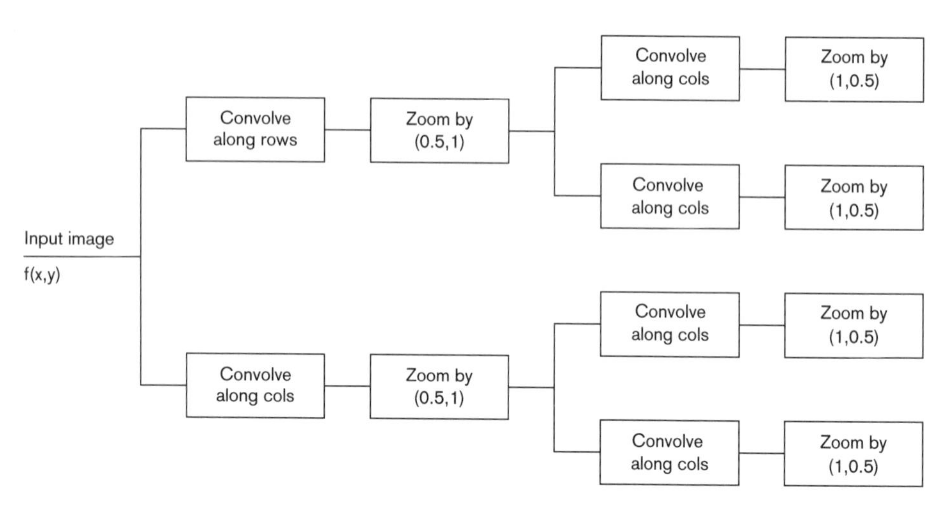

Figure 9.17 Four-band subband decomposition of an image using separable filters

It should be noted here that the decomposition of a signal using subband coding is identical to the *discrete wavelet transform*. Wavelet decomposition of a signal or an image enables analysis by localizing in time (spatial) or frequency domain. The only difference is that wavelet functions are used instead of the QMF filters. The wavelet functions are more general and do not have the symmetry properties that are part of the QMF filters.

Choice of coding technique The choice of the coding technique, including quantizer design and (variable- or fixed-length) coder design, are important steps in subband image compression.

Several types of encoders have been designed for subband coding. These include DPCM-based coders, vector quantizers, and run-length coders, as well as more sophisticated coders. Several types of uniform and nonuniform quantizers have also been used with subband compression schemata.

Like the other compression techniques we have seen, the success of a subband compression scheme depends largely on the encoding scheme used. An example of a successful compression scheme based on subband coding and using an involved quantization and coding algorithm, but having some very nice properties, is presented in reference 15.

9.2.8 *Other compression techniques*

There are numerous other compression techniques and building blocks that have not been discussed in this section. Emerging compression standards such as MPEG-4 (discussed in Section 9.3) may include compression techniques based on advances in image analysis and computer vision. These techniques, called model-based coding, rely on segmenting the image into meaningful objects and coding those objects. An example of a model-based compression scheme can be found in reference 16.

Another set of techniques uses mathematical models such as fractals to compress the image [17]. In the arena of synthetic images, algorithms such as geometry compression [18] have been developed to compress three-dimensional scenes. Geometry compression efficiently encodes the parameters defining a three-dimensional scene.

9.3 *Compression standards in imaging*

A standard is a way to ensure uniformity in implementations. It is a specification, decided by a standards committee, usually composed of representatives from interested organizations, that describes the behavior of a system or an algorithm in a predictable manner. A manufacturer claiming support for a standard must ensure that the product passes certain conformance criteria, usually specified as a part of the standard. The primary reason for standards is interoperability between products from different manufacturers.

Due to their inherent redundancy, most images can be compressed quite well by the methods described in this chapter. The need for compression standards arises because we want to be able to decompress an image on a product manufactured by, say, company A that was compressed on a product of, say, company B. Conformance to compression standards guarantees interoperability of the compression products across platforms.

The standards described in this section are decided under the auspices of two international organizations: International Telecommunications Union (ITU) and International Organization for Standardization (ISO).

In this section we consider the Group 3 standard for binary images used in facsimile machines, the JPEG standard for continuous-tone still images, and the MPEG-1 standard for motion picture compression. Group 3 was written under the auspices of the ITU. (The ITU was formerly called the CCITT.) JPEG and MPEG-1 are written under the auspices of the ISO.

These standards are summarized in Table 9.2.

Table 9.2 Summary of compression standards in imaging

Standard	Type of picture	Algorithm used	Typical compression
Group 3	Black/white	Run-length coding	5:1
JPEG	Gray scale, color	DCT, VLC	20:1
MPEG-1	Moving pictures (gray scale and color)	DCT, motion compensation, VLC	40:1

The Group 3 facsimile standard is based on compression methods for binary images. JPEG utilizes continuous-tone compression techniques, removing spectral redundancy. MPEG-1 utilizes continuous-tone as well as temporal compression across frames of a moving picture to remove motion redundancies.

Compression standards have been commercially successful. The Group 3 standard has led to the proliferation of facsimile machines and a revolution in global communications. Since the advent of JPEG, many application programs ranging from desktop publishing to desktop movies, that were once unable to use high-quality digital images, now do so routinely. There has been a sharp growth of JPEG-based products as vendors have sought to supply customers with software, printed circuit boards, and integrated circuit chips with JPEG capabilities. MPEG-1 (and its follow-on MPEG-2) promises to popularize digital video on desktop computers, as well as revolutionize television.

The discussions that follow cover the essential elements of the standards. However, standards contain many details, which cannot be covered in a book of this size. A thorough examination of the standards documents is therefore required before the standard can be implemented.

9.3.1 The Group 3 standard for binary image compression

The CCITT Group 3 standard for fax was motivated by the desire to enable the communication of binary (two-tone) documents through telephone lines. In the beginning, standards called Group 1 and Group 2, based on analog techniques, were defined. Subsequently, digital encoding of the images was incorporated into the standards known as Group 3 and Group 4. Today's fax machines almost exclusively use Group 3, the subject of the subsequent discussion. Group 3 is meant primarily for communicating A4-size documents (210 mm $\times$ 298 mm) over a telephone network.

Algorithms Group 3 is based on run-length coding followed by Modified Huffman Coding using a set of precomputed Huffman tables. The documents used to generate these tables are shown in Appendix C. These Huffman tables are fixed and are hardwired into every facsimile machine.

There are two classes of algorithms used in the fax standard. The first is *one-dimensional coding*, which compresses one line at a time. The second is *two-dimensional coding*, where vertical correlation between consecutive lines is exploited and information from the previous line may be used while encoding the current line.

One-dimensional coding In one-dimensional coding, each line of pixels—1728 pixels long and composed of alternating runs of black and white—is encoded separately. Each line is assumed to start with a white run to ensure color synchronization at the receiver. If the line starts with a black run, the code for a zero-length white run is prepended.

Following the Modified Huffman coding technique discussed in Section 9.2.1, two categories of runs for black pixels and for white pixels, for the terminating code and the makeup code, are defined. For a run of length N, the terminating code is (N modulo 64) and the makeup code is (N div 64) * 64.

Since the pixel probabilities of black runs and white runs are different, four Huffman encoding tables are used, as shown combined in Table 9.3. Note that these tables are incomplete, and complete tables may be found in Appendix C.

Table 9.3 Huffman encoding tables

Length	White run codes	Black run codes
	Terminating codes	
0	00110101	0000110111
1	000111	010
2	0111	11
3	1000	10
4	1011	011
5	1100	0011
6	1110	0010
...	...	...
62	001101011	000001100110
63	001101000	000001100111
	Makeup codes	
64	11011	0000001111
128	10010	000011001000
192	010111	000011001001
...	...	...
1664	011000	0000001100100
1728	010011011	0000001100101
EOL	00000000001	000000000001

An end-of-line (EOL) code is appended at the end of each line. This helps prevent catastrophic errors, which could be caused by mistakes in Huffman coding. The start of document is coded by six consecutive EOLs. The start of a document must also be marked by an EOL. In addition, a Fill signal consisting of a variable-length string of 0s may be used to add a pause to the message. A Fill is inserted between a line of data and an EOL.

Two-dimensional coding Also known as the *Modified READ (Relative Element Address Designate)* scheme, two-dimensional coding (an optional extension of the one-dimensional scheme) uses vertical overlap between lines to provide more compression. The line being coded is called the *current line* and the previous line is called the *reference line*. Both these lines are used in two-dimensional coding.

Initially, the first line is coded using the one-dimensional algorithm. The next $K - 1$ lines are encoded using the two-dimensional algorithms. Following this, the next line is again encoded using the one-dimensional algorithm, and so on. The parameter K is used to limit the number of lines that can be iteratively encoded in this manner. This is to minimize errors being propagated in the event of transmission errors. After a line has been encoded using one-dimensional coding, at most $(K - 1)$ lines can be encoded using two-dimensional coding. The default value of K is 2; it can be optionally set to 4 for higher resolution.

We now look at the two-dimensional coding in more detail with the aid of five delimiters, $b1$, $b2$, $a0$, $a1$, and $a2$. The first three are from the current line, and the rest are from the reference line. The relationship between these delimiters determines the mode used for encoding.

The delimiters are defined as shown in Figure 9.18. $a0$ is the first white pixel to be encoded, $a1$ is the first black pixel to the right of $a0$, and $a2$ is the first white pixel to the right of $a1$. In the reference line, $b1$ is the first black pixel after a transition, to the right of $a0$. $b2$ is the first white pixel after $b1$.

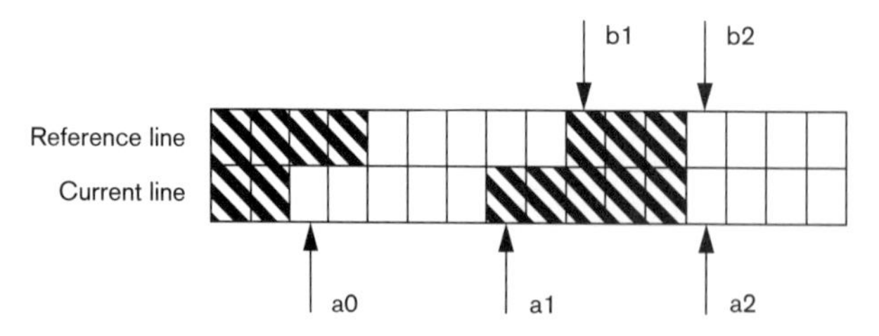

Figure 9.18 Definitions of pixel changes

Three modes of coding are possible depending on the relative positions of $a0$ and $b1$. These are called *pass* mode, *vertical* mode, and *horizontal* mode. Pass mode detects runs on the current line that have no overlap with runs of the previous line, so that encoding can move on to the next run on the current line. Vertical mode detects the situation where there is a run—already encoded—in the previous line, which is close to the current run being encoded. Horizontal mode detects the situation where there is no advantage to be gained from information from the previous line.

These modes are described in detail in the following text. Note that in Figures 9.19, 9.20, and 9.21, pixel locations in parenthesis, for example, ($a0$), indicate the new position of the pixel transition.

- *Pass mode (Figure 9.19)* When $b2$ is to the left of $a1$, pass mode is used. This is coded as 0001. $a0$ is moved right underneath $b2$. $b1$ and $b2$ are moved to the next two pixels where the color changes.

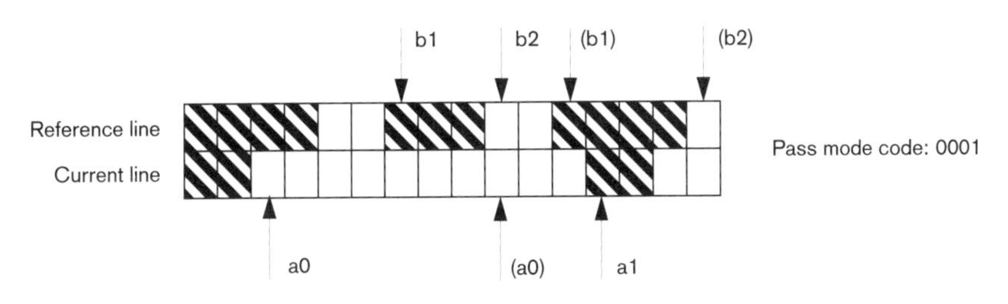

Figure 9.19 Pass mode

- *Vertical mode (Figure 9.20)* If the horizontal distance between $a1$ and $b1$ is three or less, vertical mode is used. The code word used depends on the displacement between $a1$ and $b1$, and is shown in Figure 9.18. After coding $a0$ is moved to $a1$, $a1$ to $a2$, and $b1$ to $b2$. $a2$ and $b2$ move to the next changing pixel to the right.

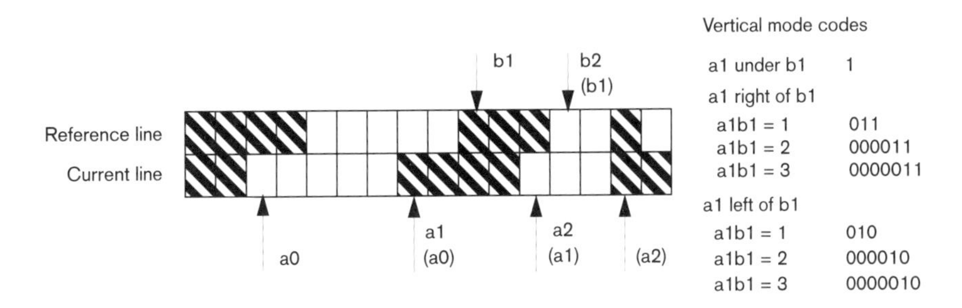

Figure 9.20 Vertical mode

- *Horizontal mode (Figure 9.21)* If the horizontal distance between $a1$ and $b1$ is more than 3, horizontal mode is used. The code word is 001 followed by the modified Huffman codes for the runs $a0a1$ and $a1a2$. After encoding these two runs, $a0$ is moved to $a2$. $b1$ and $b2$ are updated as shown.

  The coding proceeds in this manner until the EOL is reached. After this, the next line is encoded using one-dimensional or two-dimensional coding, depending on the value of K.

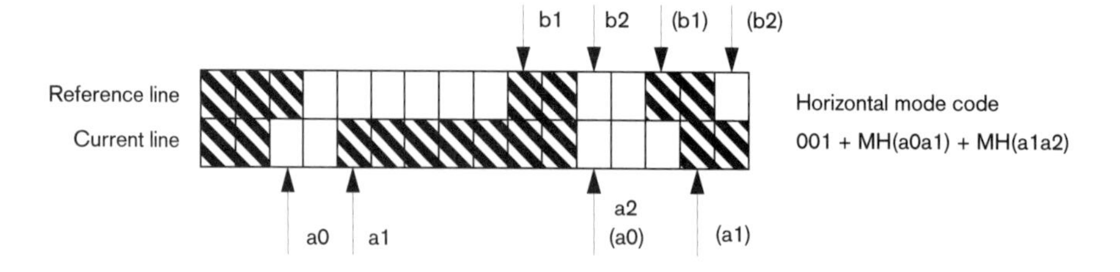

Figure 9.21 Horizontal mode

This concludes our discussion of compression standards for binary images. We now proceed to discuss compression of gray-level and color images using the JPEG compression standard.

9.3.2 JPEG standard for still picture compression

JPEG (Joint Photographic Experts Group) is a standard developed by a collaboration between the ISO and CCITT for compressing continuous-tone still pictures. It is an ensemble of high-quality compression algorithms which are flexible enough to allow trade-offs between compression ratio and picture fidelity. JPEG can handle a large variety of pictures in many different formats.

The algorithms used in JPEG were decided after competition among the proposals in 1987. A blind selection process was held where performance of different proposed algorithms was judged. Eventually, the JPEG Committee decided on four modes of compression to be standardized: *baseline*, *lossless*, *progressive*, and *hierarchical* JPEG.

The baseline mode is the basic JPEG algorithm that enables lossy compression of images. Those seeking perfect fidelity can use the lossless JPEG algorithm based on predictive coding. Progressive JPEG is intended for image transmission where the quality of the decoded picture progressively improves. In hierarchical JPEG compression the picture is successively subsampled and then compressed.

Examples of images compressed using JPEG are shown in Figures 9.22 through 9.24.

The baseline JPEG algorithm The baseline JPEG algorithm is based on the Discrete Cosine Transform and Huffman encoding. The algorithm can be broken into four operations, performed in the forward direction for compression and in the reverse direction for decompression. These operations are shown in Figure 9.25.

In the first step, the image is subdivided into 8×8 blocks. For each block $f(i,j)$, the 8×8 discrete cosine transform is calculated according to Equation 9.7. The result of this operation is an 8×8 matrix, $F(u,v)$, composed of the transform coefficients. $F(0,0)$, the top left element of

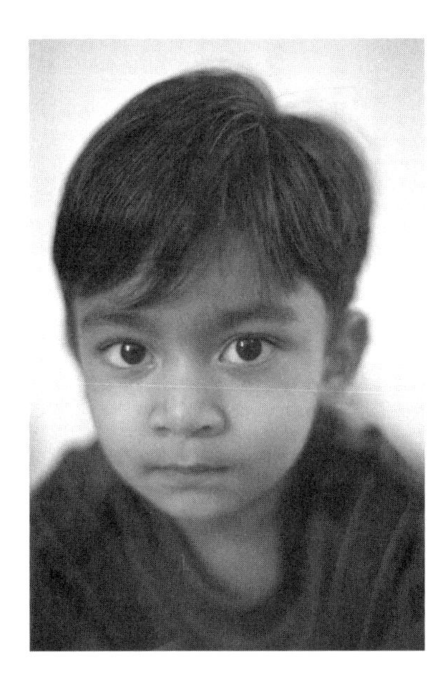

Figure 9.22 Ihsan original

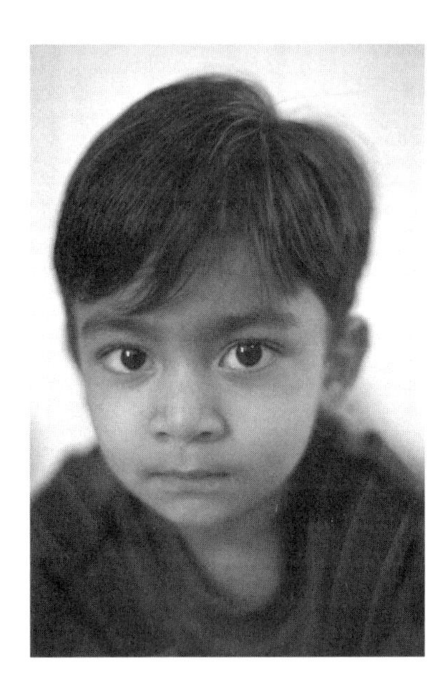

Figure 9.23 10:1 JPEG compression

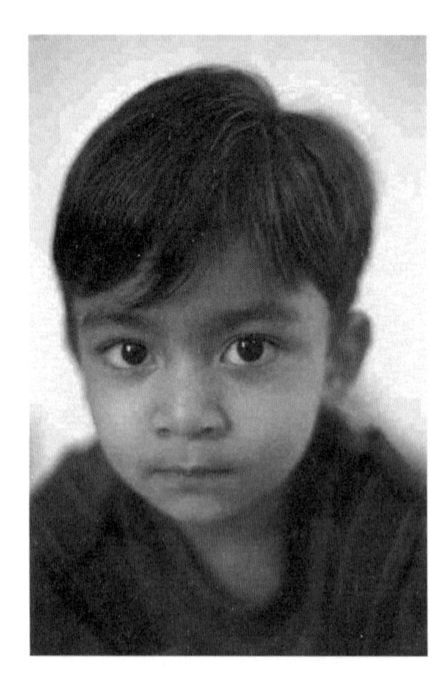

Figure 9.24 30:1 JPEG compression

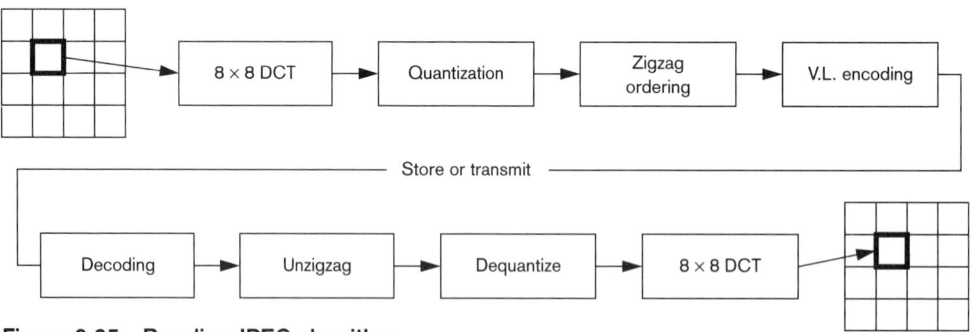

Figure 9.25 Baseline JPEG algorithm

$F(u,v)$, is proportional to the average or *DC component* of $f(i,j)$. Other values in $F(u,v)$ are proportional to the relative amounts of the corresponding frequency components present in $f(j,k)$. These are often called the *AC components*. Baseline JPEG requires that the precision of $f(i,j)$ be 8 bits and that of $F(u,v)$ be 10 bits.

The second step is quantization, where the goal is to reduce the number of different values the DCT coefficients can have. In this case, it is a normalization procedure that zeros out small, insignificant terms. This is done by the following process:

$$F^Q(u, v) = \text{Nearest Integer}\left(\frac{F(u, v)}{Q(u, v)}\right) \tag{9.17}$$

where $Q(u,v)$ is a predefined quantization (or normalization) matrix, and $F^Q(u,v)$ is the quantized DCT. A default quantization table is included in the JPEG specification, but the user may specify his or her own table. An example quantization table is shown in Figure 9.26 [1].

16	11	10	16	24	40	51	61
12	12	14	19	26	58	60	55
14	13	16	24	40	57	69	56
14	17	22	29	51	87	80	62
18	22	37	56	68	109	103	77
24	35	55	64	81	104	113	98
49	64	78	87	103	121	120	101
72	92	95	98	112	100	103	99

Figure 9.26 Example quantization table—Q(u,v) used in JPEG

After quantization, many high-frequency DCT coefficients become 0. The third step takes advantage of this by zigzag scanning the coefficients. The order of the scan is shown in Figure 9.27. This results in a string of numbers in this order: $F(0,0)$, $F(0,1)$, $F(1,0)$, $F(2,0)$, and so forth.

0	1	5	6	14	15	27	28
2	4	7	13	16	26	29	42
3	8	12	17	25	30	41	43
9	11	18	24	31	40	44	53
10	19	23	32	39	45	52	54
20	22	33	34	46	51	55	60
21	34	37	47	50	56	59	61
35	36	48	49	57	58	62	63

Figure 9.27 Zigzag scan pattern for JPEG

The final step of the compression process is variable-length encoding. The DCT coefficients are divided into DC and AC components, and separate Huffman tables are needed for encoding both classes of coefficients. Baseline JPEG allows two sets of Huffman encoding tables in order to accommodate color or other multibanded images. Each set consists of a DC Huffman table and an AC Huffman table.

The DC coefficient $F(0,0)$ is usually larger than the AC coefficients. Therefore, it is encoded differentially, using a DC Huffman table. After encoding the DC coefficient of the initial block, the difference between it and the DC coefficient of the subsequent block is computed and encoded. The DC coefficients of the subsequent blocks are encoded differentially in this manner. A typical Huffman table for encoding the differential DC components for luminance (i.e., gray-scale) images is shown in Appendix C.

The nonzero AC coefficients are coded as follows: Each coefficient is written as an 8-bit number of the form $I = NNNNSSSS$, whose Huffman code is looked up in a table. Some additional bits for sign and magnitude information are also prepended to this Huffman code. The four bits $SSSS$ define a category for the coefficient magnitude. The values in the category k are in the range $(2^{(k-1)}, 2^k - 1)$ or $(-2^k + 1, -2^{(k-1)})$, where $1 \leq k \leq 10$. These ranges are shown in Table 9.4.

Table 9.4 SSSS corresponding to coefficient magnitude

SSSS (k)	Coefficient
1	−1, 1
2	−3, −2, 2, 3
3	−7,..−4, 4..7
4	−15..−8, 8..15
5	−31..−16, 16..31
6	−63..−32, 32..61
7	−127..−64, 64..127
8	−255..−128, 128..255
9	−511..−256, 256..511
10	−1023..−512, 512..1023
11	−2047..−1024, 1024..2047
12	−4095..−2048, 2048..4095
13	−8191..−4096, 4096..8191
14	−16383..−8192, 8192..16383
15	Unused

The four bits $NNNN$ represent the number of zeros that occurred between the current and previous coefficient. If the run of zeros exceeds 16, then $I = 11110000$ denotes one run of 16, and $I = 0$ indicates end of the block. The set of values of I consists of (10 categories × 16 run lengths + 2 special symbols) = 162 symbols, which are Huffman coded.

Samples of the Huffman codes used in default baseline JPEG are shown in Table 9.5. Complete tables may be found in Appendix C for luminance AC coefficients. Note, however, that these tables are suggested tables and are not default tables.

Table 9.5 Huffman code (samples) given NNNN and SSSS

NNNN	SSSS	Code
0	1	00
0	2	01
1	1	1100
1	2	111001
4	1	111011
EOB		1010

In addition, the position (that is, relative magnitude) of the number *SSSS* within the category is represented using k bits, which is coded using a fixed-length code and appended to the Huffman code.

Using this table, the sequence *2 0 3 EOB* would be coded as 0110-11100111-1010. The 2, having $SSSS = 2$ and no zeros before it, has code 01 (from Table 9.5). To this is appended 1 for sign (positive) and 0 for position (relative magnitude) in that coefficient category. The 3, having $SSSS = 2$, and one zero before it, yields 111001. To this, 1 is appended for sign and 1 for magnitude. Finally the EOB is encoded as 1010.

During decoding, the inverse sequence of operations takes place. The bitstream is decoded first. Then the symbols are dequantized. Then they are unzigzagged. Finally the IDCT is taken.

Color images JPEG assumes no information about color spaces. The baseline JPEG algorithm can handle multiple bands, and up to 255 separate bands or components are allowed. Components can have different dimensions. The width and height of the ith component may be x_i and y_i. However, x_i and y_i must be integer multiples H_i and V_i of the smallest (x,y), and H_i, V_i must be in the range [1..4].

In common practice, the YCbCR-601 color space is often used to represent color images in JPEG. The above scheme allows for subsampling of the chrominance components, such as the 4:2:0 representation that we saw in Chapter 7.

The lossless JPEG algorithm For applications that require absolute fidelity between the original and decompressed image, lossless JPEG is a DPCM-based algorithm that guarantees pixel-by-pixel reproduction of the original image after the compression-decompression cycle. This scheme is shown in Figure 9.28.

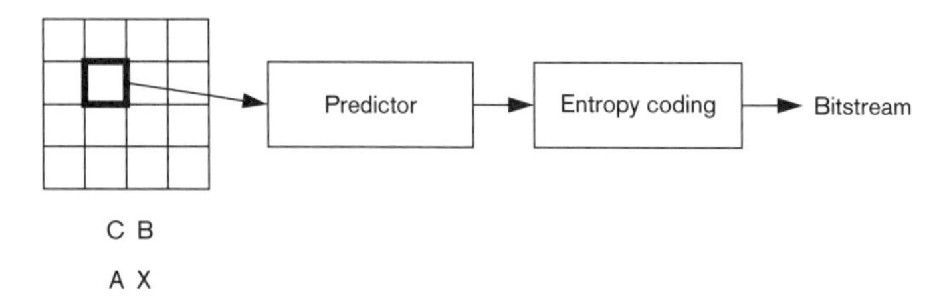

Figure 9.28 The lossless JPEG algorithm

The prediction is based on up to three past pixels, *A*, *B*, and *C*, as shown in Figure 9.28. A prediction *P* is formed for the current pixel *X* based on these past pixels. The difference between *P* and *X* is entropy coded using either Arithmetic or Huffman coding. The resulting bitstream is the compressed image.

There are eight modes of forming the prediction *P* from the past pixels. These are listed in Table 9.6.

Table 9.6 Modes of prediction in lossless JPEG

Mode	Predictor
0	No prediction
1	A
2	B
3	C
4	A + B – C
5	A + (B – C) / 2
6	B + (A – C) / 2
7	(A + B) / 2

The progressive JPEG algorithm In the progressive JPEG algorithm, shown in Figure 9.29, the goal is to encode and decode in multiple scans of the image, rather than a single scan. An application of progressive JPEG is multipass decompression of an image on a relatively slow computer, where the first pass quickly displays a rough approximation of the original image, and subsequent passes refine it. This is done so that the user is not kept waiting.

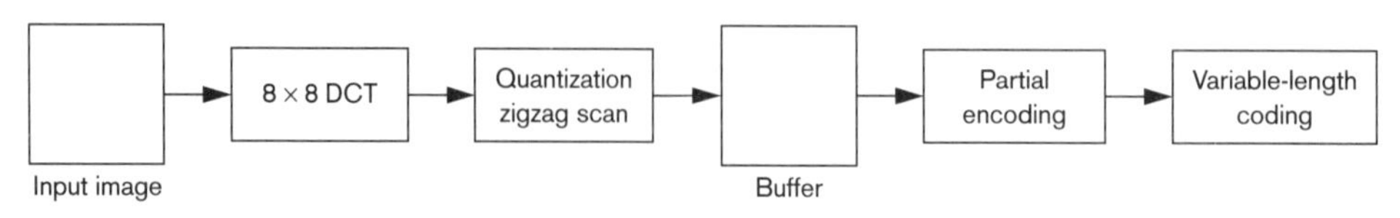

Figure 9.29 The progressive JPEG algorithm

The basic techniques used in progressive JPEG are the same as in baseline JPEG. However, the DCT coefficients are placed in a buffer before quantization. The coefficients in the buffer are partially encoded at each pass of the algorithm. There are two methods of partial encoding: *spectral selection* and *bit selection*. In spectral selection, the zigzag scanned DCT coefficients are divided into frequency bands, and a separate band is encoded during each pass. In bit selection, the first pass consists of encoding the N most significant bits of the DCT coefficients (where N is specified by the user). Subsequent passes encode the less significant bits of the coefficients until the coefficients have been encoded to their full accuracy.

It should be noted that the above two methods are complementary to each other and can be used separately or mixed. For example, the most significant N bits of the most important band may be chosen for the first pass of the encoding.

Progressive JPEG decoding consists of the inverse steps of progressive JPEG encoding.

The hierarchical JPEG algorithm The hierarchical JPEG algorithm uses multiresolution image processing techniques to deal with display devices of multiple resolutions. The algorithm consists of low-pass filtering and subsampling the image by factors of two. The reduced image is encoded by baseline, lossless, or progressive JPEG algorithms. The encoded image is then decoded and zoomed to the original resolution. Using this image as the prediction, the error between the original image and the prediction is computed. This error is encoded by using either baseline, lossless, or progressive compression. This procedure is repeated until full resolution is encoded. One use of hierarchical encoding is for high-resolution images, which must be printed at high resolution in a printer, but can be displayed at a low resolution on a computer monitor.

Implementation of JPEG The implementation of JPEG compression consists of DCT, zigzag scanning, and Huffman coding operations for gray-scale images. For color images where chromatic subsampling has been used, it may be necessary to perform an additional color conversion before the image can be displayed on an RGB color system.

In the absence of any color conversion, the most computationally expensive operation in baseline JPEG is the DCT. We have looked at several ways of implementing DCT in Section 9.2.2. Any of these may be used in the implementation of both a compressor and a decompressor. The splat algorithm discussed previously can be used in a decompressor for efficient decompression.

In many cases, images are encoded in YUV color space and subsampled chromatically. In these cases, the images must be converted into RGB color space before being displayed. If decoding is being done entirely by a microprocessor without hardware color conversion assistance, this operation may turn out to be more time consuming than any other operation in JPEG decompression.

Having looked at still picture compression, we are now ready to discuss the compression of moving pictures.

9.3.3 MPEG standard for moving picture compression

MPEG (Motion Picture Experts Group) is a set of standards defining the syntax of compressed bitstream for digital video and digital audio sequences. An MPEG encoder must produce a compressed bitstream conforming to this syntax. The MPEG decoder's function is to take the bitstream and decode it into the sequence of moving pictures from which it was generated. Like the JPEG standard discussed in the previous section, the algorithms chosen for MPEG were decided by the results of a blind competition.

There are two flavors of MPEG that are currently standardized. MPEG-1 is for lower-resolution images and encodes bitstreams at a rate of up to about 1.5 Mbits/sec. MPEG-2 supports higher-resolution images and can go up to 15 Mbits/sec. There are several other differences between these standards—the main one being that MPEG-2 is able to support motion compensation based on fields rather than frames—but they are both based on the same basic algorithms. In the next section we will look at the future of MPEG as represented by the MPEG-4 standardization activities.

A salient feature of MPEG is that it standardizes the synchronization between the video and its associated audio parts. This is done by the MPEG *systems* standard, which defines how the compressed video and compressed audio streams are multiplexed together. Thus, there are three main parts of MPEG-1 and MPEG-2. These are called *video*, *audio*, and *systems*. (MPEG-2 has some additional parts—supporting the operation of video servers over networks—that are beyond the scope of this discussion.)

Because it deals exclusively with moving pictures, and uses motion estimation, MPEG is substantially more complex than baseline JPEG or Group 3 algorithms. In this section we will be concerned with MPEG-1 video only. For a tutorial on MPEG audio, the reader should see reference 19. For information about MPEG systems, the reader should see reference 20. Because of its complexity, our review of MPEG video will not be as detailed as our review of JPEG was. Only the basic ideas will be presented, in order that we can see how the building blocks discussed earlier in this chapter can be used to build a compression system. To implement an MPEG encoder and decoder, substantially more information is needed, and the implementor must read the standards documents.

Encoder considerations An MPEG-1 video encoder compresses a sequence of digital pictures, or frames, and creates a bitstream that conforms to the MPEG-1 video bitstream syntax. Color frames are in YCbCr-601 color space, with 4:2:0 chromatic subsampling being used (discussed in Chapter 7), so that for every four *Y* pixels, there are one *Cb* and one *Cr* pixel.

MPEG-1 distinguishes between three types of pictures: *I-pictures*, *P-pictures*, and *B-pictures*, based on the method used in compression. An I-picture (intrapicture) is compressed as a stand-alone picture, without information from any other pictures in the sequence. The compression of a P-picture (predicted picture) uses the most recently compressed I- or P-picture. Finally, a B-picture (bidirectional picture) is compressed with information from both past and future, using the closest I- or P-pictures.

A typical organization of MPEG pictures is like this: *IBBPBBPBBPBBI*. A common scenario is to have 15 pictures from one I-picture to the next.

The pictures used in the prediction of P- and B-pictures are as shown in Figure 9.30. Initially, the encoder compresses the first I-picture. Then, the next P-picture is compressed. Following these, the B-pictures in between the I- and the P-picture are compressed. The second P-picture is compressed based on the first P-picture. The B-pictures between the two P-pictures are then bidirectionally compressed.

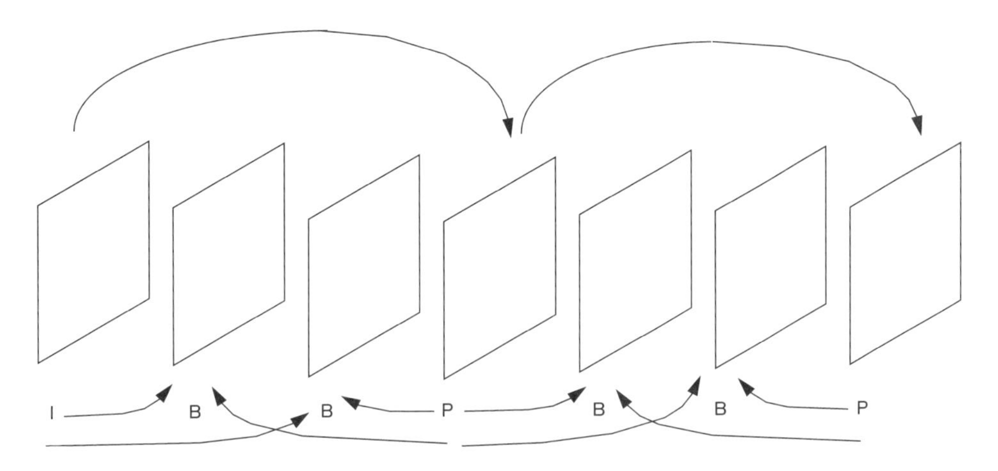

Figure 9.30 A typical sequence of MPEG pictures

In order for the decompressor to function, the order of the pictures placed in the compressed bitstream is different from the order in which they are displayed. For example, in the above case, the I- and the P-pictures are placed in the bitstream before the two B-pictures.

Compression of I-pictures is probably the simplest. It is similar to the baseline JPEG algorithm, with 8×8 DCT's followed by quantization, zigzag scanning, and variable-length coding. The quantization tables, however, are different from JPEG, with smaller values that zero out the coefficients less drastically than in JPEG. There is no motion estimation involved in the compression on I-pictures.

In compressing a P-picture, motion estimation using a block search, as discussed in Section 9.2.4, is used. The size of the block—called a *macroblock*—is 16×16 (for *Y*) and 8×8 (for *Cb* and *Cr*). Motion estimation is performed between the reference picture (that is, the last I- or P-picture) and the target picture (that is, the picture being compressed). A good match is found if the error between the target macroblock and the motion compensated macroblock from the reference picture does not exceed a predefined error threshold. In this case, the motion vector is encoded and added to the bitstream. In addition, the DCT of the "difference macroblock"—that is, the macroblock representing the difference between the block being compressed and the predicted block—is also computed and added to the bitstream after quantization and variable-length coding. If a good match is not found, then a motion vector is not used. Instead, the DCT of the macroblock is taken, quantized, and variable-length coded and added to the bitstream. This is shown in Figure 9.31.

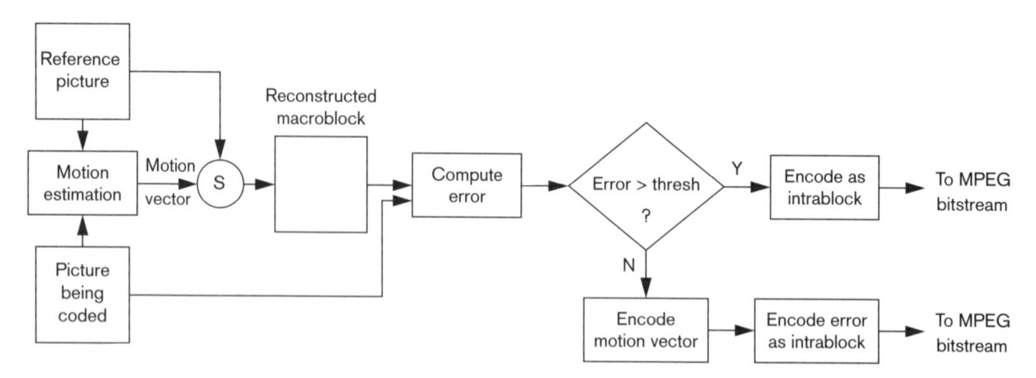

Figure 9.31 Encoding macroblocks of MPEG P-pictures

The encoding of B-pictures is the most complex of all three. Performing motion estimation with the previous I- or P-picture yields a forward motion vector. Similarly, a backward motion vector is obtained by performing motion estimation with the next I- or P-picture. The target image macroblock can now be reconstructed in one of three ways: from the previous picture using the backward motion vector, from the future picture using the backward motion vector, or averaging these two motion vectors. The "best" option, in the sense of least error, is chosen. After this, the steps followed are the same as in compressing P-pictures. Compression of B-pictures is illustrated in Figure 9.32.

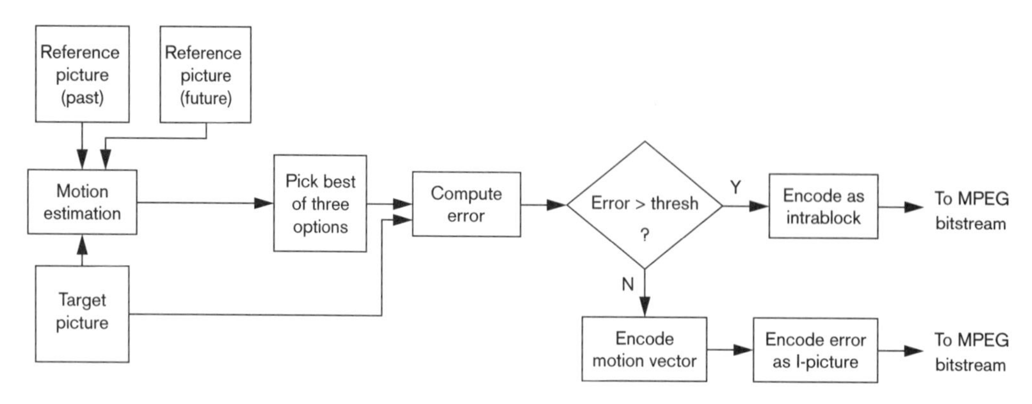

Figure 9.32 Encoding MPEG P-pictures

The above is a high-level description of MPEG-1 video compression. There are many subtleties in the standard that are not covered here.

Decoder considerations The MPEG video decoder is responsible for decompressing the compressed MPEG video bitstream. As we have seen, the order of pictures in the compressed bitstream is not necessarily the order in which they are displayed (for example, P-pictures come before B-pictures). To decompress a B-picture, a future I- or P-picture and a past I- or P-picture are needed. Thus, the decompressor needs to buffer two pictures for motion compensation: one for the relevant previous picture and one for the relevant future picture.

For example, if the sequence $I_1P_4B_2B_3P_7B_5B_6P_{10}\ldots$ arrives at the decompressor, first I_1 is decompressed and placed in the previous buffer. Then P_4 is decompressed using I_1 and placed in the future buffer. Then B_2 and B_3 are decompressed using these buffers. Display of pictures can begin as soon as the two buffers are full, and the decompressor can display one picture for each picture that is being decompressed. Continuing the process, when P_7 comes in, P_4 is copied into the previous buffer and P_2, after decompression, is copied into the future buffer, after which B_5 and B_6 are decompressed.

The decompression of I-pictures is similar to the baseline JPEG decompression shown in Figure 9.33. The decompressed image must be placed in the buffer instead of being displayed.

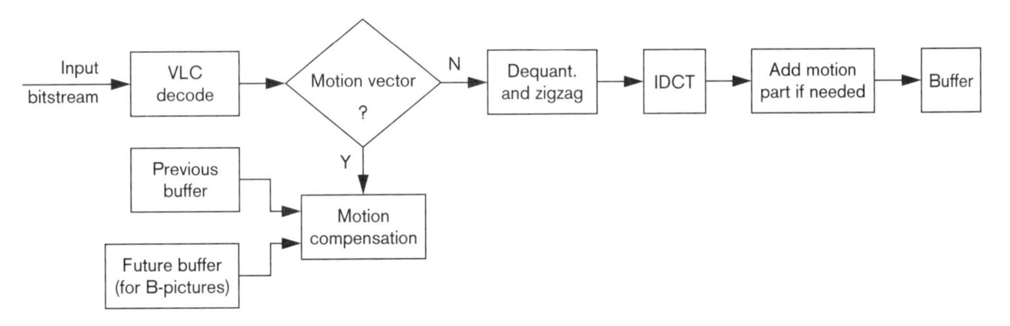

Figure 9.33 Decoding MPEG P- and B-pictures

The decompression of P-pictures is shown in Figure 9.33. The decompressor first performs a variable-length decoding. Then, if a motion vector is present in the bitstream, motion compensation must be done. In addition, an IDCT must be performed. If motion compensation was done, then the IDCT results are added to the motion compensated block and placed in the buffer. Otherwise, the IDCT results are directly placed in the buffer.

The decompression of B-pictures is similar to the decompression of P-pictures. The main difference is that two buffers are used to perform the motion compensation.

Computational complexity Several attempts have been made to measure the computational complexity of MPEG-1 and MPEG-2. The processing requirements are often expressed in Million Operations per Second (MOPS), where each operation could be a load, store, addition, or multiplication. In reference 21, it is estimated that the computational requirement for MPEG-1 compression of CCIR-601 resolution images varies between 3020 and 4585 MOPS, whereas MPEG-1 decompression requires 395 to 466 MOPS. A similar estimate of 400 MOPS for MPEG-2 decompression at CCIR-601 resolution has been made in reference 22. While these numbers provide a rough estimate of the demands placed on the processor by MPEG video, the lessons learned in this book show that choice of clever algorithms and implementation tricks can substantially reduce the processing requirements.

However clever the implementation, MPEG-1 and MPEG-2 video compression are demanding algorithms, which stretch the performance capabilities of any piece of hardware. The computations needed by the compressor are much greater than those needed by the decompressor because of the motion estimation. For this reason, real-time MPEG-1 and MPEG-2 compression usually require dedicated hardware in the form of integrated circuit chips, add-on boards, or even turn-key systems. However, we anticipate that general-purpose microprocessors in the near future will have the processing power needed for real-time MPEG encoding.

Using clever IDCT and color conversion algorithms, the decompression of MPEG-1 video and systems is possible on desktop computers. Several software solutions are available for decompressing MPEG-1 system streams on desktop computers. More recently, it has been shown [22] that MPEG-2 video can be decompressed using the UltraSPARC microprocessor and the Visual Instruction Set described in Chapter 3.

9.3.4 MPEG-4: the future of MPEG

A newer flavor of MPEG, called MPEG-4, is being standardized as this book is being written. (There is no MPEG-3.) MPEG-4 emphasizes content-based coding and high-compression ratio for emerging multimedia applications. Many of the ideas behind MPEG-4 originate in a pan-European project called Mobile Audio-Visual Terminal (MAVT) in which several European companies and governments participated.

The application areas that MPEG-4 addresses are shown in Figure 9.34. The underlying assumption is that the traditional boundaries between telecommunications, computer, and TV/film industries are no longer well defined and there is a convergence taking place between these technologies. MPEG-4 addresses the shaded areas represented by this convergence.

A key feature of MPEG-4 is the support of *content-based* interactivity in multimedia applications. This allows the user to interact with the individual objects present in an audio-visual scene. While this level of interaction has been available in the area of synthetic images (computer graphics) for many years, it is only recent advances in computer vision and model-based compression that enables such capabilities when dealing with natural images. Content-based compression—as distinct from the other types of compression studied in this chapter—relies on being able to understand the objects present in an image, and compresses the image based on this understanding.

A second key feature of MPEG-4 is the support of both natural and synthetic imagery, as well as hybrid scenes, which are made up of natural and artificially generated images.

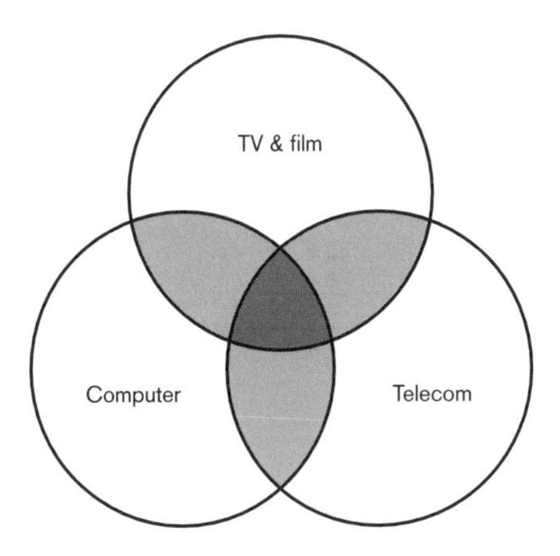

Figure 9.34 MPEG-4 areas of interest

The developers of MPEG-4 hope that it will be used in such applications as content-based retrieval of information from on-line libraries and travel information databases, interactive home shopping, communicating from a mobile terminal, and multimedia entertainment.

MPEG-4 is composed of four different elements: *tools, algorithms, profiles,* and *MPEG-4 syntax. description language (MSDL).*

A tool is a building block of the MPEG-4 system. For example, motion compensation, or segmentation of the objects in an image, is a tool. An algorithm is a collection of tools that provides a compression functionality. An example of an algorithm is MPEG-1 video decompression. A profile is an algorithm, or a set of algorithms, which is chosen to address a specific class of applications. MSDL is a programming language which allows the description, selection, and downloading of tools, algorithms, and profiles.

The relationship between these elements is shown in Figure 9.35.

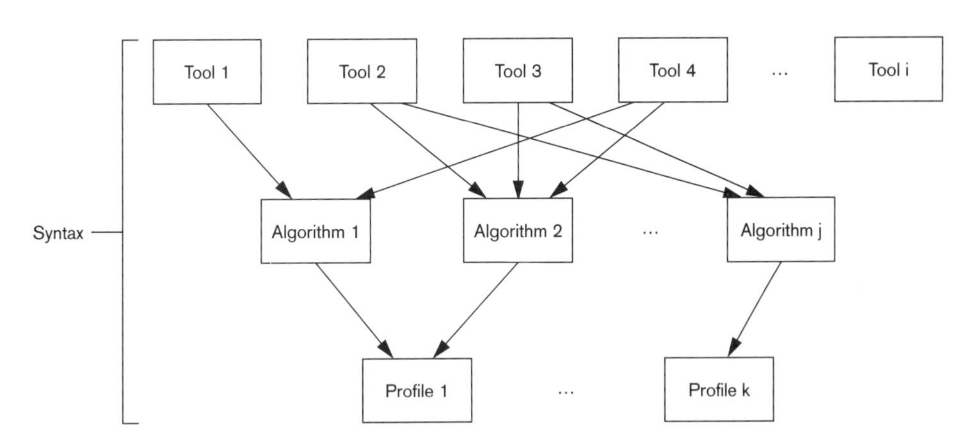

Figure 9.35 Structure of MPEG-4

The downloading feature of the MSDL merits further discussion. The idea is very similar to that of downloading of executable content over the Internet using Java, as discussed in Section 4.10. Recognizing the diverse nature of the applications it intends to capture, MPEG-4 allows a large amount of flexibility in the choice of compression techniques. It assumes that not all decoders will have the built-in capability for understanding every compression technique being used. If the decoder is unable to decode a particular compressed sequence, it uses the MSDL to download the decoding program as part of the compressed bitstream. This is similar to the Java programming model shown in Figure 4.15, with the decoder taking on the role of the client and the encoder taking the role of the server.

Work on the definition of the elements of MPEG-4 proceeds as of this writing (see reference 23). While no product based on MPEG-4 has arrived on the market, the developers of MPEG-4 have solved a number of difficult problems and are closing in on a viable standard.

9.4 Conclusion

Image data compression is a thriving area of computer imaging. In this chapter we have looked at several building blocks useful for building compression and decompression systems. We have also considered compression and decompression systems based on standards—fax, JPEG, and MPEGvideo—and have seen how the building blocks of image compression can be used to build standards-based systems. Finally, we have looked at the emerging MPEG-4 standard.

A lucid and complete account of image compression building blocks can be found in reference 24. To probe further on the compression standards, the reader can look at reference 1 for a detailed look at JPEG, and reference 21 for an abbreviated look at MPEG and H.261.

9.5 References

1. J. Pennebaker and J. Mitchell, *JPEG Still Image Data Compression Standard*, New York: Van Nostrand Reinhold, 1993.
2. N. Abramson, *Information Theory and Coding*, New York: McGraw-Hill, 1963.
3. D. A. Huffman, "A Method for the Construction of Minimum Redundancy Codes," *Proceedings of the IRE*, Vol. 40, 1952, pp. 1098–1101.
4. R. Sedgewick, *Algorithms*, New York: Addison-Wesley, 1988.
5. A. K. Jain, *Fundamentals of Digital Image Processing*, New York: Prentice Hall, 1988.
6. E. Feig, "A Fast-Scaled DCT Algorithm,"*Proceedings of the SPIE, Image Processing Algorithms and Techniques*, Vol. 1244, 1990.
7. L. Macmillan and L. Westover, "A Forward Mapping Realization of the Inverse Discrete Cosine Transform," *Data Compression Conference*, Los Alamitos, CA: IEEE Computer Society Press, 1992, pp. 219–228.

8 H. G. Musmann, "Predictive Image Coding," *Advances in Electronics and Electron Physics*, Suppl. 12, W. K. Pratt, ed., New York: Academic Press, 1979, pp. 73–112.

9 J. W. Woods and S. D. O'Neil, "Sub-band Coding of Images," *IEEE Transactions on Acoustics, Speech, and Signal Processing*, Vol. 34, No. 5, May 1986, pp. 1278–1288.

10 A. N. Netravali and B. G. Haskell, *Digital Pictures*, New York: Plenum Press, 1988.

11 E. Dubois and J. Konrad, "Review of Techniques for Motion Estimation and Motion Compensation," colloquium on HDTV, 1990.

12 Y. Linde, A. Buzo, and R. M. Gray, "An Algorithm for Vector Quantizer Design," *IEEE Transactions on Communications,* Vol. COM-28, No. 1, 1980, pp. 84–95.

13 R. E. Crochiere, S. A. Webber, and J. L. Flanagan, "Digital Coding of Speech in Subband Signals," *Bell System Technical Journal*, Vol. 55, No. 8, October 1976, pp. 1069–1085.

14 J. D. Johnston, "A Filter Family for Use in Quadrature Mirror Filter Banks," *Proceedings IEEE ICASSP*, Denver, CO, 1980, pp. 291–294.

15 J. Shapiro, "Embedded Image Coding Using Zerotrees of Wavelet Coefficients," *IEEE Transactions on Signal Processing*, Vol. 41, No. 12, December 1993.

16 M. Kaneko, A. Hike, and Y. Hatori, "Coding of Facial Image Sequence Based on a 3-D Model of the Head and Motion Detection," *Journal of Visual Communications and Image Representation*, Vol. 2, No. 1, March 1991, pp 39–54.

17 M. F. Barnsley and L. P. Hurd, *Fractal Image Compression*, Wellesley, MA: AK Peters, 1993.

18 M. Deering, et al., "Geometry Compression," *Proceedings of SIGGRAPH*, 1995.

19 D. Pan, "A Tutorial on MPEG/Audio Compression," *IEEE Multimedia*, Summer 1995.

20 ISO/IEC International Standard IS 11172-2, "Information Technology—Coding of Moving Pictures and Associated Audio for Digital Storage Media at up to about 1.5 Mbits/s—Part 2: Systems."

21 V. Bhaskaran and K. Konstantinides, *Image and Video Compression Standards*, Boston: Kluwer Academic Publishers, 1995.

22 C-G Zhou, L. Kohn, D. Rice, I. Kabir, A. Jabbi, "MPEG Video Decoding Using the UltraSPARC Visual Instruction Set," *COMPCON*, Spring 1995.

23 ISO/IEC JTC1/SC29/WG11 MPEG 95/027 MPEG-4 Proposal Package Description—Revision 2.0.

24 M. Rabbani and P. Jones, *Digital Image Compression Techniques*, Bellingham, WA: SPIE Press, 1991.

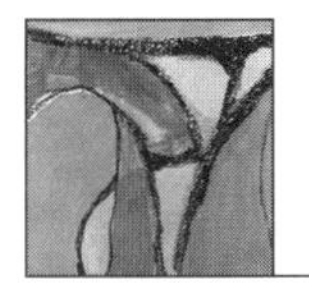

appendix A

Benchmarking and evaluation of imaging products

A.1 Benchmarking of imaging products

With the all-important role that performance plays in imaging products, one would think that there would be well-established imaging benchmarks that measure the performance of imaging products in the same way that SPEC measures the performance of workstations, X11PERF measures the performance of implementations of the X Windows system, and GPC measures the performance of graphics software. Unfortunately, as we have seen in Chapter 3, this is not the case: There is no well-established benchmark for imaging products.

There are several reasons for this deficiency. Until recently, due to the inability of general-purpose computers to handle the computational demands of imaging, most imaging systems were based on special-purpose hardware. Hardware imaging computers excelled at performing a (usually small) set of primitive imaging operations. Thus, it was difficult for the manufacturers to agree on a broad enough set of imaging functions to benchmark, since a given machine performed poorly on unaccelerated functions. In addition, the sheer diversity of imaging operations made it difficult to find a useful set of functions on which to measure performance. Finally, the lack of a standard set of imaging functions—soon to be filled by the PIKS imaging standard—also contributed to this deficiency.

A.1.1 The Abingdon Cross

With these difficulties in mind, a test called the Abingdon Cross [1] was proposed in the early 1980s to benchmark imaging computers. The Abingdon Cross's goal was to find the skeleton (or the medial axis transform) of a white cross in a black background, in the presence of noise.

Unfortunately, there were several serious problems with the Abingdon Cross. First, there were many ways to solve the problem. In principle, the idea was to filter out the noise, threshold the image, and then apply a morphological filter to the image to extract the medial axis transform. However, the choice of the filter (median or linear?), the method for setting the threshold, and the algorithms to extract the medial axis transform were left to the implementor. Therefore, when comparing the results of the Abingdon Cross test on different machines, one was never quite sure if one were comparing apples to apples.

Another problem with the Abingdon Cross was that it omitted some fundamental imaging operations. For example, an imaging product could do very well on the Abingdon Cross test without doing so well in copying images, or lookup table operations, or convolutions. Conversely, a mediocre score on the test did not necessarily mean that the product was slow. For example, the Vicom image processor scored average on the Abingdon test. However, for many years, the Vicom was the only imaging machine capable of performing real-time convolutions, a fact that was never reflected in the test.

Benchmarking suggestions All of the above means that there is no standard benchmark in imaging. However, the manufacturer of the product needs to provide benchmark numbers in order to distinguish the product from its competitors. Wherever possible, benchmark numbers should be provided for a variety of image sizes, since the processing rate can vary so much with the image size.

Some questions that the benchmark should answer (to some degree of precision if not exactly) are:

- *What is the copy rate in megapixels/second?* The copy rate of the product, that is, the number of pixels that it can copy per second from one location to another, tells us about the dataflow bottlenecks, and gives us some idea of what to expect from the product. In the case of imaging hardware with dedicated image memory, the rates for copying between this hardware memory and program memory (i.e., rates of copying between accelerated memory and unaccelerated memory) should also be provided.

- *What is the convolution rate in megapixels/second?* In addition to being a fundamentally important imaging operation, convolution is a moderately complex function that exercises the compute horsepower of the product under evaluation. Therefore, this figure gives us a quick idea of how fast the product crunches pixels. Many hardware devices have dedicated hardware to perform convolution. If a "special-case" implementation of convolution (such as one based on table lookups to avoid multiplications) is used for benchmarking, then it should be stated and figures for a general-case implementation should also be provided.

- *What are the rates, in megapixels/second, for some other functions useful for the target market?* The choice of these functions can vary greatly with the intended application of the product. Typically, table lookup and resampling functions are important. The figures should be chosen such that they convey an idea of how well, in general, the product is expected to perform (i.e., few special cases.)

- *What are the largest and smallest images and pixel sizes that can be handled?* This is an important question because the size of the images varies a good deal with applications, and not all architectures are able to handle different sizes equally well. Some graphic arts applications require images that are a few hundred megabytes, whereas many other applications require images between 256 KB and 1 MB.

A.2 Evaluation of imaging products

Evaluation of any product is a complicated affair. Many factors involved in an evaluation are beyond the scope of this discussion. In this section, we will attempt to focus on some technical issues that should be addressed while evaluating an imaging product.

In general, there are two ways to position an imaging product in the market: as an Original Equipment Manufacturer (OEM) product, or as an end-user product. The OEM uses the imaging product as a "part" in an end-user product. For example, a hardware imaging accelerator and its associated software libraries can be used internally in a photocopier to perform fast image enhancement. An end-user product, as the name suggests, is a product that is sold to the user directly. Examples are an image -editing application program or a turn-key document management system.

When an imaging product is being evaluated, its proposed use—that is, OEM or end-user—should be used to guide the evaluation. We now discuss the general principles of evaluation of an imaging product in these two categories.

A.2.1 Evaluation of an imaging product for OEM purposes

In general, this kind of an imaging product performs a subtask which is one of possibly many subtasks making up the larger overall task that the product performs. These functions can be repetitive, as in a photocopier, or one-time, as in an MRI scanner. Often, there are timing constraints on the entire task, such as, "60 photocopies per minute," or "one MRI scan in five minutes." Depending on how long the other subtasks take, there will be timing constraints on the imaging part.

Some preliminary questions about the product must be asked to eliminate products that are completely unsuitable:

- *Does it fit in with the rest of the final product being envisioned?* This question should take into account issues such as input/output and bandwidth compatibility, as well as other related issues about how well it fits into the space allotted for the imaging product.
- *Does it have the minimum acceptable functionality?* Clearly, the imaging product being evaluated must be able to perform the minimum functions necessary for the product.
- *Does it meet minimum performance requirements?* The product must be able to perform at an "interesting" speed to be considered.
- *Is the price/performance acceptable?* The price of the product, given its performance, must be competitive and within the limits set for the cost of the final product.

Once the product has passed these tests, a more detailed analysis of the product can be performed. The following is a possible set of questions:

- *If the product is programmable, how difficult is it to program?* The cost of programming varies with the complexity of the code that must be written. If microcode needs to be written, then the programming cost is much higher than if a software imaging library is provided.

- *What is the overall timing requirement of the end product? What is the timing share of the imaging product?* The timing of the finished product, as well as the maximum time allotted to the imaging product must be known.
- *Is it able to meet this timing requirement for the "default" operations and image size?* Often, we need a product to perform well in the common case and adequately in the uncommon cases. Clearly, the product must hold the promise of meeting or beating the timing requirements for the common cases.
- *Is it able to meet this requirement for all the operations and image sizes?* This is a harder requirement to meet. Most imaging products do not perform all functions equally well. The implementor chooses to optimize some functions based on the perceived priority of the functions.

A.2.2 Evaluation of an imaging product for end-user purposes

Evaluation of an end-user imaging product is more focused on the characteristics of the entire system as a whole. Naturally, more attention must be paid to how the system appears to the user. The following list of questions should be addressed while evaluating an imaging product for the end user:

- *What is the display capability of the product?* Most imaging products have displays associated with them. Some of these are monochrome, some gray level, and some color. In color, one can have 8-bit color which gives 256 possible colors, or 24-bit color, which means 4 million colors. In addition, the resolution of the display must be taken into consideration.
- *What is the size of images and pixels it can handle?* Some application areas, such as prepress and graphic arts, need to handle very large images. Other areas, such as medical imaging, require high dynamic range pixels (e.g., 12- to 16-bit precision).
- *How easy is it to use?* Ease-of-use is critical for end-user products, especially because we cannot expect that the users will be technically proficient engineers.
- *Is the functionality adequate?* The product must provide adequate functionality for the job for which it is intended.
- *What is the response time of the product?* The response time is important for applications where the user of the product uses it to process large amounts of images.

A.3 References

1 K. Preston, Jr., "Abingdon Cross," *IEEE Computer*, July 1989, pp. 9–20.

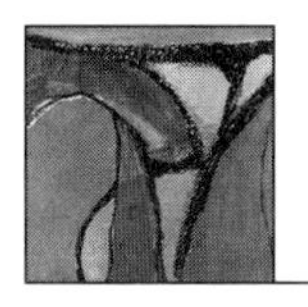

appendix B

Imaging resources

B.1 General image processing textbooks

The following books are general image processing textbooks.

Baxes, G., *Digital Image Processing*, 2nd edition, New York: Wiley, 1994. A basic introduction to image processing including some hands-on examples.

Castleman, K. R., *Digital Image Processing*, Englewood Cliffs, NJ: Prentice Hall, 1979. This was one of the first books to appear on image processing. Lucid and motivating. A second edition, including a new chapter on wavelets, appeared in 1996.

Gonzalez, R., and R. E. Woods, *Digital Image Processing*, 2nd edition, New York: Addison-Wesley, 1993. Second edition of another classic book. Covers basic image processing. Somewhat easier to understand than Jain or Pratt.

Jain, A. K., *Fundamentals of Image Processing*, Englewood Cliffs, NJ: Prentice Hall, 1988. Classic graduate-level textbook. Formal treatment of the subject. Particularly strong on probablistic mathematical models and compression.

Pratt, W. K., *Digital Image Processing*, 2nd edition, New York: Wiley, 1991. Second edition of another classic graduate-level textbook. Encyclopaedic treatise on the subject. Second edition adds chapters on image analysis and mathematical morphology but removes chapter on compression.

Rosenfeld, A., and A. C. Kak, *Digital Picture Processing*, 2nd edition, New York: Academic Press, 1982. Good comprehensive coverage, with emphasis on image analysis.

B.2 How-to books on image processing

Dougherty, E., and P. Laplante, *Introduction to Real Time Image Processing,* Bellingham, WA: SPIE Press, 1995. Course notes turned into a book. Emphasizes technical image processing.

Lindley, C., *Practical Image Processing in C.* The first published book on C programming for image processing. It has an interesting section on frame grabber design.

Myler, H., and A. Weeks, *Imaging Recipes in C*, Englewood Cliffs, NJ: Prentice Hall, 1993. More image processing recipes for straightforward implementations of image processing functions.

Pitas, I., *Digital Image Processing Algorithms*, Englewood Cliffs, NJ: Prentice Hall, 1993. Good coverage of technical image processing. Has many C programming examples, with emphasis on straightforward implementations.

B.3 Books on particular topics

Bhaskaran, V., and K. Konstantinides, *Image and Video Compression Standards*, Boston: Kluwer Academic Publishers, 1995. Covers JPEG, MPEG, and H.261, as well as various compression techniques. Also covers hardware solutions.

Hussain, Z., *Parallel Image Processing,* New York: Ellis Horwood, 1991. Imaging algorithms for parallel computers. Interesting architectures, developed mostly for academic use.

Mitchell, J., and M. Pennebaker, *JPEG Still Image Data Compression Standard*, New York: Van Nostrand Reinhold, 1991. *The* reference book on JPEG compression. Includes discussion of various DCT algorithms, as well as all the flavors of JPEG.

Rabbani, M., and P. Jones, *Image Data Compression*, Bellingham, WA: SPIE Press, 1991. Excellent introduction to several types of compression algorithms. Includes a worked example of JPEG compression.

Wolberg, G., *Digital Image Warping*, Los Alamitos, CA: IEEE Computer Society Press, 1990. Excellent book on geometric processing of images. Lays out the foundation of the material covered in Chapter 8.

B.4 Other books of interest

Graphics Gems, Vols. 1–4, Boston: Academic Press, 1990–1994. This series of books looks at 2-D and 3-D graphics and imaging algorithms and their fast implementation. Good for ideas on how to optimize pixel processing.

Hennessy, J., and D. Patterson, *Computer Architecture: A Quantitative Analysis*, San Mateo, CA: Morgan Kaufmann. Invaluable reference on various aspects of computer design for performance.

Pratt, W. K., *PIKS Foundation C Programmer's Guide*, Greenwich, CT: Manning Publications, 1995. PIKS is the only ANSI standard for image processing and this is the only reference book on PIKS.

———, *The XIL Book*, in press. This book is devoted to the use of the XIL library for performing image processing operations.

B.5 Internet resources

There are several newsgroups for information on image processing. These include:

- *comp.ai.vision* a moderated group for discussing computer vision
- *comp.compression* discussion on compression of images, text, video, and so forth.
- *comp.graphics* general graphics discussions (such topics as color quantization are more likely to show up here than in the other groups)
- *sci.image.processing* image processing discussion

B.6 Free imaging and video software

There are numerous imaging and related software packages available free of charge on the Internet. These include:

- *gimp* This is a freely available image manipulation software package from http://www.xcf.berkeley.edu/~gimp/gimp.html. It is currently available on UNIX systems only.
- *JPEG software* The Independent JPEG group has published a free JPEG codec. This can be obtained from ftp.uu.net in the directory graphics/jpeg.

- *Khoros* Khoros was originally developed at the University of Arizona; it is currently owned by Khoral Research International. It is a comprehensive image processing package, available from ftp.khoral.com. Khoros runs on a variety of hardware platforms, including several models of UNIX workstations, Macintosh, and IBM-compatible PCs, and lets you perform a variety of image processing operations.
- *MPEG software* There are several public-domain software MPEG codecs, including one from the University of California at Berkeley (s2k-ftp.cs.berkeley.edu:/multimedia/mpeg) and one from Stanford University (havefun.stanford.edu:/pub/mpeg).
- *nih-image* Image processing software for the Macintosh platform only, available via ftp from zippy.nimh.nih.gov.
- *pbmplus (or netpbm)* Written by Jef Poskanzer. This is primarily a file conversion program, but with basic image processing functions built in, including convolution, scale, and color quantization. It can be obtained from http://www.arc.umn.edu/GVL/Software/pbmplus.html.
- *The Utah Raster Toolkit* This is a free software package for performing basic imaging operations as well as conversion and display of images. It is available at http://www.arc.umn.edu/GVL/Software/urt.html. Works on a variety of hardware platforms with X11 and SGI displays.

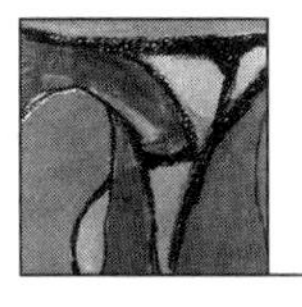

appendix C

Compression tables

C.1 Modified Huffman tables for fax Group 3

Table C.1 shows the terminating codes for Group 3 modified Huffman coding, as discussed in Chapter 9.

Table C.1 Group 3 terminating codes

Run length	Code word (white)	Code word (black)
0	00110101	0000110111
1	000111	010
2	0111	11
3	1000	10
4	1011	011
5	1100	0011
6	1110	0010
7	1111	00011
8	10011	000101
9	10100	000100
10	00111	0000100
11	01000	0000101
12	001000	0000111
13	000011	00000100
14	110100	00000111
15	110101	000011000
16	101010	0000010111
17	101011	0000011000
18	0100111	0000001000
19	0001100	00001100111
20	0001000	00001101000
21	0010111	00001101100
22	0000011	00000110111
23	0000100	00000101000
24	0101000	00000010111
25	0101011	00000011000
26	0010011	000011001010
27	0100100	000011001011
28	0011000	000011001100
29	00000010	000011001101

Table C.1 Group 3 terminating codes (continued)

Run length	Code word (white)	Code word (black)
30	00000011	000001101000
31	00011010	000001101001
32	00011011	000001101010
33	00010010	000001101011
34	00010011	000011010010
35	00010100	000011010011
36	00010101	000011010100
37	00010110	000011010101
38	00010111	000011010110
39	00101000	000011010111
40	00101001	000001101100
41	00101010	000001101101
42	00101011	000011011010
43	00101100	000011011011
44	00101101	000001010100
45	00000100	000001010101
46	00000101	000001010110
47	00001010	000001010111
48	00001011	000001100100
49	01010010	000001100101
50	01010011	000001010010
51	01010100	000001010011
52	010101	000000100100
53	00100100	000000110111
54	00100101	000000111000
55	01011000	000000100111
56	01011001	000000101000
57	01011010	000001011000
58	01011011	000001011001
59	01001010	000000101011
60	01001011	000000101100
61	00110010	000001011010
62	00110011	000001100110
63	00110100	000001100111

Table C.2 shows the makeup code for modified Huffman coding, as discussed in Chapter 9.

Table C.2 Group 3 makeup codes

Run length	Code word (white)	Code word (black)
64	11011	0000001111
128	10010	000011001000
192	010111	000011001001
256	0110111	000001011011
320	00110110	000000110011
384	00110111	000000110100
448	01100100	000000110101
512	01100101	0000001101100
576	01101000	0000001101101
640	01100111	0000001001010
704	011001100	0000001001011
768	011001101	0000001001100
832	011010010	0000001001101
896	011010011	0000001110010
960	011010100	0000001110011
1024	011010101	0000001110100
1088	011010110	0000001110101
1152	011010111	0000001110110
1216	011011000	0000001110111
1280	011011001	0000001010010
1344	011011010	0000001010011
1408	011011011	0000001010100
1472	010011000	0000001010101
1536	010011001	0000001011010
1600	010011010	0000001011011
1664	011000	0000001100100
1728	010011011	0000001100101
EOL	000000000001	000000000001

C.2 Images used for fax Group 3

The following eight images were used to generate the tables for Groups 3 and 4 fax standards.

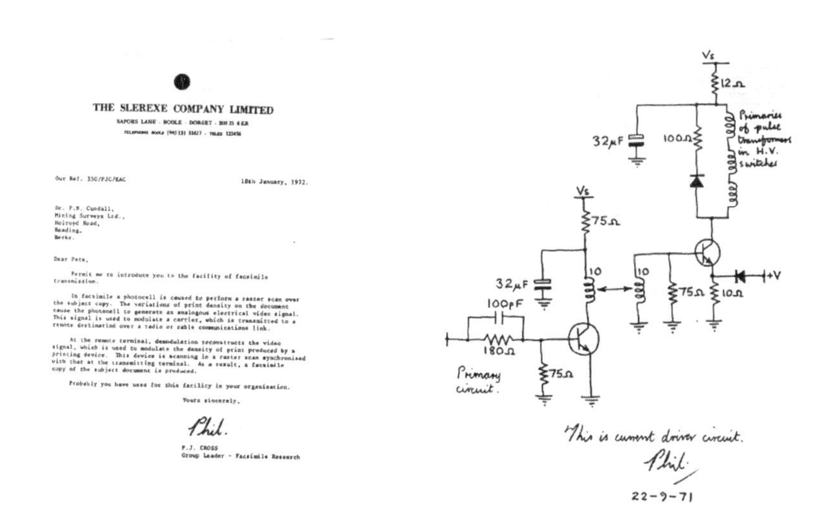

THE SLEREXE COMPANY LIMITED

SAPORS LANE · BOOLE · DORSET · BH25 8ER

TELEPHONE BOOLE (945 13) 51617 · TELEX 123456

Our Ref. 350/PJC/EAC

18th January, 1972.

Dr. P.N. Cundall,
Mining Surveys Ltd.,
Holroyd Road,
Reading.
Berks.

Dear Pete,

Permit me to introduce you to the facility of facsimile transmission.

In facsimile a photocell is caused to perform a raster scan over the subject copy. The variations of print density on the document cause the photocell to generate an analogous electrical video signal. This signal is used to modulate a carrier, which is transmitted to a remote destination over a radio or cable communications link.

At the remote terminal, demodulation reconstructs the video signal, which is used to modulate the density of print produced by a printing device. This device is scanning in a raster scan synchronised with that at the transmitting terminal. As a result, a facsimile copy of the subject document is produced.

Probably you have uses for this facility in your organization.

Yours sincerely,

Phil.

P.J. CROSS
Group Leader - Facsimile Research

Figure C.1

Figure C.2

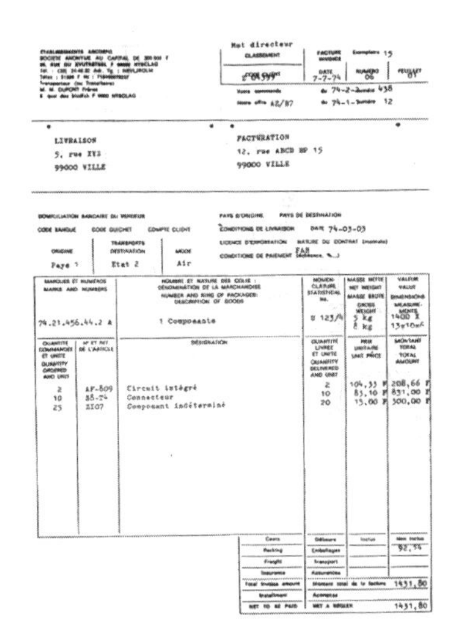

Mot directeur

LIVRAISON
5, rue XYZ
99000 VILLE

FACTURATION
12, rue ABCD BP 15
99000 VILLE

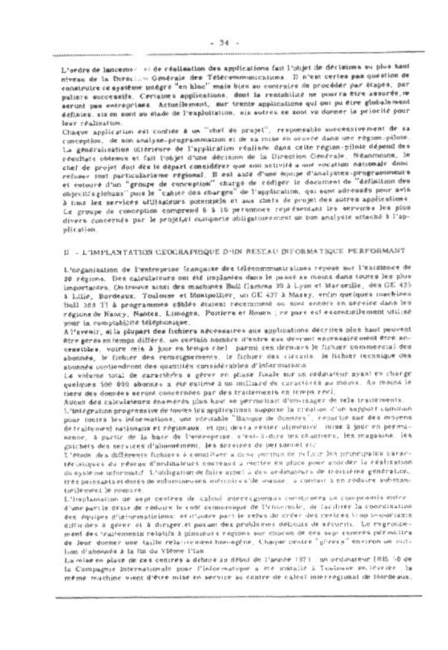

Figure C.3

Figure C.4

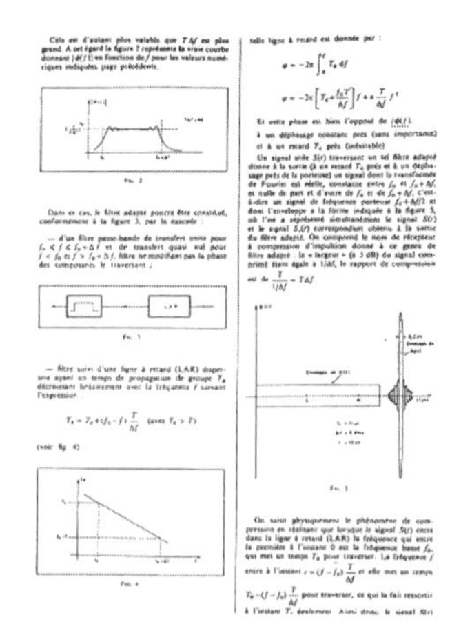

Cela est d'autant plus valable que $T\Delta f$ est plus grand. A cet égard la figure 2 représente la vraie courbe donnant $|\Phi(f)|$ en fonction de f pour les valeurs numériques indiquées page précédente.

Fig. 2

Dans ce cas, le filtre adapté pourra être constitué, conformément à la figure 3, par la cascade :

— d'un filtre passe-bande de transfert unité pour $f_0 \leq f \leq f_0 + \Delta f$ et de transfert quasi nul pour $f < f_0$ et $f > f_0 + \Delta f$, filtre ne modifiant pas la phase des composants le traversant ;

Fig. 3

— filtre suivi d'une ligne à retard (LAR) dispersive ayant un temps de propagation de groupe T_R décroissant linéairement avec la fréquence f suivant l'expression

$$T_R = T_0 + (f_0 - f)\frac{T}{\Delta f} \quad (\text{avec } T_0 > T)$$

(voir fig. 4)

Fig. 4

telle ligne à retard est donnée par :

$$\varphi = -2\pi \int_0^f T_R \, df$$

$$\varphi = -2\pi \left[T_0 + \frac{f_0 T}{\Delta f}\right] f + \pi \frac{T}{\Delta f} f^2$$

Et cette phase est bien l'opposé de $|\Phi(f)|$,

à un déphasage constant près (sans importance) et à un retard T_0 près (inévitable).

Un signal utile $S(t)$ traversant un tel filtre adapté donne à la sortie (à un retard T_0 près et à un déphasage près de la porteuse) un signal dont la transformée de Fourier est réelle, constante entre f_0 et $f_0 + \Delta f$, et nulle de part et d'autre de f_0 et de $f_0 + \Delta f$, c'est-à-dire un signal de fréquence porteuse $f_0 + \Delta f/2$ et dont l'enveloppe a la forme indiquée à la figure 5, où l'on a représenté simultanément le signal $S(t)$ et le signal $S_1(t)$ correspondant obtenu à la sortie du filtre adapté. On comprend le nom de récepteur à compression d'impulsion donné à ce genre de filtre adapté : la « largeur » (à 3 dB) du signal comprimé étant égale à $1/\Delta f$, le rapport de compression est de $\frac{T}{1/\Delta f} = T\Delta f$

Fig. 5

On saisit physiquement le phénomène de compression en réalisant que lorsque le signal $S(t)$ entre dans la ligne à retard (LAR) la fréquence qui entre la première à l'instant 0 est la fréquence basse f_0, qui met un temps T_R pour traverser. La fréquence f entre à l'instant $t = (f - f_0)\frac{T}{\Delta f}$ et elle met un temps $T_R - (f - f_0)\frac{T}{\Delta f}$ pour traverser, ce qui la fait ressortir à l'instant T_r également. Ainsi donc, le signal $S(t)$

Figure C.5

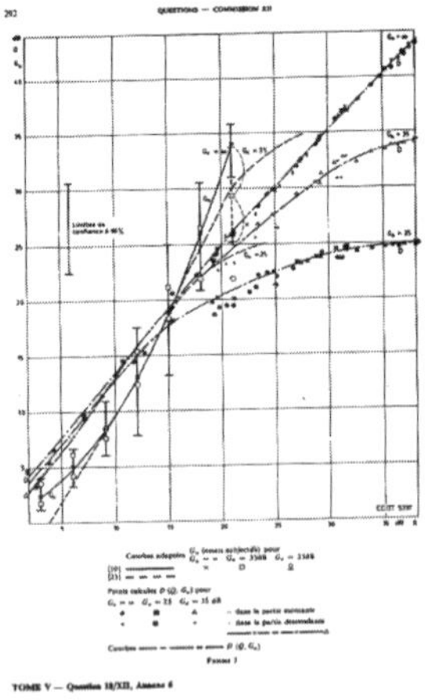

Figure C.6

CCITTの概要

Figure C.7

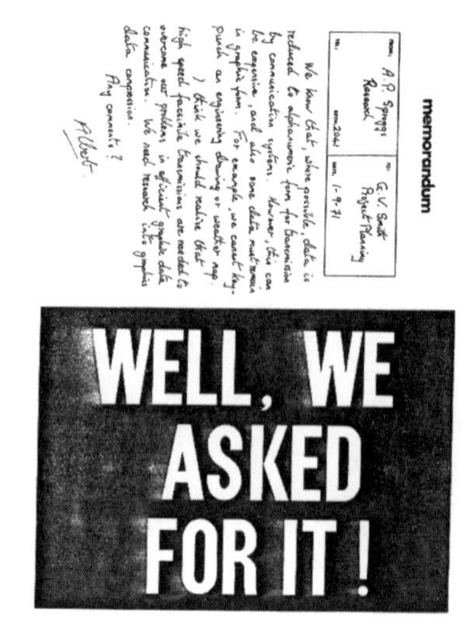

memorandum

FROM: A.P. Spraggs
Research

TO: G.V. Smit
Project Planning

EXTN. 2041

DATE: 1-9-71

We know that, where possible, data is reduced to alphanumeric form for transmission by communication systems. However, this can be expensive, and also some data must remain in graphic form. For example, we cannot key-punch an engineering drawing or weather map.

I think we should realize that high speed facsimile transmissions are needed to overcome our problems in efficient graphic data communication. We need research into graphic data compression.

Any comments?

Albert.

Figure C.8

C.3 JPEG Huffman tables

Table C.3 shows typical Huffman codes needed for coding the luminance DC coefficients differentially. Note that a separate table, not listed here, is suggested for chrominance coefficients.

Table C.3 Table for luminance DC difference

Category	Code length	Code word
0	2	00
1	3	010
2	3	011
3	3	100
4	3	101
5	3	110
6	4	1110
7	5	11110
8	6	111110
9	7	1111110
10	8	11111110
11	9	111111110

Table C.4 shows typical Huffman codes needed to encode the luminance AC coefficients. A separate table, not listed here, is suggested for encoding the chrominance coefficients.

Table C.4 Huffman table for luminance AC coefficients

NNNN/SSSS	Code length	Code word
0/0 (EOB)	4	1010
0/1	2	00
0/2	2	01
0/3	3	100
0/4	4	1011
0/5	5	11010
0/6	7	1111000
0/7	8	11111000
0/8	10	1111110110
0/9	16	1111111110000010
0/10	16	1111111110000011
1//1	4	1100

Table C.4 Huffman table for luminance AC coefficients (continued)

NNNN/SSSS	Code length	Code word
1/2	5	11011
1/3	7	1111001
1/4	9	111110110
1/5	11	11111110110
1/6	16	1111111110000100
1/7	16	1111111110000101
1/8	16	1111111110000110
1/9	16	1111111110000111
1/10	16	1111111110001000
2/1	5	11100
2/2	8	11111001
2/3	10	1111110111
2/4	12	111111110100
2/5	16	1111111110001001
2/6	16	1111111110001010
2/7	16	1111111110001010
2/8	16	1111111110001100
2/9	16	1111111110001101
2/10	16	1111111110001110
3/1	6	111010
3/2	9	1111101111
3/3	12	1111111110101
3/4	16	11111111110001111
3/5	16	11111111110010000
3/6	16	11111111110010001
3/7	16	11111111110010010
3/8	16	11111111110010011
3/9	16	11111111110010100
3/10	16	11111111110010101
4/1	6	111011
4/2	10	1111111000
4/3	16	1111111110010110
4/4	16	1111111110010111
4/5	16	1111111110011000
4/6	16	1111111110011001
4/7	16	1111111110011010
4/8	16	1111111110011011
4/9	16	1111111110011100
4/10	16	1111111110011101

Table C.4 Huffman table for luminance AC coefficients (continued)

NNNN/SSSS	Code length	Code word
5/1	7	1111010
5/2	11	11111110111
5/3	16	1111111110011110
5/4	16	1111111110011111
5/5	16	1111111110100000
5/6	16	1111111110100001
5/7	16	1111111110100010
5/8	16	1111111110100011
5/9	16	1111111110100100
5/10	16	1111111110100101
6/1	7	1111011
6/2	12	111111110110
6/3	16	1111111110100110
6/4	16	1111111110100111
6/5	16	1111111110101000
6/6	16	1111111110101001
6/7	16	1111111110101010
6/8	16	1111111110101011
6/9	16	1111111110101100
6/10	16	1111111110101101
7/1	8	11111010
7/2	12	111110100111
7/3	16	1111111110101110
7/4	16	1111111110101111
7/5	16	1111111110110000
7/6	16	1111111110110001
7/7	16	1111111110110010
7/8	16	1111111110110011
7/9	16	1111111110110100
7/10	16	1111111110110101
8/1	9	111111000
8/2	15	111111111000000
8/3	16	1111111110110110
8/4	16	1111111110110111
8/5	16	1111111110111000
8/6	16	1111111110111001
8/7	16	1111111110111010
8/8	16	1111111110111011
8/9	16	1111111110111100

Table C.4 **Huffman table for luminance AC coefficients (continued)**

NNNN/SSSS	Code length	Code word
8/10	16	1111111110111101
9/1	9	111111001
9/2	16	1111111110111110
9/3	16	1111111110111111
9/4	16	1111111111000000
9/5	16	1111111111000001
9/6	16	1111111111000010
9/7	16	1111111111000011
9/8	16	1111111111000100
9/9	16	1111111111000101
9/10	16	1111111111000110
10/1	9	111111010
10/2	16	1111111111000111
10/3	16	1111111111001000
10/4	16	1111111111001001
10/5	16	1111111111001010
10/6	16	1111111111001011
10/7	16	1111111111001100
10/8	16	1111111111001101
10/9	16	1111111111001110
10/10	16	1111111111001111
11/1	10	1111111001
11/2	16	1111111111010000
11/3	16	1111111111010001
11/4	16	1111111111010010
11/5	16	1111111111010011
11/6	16	1111111111010100
11/7	16	1111111111010101
11/8	16	1111111111010110
11/9	16	1111111111010111
11/10	16	1111111111011000
12/1	10	111111010
12/2	16	1111111111011001
12/3	16	1111111111011010
12/4	16	1111111111011011
12/5	16	1111111111011100
12/6	16	1111111111011101
12/7	16	1111111111011110
12/8	16	1111111111011111

Table C.4 Huffman table for luminance AC coefficients (continued)

NNNN/SSSS	Code length	Code word
12/9	16	1111111111100000
12/10	16	1111111111100001
13/1	11	11111111000
13/2	16	1111111111100010
13/3	16	1111111111100011
13/4	16	1111111111100100
13/5	16	1111111111100101
13/6	16	1111111111100110
13/7	16	1111111111100111
13/8	16	1111111111101000
13/9	16	1111111111101001
13/10	16	1111111111101010
14/1	16	1111111111101010
14/2	16	1111111111101100
14/3	16	1111111111101101
14/4	16	1111111111101110
14/5	16	1111111111101111
14/6	16	1111111111110000
14/7	16	1111111111110001
14/8	16	1111111111110010
14/9	16	1111111111110011
14/10	16	1111111111110100
15/0 (ZRL)	11	11111111001
15/1	16	1111111111110101
15/2	16	1111111111110110
15/3	16	1111111111110111
15/4	16	1111111111111000
15/5	16	1111111111111101
15/6	16	1111111111111010
15/7	16	1111111111111010
15/8	16	1111111111111100
15/9	16	1111111111111101
15/10	16	1111111111111110

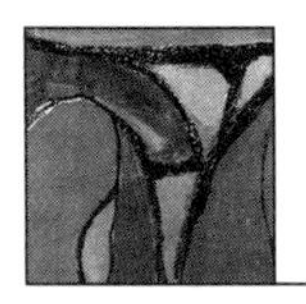

appendix D

Utility and header files for libpci

The following is a listing of utility routines that have been used by some of the example programs in the book.

```
/*
        Utility Routines

        pci_print_pixels     prints out a block of pixels
        pci_printblock_32    prints out a block of 32-bit integers
        pci_rect             creates an image of constant value
        pci_threshold        thresholds an image
        pci_bright           converts a binary image into a 0/255 image
        pci_alloc_image      allocates memory for a PCI image
*/

#include <stdio.h>
#include <stdlib.h>
#include <math.h>
#include <sys/time.h>
#include "pci.h"


/*
        Print out a block of pixels
*/

#define MAX_SPAN 30
void
pci_print_pixels(pci_image *image, int row, int col, int w, int h)
{
        int i,j;
        unsigned char *data, *sdata;
        int lb;

        if (w>MAX_SPAN) {
               printf ("pci_print_pixels: block too large\n");
               exit(1);
        }

        lb = image->linebytes;
        data  = image->data + row *lb + col;

        printf ("col  ");
        for (i=0; i<w; i++)
             printf ("%5d", (col+i));
        printf ("\n");

        for (j=0; j<h; j++) {
             printf ("%5d ", (j+row));
             for (i=0; i<w; i++) {
                  printf ("%5d", *(data + j*lb + i));
             }
             printf ("\n");
        }
}
```

```
/*
        Print out a block of 32-bit pixels
*/

void
pci_printblock_32 (int *da, int w,int h,int lb,char *s)
{
        int *p;
        int i,j;
        int lp = lb>>2;         /* pixels per line */

        printf ("%s\n",s);

        p = (int *)da;

        printf("             ");
        for (j=0;j<w;j++)
                printf ("   %d    ",j);
        printf ("\n\n");

        if (w>MAX_SPAN) {
                printf ("pci_printblock_32: block too large\n");
                exit(1);
        }

        for (i=0;i<h;i++) {
                printf ("  %d     ",i);
                for (j=0;j<w;j++,p++)
                        printf ("%8d",*p);
                printf ("\n");
                p = (da += lp);
        }
}


/*
        Create an image of constant value
*/

void
pci_rect(pci_image *image, int x, int y, int w, int h, unsigned char val)
{
        unsigned char *pixel;
        int im_w, im_h;
        int i,j;

        pixel = image->data;
        im_w = image->width;
        im_h = image->height;

        for (i=0; i<im_w*im_h; i++)
             *pixel++ = 0;
```

```
        for (i=0; i<h; i++) {
            pixel = image->data + im_w*(y+i) + x;
            for (j=0; j<w; j++)
                *pixel++ = val;
        }
}


/*
        Thresholds an image
*/

void
pci_threshold(pci_image *src, pci_image *dst)
{
        register unsigned char *spixptr, *dpixptr, *spix, *dpix, *sptr;
        int rows, cols;
        int slb, dlb;
        int dw, dh;
        int lw, lh;
        int i;
        int ir,result;

        /* get the info from the images */
        spixptr = spix = src->data;
        slb = src->linebytes;
        dpixptr = dpix = dst->data;
        dlb = dst->linebytes;
        dw = dst->width; /* dest width/height better be same as src's */
        dh = dst->height;

        while (dh--) {
                lw = dw;
                while (lw--) {
                        ir = (int) *spixptr;
                        if (ir > THRESH)
                                *dpixptr = 1;
                        else
                        *dpixptr = 0;

                        dpixptr++; spixptr++;
                }
                spixptr = (spix += slb);
                dpixptr = (dpix += dlb);
        }
}
```

```
/*
   Convert an image with values 0/1 to image with values 0/255
   Useful for displaying the results of morphological processing
*/

void
pci_bright(pci_image *src, pci_image *dst)
{
        register unsigned char *spixptr, *dpixptr, *spix, *dpix, *sptr;
        int rows, cols;
        int slb, dlb;
        int dw, dh;
        int lw, lh;
        int i;
        int ir,result;

        /* get the info from the images */
        spixptr = spix = src->data;
        slb = src->linebytes;
        dpixptr = dpix = dst->data;
        dlb = dst->linebytes;
        dw = dst->width; /* dest width/height better be same as src's */
        dh = dst->height;

        while (dh--) {
                lw = dw;
                while (lw--) {
                        ir = (int) *spixptr;
                        if (ir == 1)
                                *dpixptr = 255;
                  else
                       *dpixptr = 0;

                        dpixptr++; spixptr++;
                }
                spixptr = (spix += slb);
                dpixptr = (dpix += dlb);
        }
}


/*
        Allocate memory for a pci image
*/

pci_alloc_image(pci_image *image,  int rows, int cols, int bands)
{

        image->height = rows;
        image->width = cols;

        image->data = (unsigned char *)
             malloc(rows*cols*sizeof(unsigned char)*bands);
```

```
        image->bands = bands;
        image->pixel_stride = bands;
        image->linebytes = cols*bands;
}
```

The following is a listing of the include file `pci.h`.

```
/*
        File Name: pci.h
        include file for libpci
*/

#define MAXSHORT 65535
#define MAXBYTE 255
#define PI 3.14159
#define ZERO 0.00000000001
#define THRESH 120

/* #defines related to bicubic scaling algorithm */

#define SPACC   5               /* SubPixelACCuracy = 5 */
#define SUPPORT 5               /* region of support for filter */
#define SCA     4096            /* filter coeffs scaled to 12 bits */
#define SHIFT   12
#define HALF    (1<<11)

#define ROUNDDOWN(IX,X)\
        IX = (int) X;\
        if ((X<ZERO) && (X!=(float)IX))\
             IX -= 1;

typedef struct pci_image {
        unsigned char *data;
        unsigned int width;
        unsigned int height;
        unsigned int bands;
        unsigned int pixel_stride;
        unsigned int linebytes;
} pci_image;

typedef struct {
        int src_off;        /* offset of the src pixel at center of conv */
        int coeff[SPACC];   /* five pixels in region of support */
} pci_ftab;


/* FUNCTION PROTOTYPES */

/* table lookup related routine from Chapter 4 */
void pci_lookup(pci_image *src, pci_image *dest, unsigned char *lut);
void pci_lookup2(pci_image *src, pci_image *dest, unsigned char *lut);
void pci_init_lookup_8(unsigned char *lut);
```

```
/* point operations from Chapter 5 */
void pci_histogram(pci_image *src, int *histo);

void pci_histogram2(pci_image *src, int *histo);

void pci_histeq(pci_image *src, pci_image *dest);

void pci_blend(pci_image *src1, pci_image *src2,
             pci_image *dest, pci_image *alpha);

void pci_lookup (pci_image *src, pci_image *dest, unsigned char *lut);


/* convolution routines  from Chapter 6*/
void pci_cnv_3x3(pci_image *in, pci_image *out, float *kernel);
void pci_cnvmxn(pci_image *src, pci_image *dst,
             int height, int width, float *kernel);


/* median filter from Chapter 6 */
void pci_mdn3x3_histo(pci_image *src, pci_image *dst);


/* morpho routines from Chapter 6 */
void pci_morph(pci_image *src, pci_image *dst, unsigned short ^m_table);
void pci_gen_erode(unsigned short *table);
void pci_gen_dilate(unsigned short *table);
void pci_gen_ipr(unsigned short *table);


/* color conversion routines from Chapter 7 */
void pci_gentab_l2nl();
void pci_gentab_nl2l();
void pci_ycc601_4202rgb709(pci_image *iny,pci_image *incb,
                     pci_image *incr,pci_image *outrgb);
void pci_rgbl2rgb709(pci_image *in,pci_image *out);
void pci_ycc6012rgb709(pci_image *in,pci_image *out);
void pci_ycc6012rgbl(pci_image *in,pci_image *out);
void pci_rgbl2photoycc(pci_image *in,pci_image *out);
void pci_rgbl2cmyk(pci_image *in,pci_image *out);


/* nearest color quantization */
void pci_nearest_color(pci_image *src,pci_image *dest,
                  int *dimensions, int *multipliers);


/* scale routines for Chapter 8 */
void pci_scale_nn(pci_image *src, pci_image *dest);
void pci_scale_nn_bres(pci_image *src, pci_image *dest);
void pci_scale_bl(pci_image *src, pci_image *dest);
void pci_scale_bl2(pci_image *src, pci_image *dest);
void pci_scale_bc(pci_image *src, pci_image *dest);
void pci_scale_bc_filter(int *filter);
```

```
float pci_cubic(float x);
void pci_scale_bc_htab(pci_image *src, pci_image *dest, pci_ftab *htable);
void pci_scale_bc_vtab(pci_image *src, pci_image *dest, pci_ftab *vtable);
void pci_scale_bc_h(pci_image *src, pci_image *inter, pci_ftab *htable);
void pci_scale_bc_v(pci_image *inter, pci_image *dest, pci_ftab *vtable);


/* rotate routines for Chapter 8 */
void pci_rotate (pci_image *src, pci_image *dest, int xorig, int yorig,
               float degrees);
void pci_rotate_2pass (pci_image *src, pci_image *inter, pci_image *dest,
              int xorig, int yorig, float degrees);
void pci_sfx(pci_image *src, pci_image *dest, double R);
void
pci_sfx_gentab(pci_image *src, pci_image *xtab, pci_image *ytab, float R);
void
pci_sfx2(pci_image *src, pci_image *dest, pci_image *xtab, pci_image *ytab);


/* DCT routines from Chapter 9 */
void pci_dct(int pixels[8][8], int dct[8][8]);
void pci_fast_dct(int *in, int *out, int delta);


/* utility routines */
void pci_bar(pci_image *image, int numbars);
void pci_gray(pci_image *image);
void pci_print_pixels(pci_image *image, int row, int col, int w, int h);
void pci_printblock_32(int *da, int w, int h, int lb, char *s);
void pci_rect(pci_image *image, int x, int y, int w, int h, unsigned char
val);
void pci_threshold (pci_image *in, pci_image *out);
void pci_bright(pci_image *src, pci_image *dst);
```

index

T

U

V

W

X

Y

Z